Soil Microbiology and Soil Biotechnology

Volume 2

by
Prof. Dr. S. Kannaiyan
Former Vice-Chancellor
TNAU, Coimbatore - 641 003
and
Former Chairman
National Biodiversity Authority (GoI)
Chennai - 600 041

2021
Associated Publishing Company®
A Division of
Astral International Pvt. Ltd.
New Delhi – 110 002

ISBN 9788185211855 (SET)

Published by : **Associated Publishing Company®**
A Division of
Astral International Pvt. Ltd.
– ISO 9001:2015 Certified Company –
4736/23, Ansari Road, Darya Ganj
New Delhi-110 002
Ph. 011-43549197, 23278134
E-mail: info@astralint.com
Website: www.astralint.com

Digitally Printed at : Replika Press Pvt. Ltd.

Dedicated

To

My Beloved

Wife

Mrs. Banumathi Kannaiyan

Contents

(Vol. 2)

Preface

The characteristics of the soil environment vary with climate. Soils differ in depth, physical properties, chemical composition and origin. There are five major categories of soil constituents such as mineral particles, organic residue, water, gases and biological system. The physical structure, aeration, water holding capacity and awailability of nutrients are determined by the proportion of the mineral particles which are formed by the weathering of rock and the degradative metabolic activities of microorganisms. The vast differences in the composition of soils together with differences in their physical characteristics and agricultural practices by which they are cultivated, reflect equally large differences in their microbial population.

Variations in climatic conditions may selectively favour certain physiological types. Interactions between and among microbiol species no doubt have an important effect on the members of the population. This situation is an extemely complex condition. Predatory protozoa and antibiotic producing actinomycetes may eliminate a segment of the flora whereas cellulolytic and proteolytic organisms may provide nutrients for less versatile biochemical species. Plant roots penetrate the soil, breaking up mineral particles, removing nutrients, converting them to organic compounds and secreting metabolic products. When plants and other soil inhabitants die, their organic components are mineralized by microorganisms, which likewise contribute metabolic excretions and dead cells. In the soil there is a constant and complex cycling of organic and inorganic materials. The most important role of soil microorganisms is their function as biogeochemical agents for the mineralization of organic carbon, nitrogen, phosphorus and sulphur compounds so that these biologically important elements are available for reuse.

Soil Microbiology and Soil Biotechnology covers in a major way on Soil Bio-resources, Soil organic matter and organic manure decomposition, Nitrification and Denitrification process in Soil, Sulphur bacteria and the environment. It also covers Rhizosphere Microflora and Plant Microbe interaction, Microbial Ecology of Arid Soil, Industrial potentials of Microbial pigments, Beneficial applications of the fungal antagonists - *Trichoderma* and Microrganisms and Environment. It also covers Biological fertilizers, *Azospirillum* in Soil and its colonization in plant roots, Liquid Bioinoculants, Mineral Phosphate Solubilization by blue green algae, AM fungi, Role of Host genes in Symbiotic Nitrogen Fixation, Bioecology and *Biotechnological* applications of *Trichoderma*, Transformation Technology against Pests and Diseases, Production Technology of Paddy Straw Mushroom, Oyster Mushroom and Biodiversity of Oyster, Paddy Straw and Milky Mushroom. The book *Soil Microbiology and Biotechnology Volume 1 and 2* covers both basic aspects and applied aspects of Soil Microorganisms.

I am grateful to the contributors who readily agreed to share their rich experience and knowledge in their respective fields of specialization to the important volumes meant for scholars and P.G. students. I profoundly thank M/s Associated Publishing Company, New Delhi and especially its owner and publisher Shri S.K. Dutta for all his sincere efforts in bringing out the volumes expeditiously.

I wish to place on record my special thanks to my beloved wife Mrs. Banumathi Kannaiyan, my son Dr. Lenin Kannaiyan, Ph.D. (UK) and daughter-in-law Dr. (Mrs) Hemavathy Lenin and daughter Er. (Mrs) Curie Pradeep, M.S. (USA) and son-in-law Er. D. Pradeep, M.S. (USA), M.S. (Food Engg. USA) who are always with me by extending their full support and cooperation in all my academic and scientific endeavours.

Thanjavur **Prof. Dr. S. Kannaiyan**

November, 7, 2009

Soil Microbiology and Soil Biotechnology : Vol. 2
Editor : S. Kannaiyan
Associated Publishing Co., New Delhi, 2010, pp. 1-26

1

Integrated Nutrient Management Strategy in Wetland Rice Ecosystem

*S. Kannaiyan**

Introduction

With the advent of modern crop varieties, more land under irrigation and greater use of fertilizers and other inputs, Asia has changed within the last 20 years from a level of food scarcity, to a level of food sufficiency. Increased fertilizer use has contributed to about one-fourth of the increased rice production in Asia. In some countries, fertilizer prices were subsidised and thereby enabling the farmers to apply maximum levels of fertilizers for their crops. But, during the same period, use of organic fertilizers including green manures for rice production declined substantially.

Under natural conditions, plant nutrients enter the soil, through the decomposition of dead plant materials. In agriculture, they are added by the farmers as residues of the harvest or in the form of stable manures and other organic materials. All these materials and the roots of the harvested plants are transformed during humification, the process by which inorganic plant nutrients become available for next crop.

For agriculture to be productive and sustainable, (i) renewable resources must be maintained, (ii) non-renewable resources must be used with foresight, (iii) the intrinsic value of the natural environment must be recognized, (iv) farm families must be able to make a decent living, and (v) increasing and changing demands for agricultural products must be satisfied at affordable prices. The combination of these commandments harbours conflicting productivity and sustainability demands that have to be satisfied at the same time, forming a continuing source of political debate (De Wit, 1990).

Land resources are broadly put into uses in (i) agricultural sector consisting of net cultivated area, current fallows, and cultivable waste lands (ii) ecological sector comprising of forests, grazing lands and barren lands and (iii) Non-agricultural sectors. During the last three decades, the non-agricultural sectors, increased at the rate of one per cent per annum. An expansion of non-agricultural sectors is an indication of the growth and development of the country but, it exerts enormous pressure on land resources which would certainly affect the agricultural production of the country. The pressure on land resources in the Indian

* Former Chairman, National Biodiversity Authority, Chennai - 600041

context could be measured from the fact that it shares only two per cent of the world's geographic area, but supports around 18 per cent of the world's human population and over 15 per cent of the world's livestock (Kannaiyan *et al.*, 1998). The resultant overstraining of land resources, by a disproportionate human and animal population led to the widespread land degradation and non-sustainable ecosystem. Therefore, land resources management should aim at sustainable crop yield.

To match the food requirements of increasing population, the developing countries are left with no alternatives than to intensify their agricultural systems rapidly to increase production. Any system of intensive cropping drains from the soil very heavily the available plant nutrients and, therefore, increased and judicious use of mineral fertilizers is the key factor in the systems aiming at an intensification of crop production. It is generally recognized that in an input package, fertilizer contributes for 50 percent to yield increases. It is also estimated that to produce the food required, developing countries as a group will have to increase their consumption of fertilizers three to four fold for better production. However, rising costs and heavy demand for fossil energy by other sectors of economy will be a serious constraint to reach this target. Therefore, the highest possible efficiency in the use of fertilizers as well as maximum use of available alternative renewable sources of plant nutrients like organic materials and biological nitrogen fixation will be utilized. The concept of integrating all sources of plant nutrients into a productive agricultural system has therefore received increasing attention during these years.

Sustainable agriculture

The basic thrust of sustainable agriculture is the improvement of the quality of life in the context of environmentally sound approach so that the resource base is maintained or enhanced for future generations. Sustainable agriculture involves a successful management of resources for agriculture to satisfy changing human needs while maintaining or enhancing the environment and conserving natural resources that provides food, income and livelihood for the present and future economic productivity and eco-system services for these natural resources (Kannaiyan, 1999).

The food originates from efficient and environmentally safe production technologies that conserve and enhance natural resource base of crops, animal husbandry and forestry. The physical dimension of food and nutrition security should involve on knowledge intensive and eco-friendly farming technologies for achieving not only the food production but also the quality food.

The stagnation in maximum yield potential during the recent years in crops like rice and wheat is a real concern for many Agricultural Scientists and Professional Technocrats. Productivity increasing technology and food loss reduction strategies such as improved storage, processing and marketing should be ecologically sustainable, economically viable and socially equitable. Cost and risk must be low and return must be attractive if the poor are to derive the benefit from the eco-friendly bio-resources technologies. Sustainable Agriculture has been defined by various authors and international agencies as follows:

> "Sustainable agriculture should involve successful management of resources, for agriculture to satisfy changing human needs while maintaining, or enhancing the environment and conserving natural resources" (FAO, 1989).

"Sustainable agriculture is a management system for renewable natural resources that provides food, income and livelihood for present and future economic productivity and ecosystem services of these resources" (TAC, 1988).

"Sustainable agriculture systems are those that are economically viable and meet society's need for safe and nutrients food while converting or enhancing natural resources and the quality of the environment for future generations" (Baier, 1990).

"Sustainable system is one with non-negative trend in measured outputs" (Lynam and Herdt, 1988).

"Sustainable agriculture system (SAS) is an efficient agricultural management option, by fully exploiting the biopotential of biological resources including the beneficial microbial systems, by protecting the environment, for improving the quality of human life" (Kannaiyan, 1998).

Integrated nutrient management

INM advocates the judicious use of all available nutrient sources (Soil, biological, organic, mineral fertilizer) to maintain nutrient balance in soils (Soil health) and to promote balanced fertilization of crops. **FAO** uses the term **"Integrated Plant Nutrient System" (IPNS)** to emphasize the need for balanced nutrient application to cropping system to maintain soil quality, reduce fertilizer cost to farmers, and improve crop yields and profitability in rice farming (Balasubramanian, 1999).

The required productivity and sustainability can be achieved by integrated nutrient management (**INM**) which is the best possible combination of available nutrient management technologies which are biophysically powerful, economically attractive and socially acceptable.

INM should be considered as a cultivation system that integrates efficient use of soil and applied plant nutrients with appropriately modified agronomic and pest management practices that help to increase and sustain crop production over long time while conserving the natural resources and protecting the environment.

We must also remember that **"crops do not respond to a fertilizer application *per se*, but rather to the soil's response to that application"** as stated by Summer and Farina (1986).

Various components of INM and their perspective importance

Various major sources of plant nutrients are soil, mineral fertilizers, organic matters and atmospheric nitrogen fixed by micro-organisms and carried through precipitation. The main aim of the integrated approach is to tap all the sources in a judicious way and ensure their efficient use.

Soil sources

The nutrient supplying capacity of the soil is well known. Due to continuous and intensive cultivation, the supplying capacity of soil is getting limited and therefore, they are becoming deficient in some or other nutrients (Roy and Braun, 1983). To enhance the soil nutrient supply the following should receive greater attention.

- appropriate soil management and conservation practices to reduce losses of nutrients,
- amelioration of problem soils for mobilising unavailable nutrients,
- appropriate crop varieties, cultural practices and cropping systems for maximizing utilization of available nutrients,
- other practices, like inoculation with insoluble phosphate dissolving bacteria and fungi, which have shown promise in mobilising unavailable soil phosphorus and making it available to the plants during their growth.

Mineral fertilizers

Major role in plant nutrition played by mineral fertilizers for sustaining and increasing agricultural production needs no emphasis.

With a view to energy and economic considerations, increased efficiency in their use is of utmost importance. At a time when we are posed with the problem of feeding increasing population in developing countries and the meager level of mineral fertilizers being actually used at farm level, any approach of further reducing the dose of mineral fertilizers and supplementing with alternative sources should be advocated with great caution and depending on current level of its use. The direction should be to maximise agricultural production per unit area and unit time by optimising its use efficiency through the complementary use of organic and other alternative sources of plant nutrients. Any additional nutrients applied through other sources should be accounted for making up the gap between the recommended and actual level of fertilizer application.

The **FAO** Fertilizer Programme in Africa and Asia is fully geared to increasing fertilizer use efficiency through the following:

- scheduling recommendations based on a cropping system rather than for a single crop in the system;
- by improving all other production factors and eliminating limiting factors, including secondary and micronutrients;
- minimising losses in the field through appropriate time and methods of application;
- minimising losses in the transport chain; and
- use of appropriate products including supergranule and coated urea, direct use of locally available phosphate rocks, *etc.*

Organic sources

The organic manures are the available by-products of farming and allied industries, derived from plant and animal resources. The available organic resources can be classified in their order of importance as follows:

- cattle-shed wastes like dung, urine and litter (farmyard manure);
- crop-wastes/residues like sugarcane trash, stubbles, weeds, straws and spoilt fodder;
- human habitation wastes like night soil, urine, town refuse, sewage sludge and sullage;
- slaughter--house wastes, animal carcasses and by-products such as blood, meat, bones, fishery wastes, leather and wool wastes, *etc.*

- sheep and goat-dropping;
- by-products of agriculture-based industries such as : oil cakes, wastes from fruit and vegetable processing, bagasse and press-mud from sugar factories, saw dust, tobacco wastes and seeds, rice husk and bran, tea-waste, and cotton-dust from textile industries;
- forest-litter;
- wastes from marine algae and seaweeds;
- miscellaneous products such as water hyacinth, tank silt, *etc.*

It is estimated that about 130 million tonnes of plant nutrients are theoretically available in this form, although the quality actually usable are much lower. Problems in their use are: low content of nutrients, slow release of the nutrients due to slow rate of mineralization, transportation difficulty and cost owing to their bulky nature, and widespread use as fuel and feed for cattle due to unavailability of alternate sources.

The nutrients present in animal dung, crop residues and other organic materials can be recycled in soil either by composting, direct incorporation of organic residues in soil or by mulching.

Rural composting

Though improved technology is now available, yet in most cases it is not followed by the farmers. It is therefore important to motivate farmers not to waste the farm residues but to conserve and compost them in correct ways, like, to select an appropriate place for composting, chopping the crop residues, adding fungal cultures (*Aspergillus* sp., *Penicillium* sp., *etc.*) or fresh dung slurry for accelerating decomposition, adding a starter dose of nitrogen (0.25 to 0.5 percent) to lower the **C/N** ratio and enriching with low grade rock phosphate (1 to 2 percent) to improve the quality and to avoid loss of valuable ammonia through volatilization.

Mechanical composting

Establishing compost plants in the urban areas to process the refuse for agricultural purposes has been successful in many developed countries. However, this has been of limited success in developing countries due to higher cost of production and transportation of a low grade nutrient source. Experience has shown that reduction of civil works and mechanization can help in considerably reducing the capital cost. The cost of production of compost should therefore be reduced by only essential mechanization and by reducing its bulkiness and by enrichment so as to increase the plant nutrient content per unit of biomass.

Biogas technology

An obstacle to greater use of organic materials, particularly animal waste, as plant nutrients is their widespread use as fuel. All their plant nutrient value is lost through burning, but can be remedied if the animal wastes (or other suitable wastes) are used to produce methane (biogas) by the anaerobic fermentation process. The methane can be used for cooking, lighting and running engines and the digested slurry can be used directly with irrigation water, composted with farm residues or sun-dried and used as manure. However, the removal of several constraints like the still high cost of the digesters, the lack of sound infrastructure for installation, operation and servicing are pre-requisites for large-scale adoption by the farmers.

Incorporation of crop residues and mulching

Crop residues are important renewable scattered organic sources and are readily available to farmers. When composting cannot be practiced, they could be directly applied to the field. Their effect on improving soil properties, conserving soil moisture and controlling weeds are well recognized. However, their availability is very often constrained as they are used by farmers as a source of fuel and feed for the cattle. While using wide C/N ratio materials, care should be taken to supplement with nitrogen and phosphorus. Otherwise crop growth will suffer due to their deficiency caused by microbial immobilization during the initial period of decomposition.

Sewage/sullage

It is equally important to recycle liquid wastes and use profitably for agricultural purpose. However, much further progress is needed in the development of suitable infrastructures including engineering technologies for sewage treatment and measures to check health hazards. According to a recent report of the Association of Germen (**FRG**) Agricultural Experimental and Research Institutes (**LUFA**), out of 306 samples of sewage sludge tested, 40 percent samples were designated unfit for agricultural use on account of their high heavy metal content (especially in zinc and copper).

FAO's role in organic recycling assistance programme has made considerable impact in the Asia region and similar activities are now being extended in the Africa and Latin America regions.

Biological sources

Legumes contribute to soil fertility directly through their unique ability to fix atmospheric molecular nitrogen in association with rhizobia. There are two modes of nitrogen fixation that are of considerable importance, one is the recycling of nitrogen fixed by select leguminous plants/trees through their incorporation in the soil as green manures thus making the nutrients in it, particularly nitrogen, available to the succeeding crop. The other is the nitrogen fixation through some symbiotic and non-symbiotic micro-organisms and making it available to the associated field crop.

INM for intensive irrigated rice systems

Rice being the most important crop for us, a special mention is required on **INM** for rice. **Nitrogen (N)** management is one of the keys in rice production. Studies have shown that a dynamic soil-and plant-based management is needed for **N**. Quantity of applied **N** is to be adjusted to variation in the indigenous **N** supply. This is as important as the timing, placement and source of applied **N**.

Estimation of crop demand, potential indigenous nutrient supply, and recovery from applied inorganic and organic sources over time is required to predict the total amount of applied **N** that is needed. Estimation of soil **N** release during early growth stages to identify the need for a basal **N** application and monitoring plant **N** status to optimize timing of split applications in relation to crop demand and soil **N** supply are required (Dobermann and White, 1999).

As far as **P** and **K** are considered, a long term strategy is required. **P** and **K** are not lost or added to the root zone by the biological and chemical processes affecting **N.** Therefore, maximizing the recovery efficiency of **P** and **K** is less important than predicting the need for applied nutrients and the amount to supply. Estimation of crop demand, potential indigenous nutrient supply, and recovery from applied inorganic and organic sources over time to predict the amount of applied **P** and **K** are required. Knowledge of the relationship between the **P** and **K** balance and changes in potential indigenous nutrient supply over time are also needed. The results from **RTDP** (Reversing Trends in Declining Productivity) project of **IRRI,** have shown that there was wide occurrence of nutrient-budget deficits of **K** induced by continued production of high rice yields. Correction of this would increase the efficiency of use of other elements, particularly **N** and **P** (Anonymous, 2000).

Diagnosis of potential deficiencies in the key tool for the management of Mg, Zn and S. Once identified as a problem, deficiencies can be alleviated by regular or one-time measures as part of general fertilizer/soil use recommendation.

Diagnostic criteria are needed to identify other nutrient disorders such as salinity, iron toxicity, or boron toxicity to make adjustments in the **N**, **P**, and **K** management.

Surveys throughout South and Southeast Asia during 1994-97 showed that indigenous nutrient supply varies widely among rice fields (Dobermann *et al.,* 1996). This means that rice crops require different amounts of nutrients in different fields, depending on indigenous nutrient supply and crop demand. When blanket recommendations are practiced, nutrients are managed poorly. This practice leads to declining productivity, higher fertilizer cost, reduced profitability, wastage of nutrients, and fertilizer-related pollution of the environment.

Impeded drainage in the lowland ecosystems, especially during the monsoon season, is a serious issue affecting nutrient use efficiencies and drastically limiting the potential yields. Recent results showed that inspite of soils testing for higher available **K** status, **K** deficiencies occur because of impeded drainage leading to reduced root oxidizing power. This shows that existing **K** availability index requires reconsideration for such situations.

Role of biofertilizers in INM

Nitrogen fixing microorganisms convert about 139-170 million tonnes of nitrogen every year into fertilizer nitrogen for which the energy bill is paid by Nature. It is atleast several fold greater than the inputs of nitrogen from chemical fertilizer factories to the tune of 65 million tonnes per year. The fact, that the total world biological nitrogen fixation is three times that of industrially produced nitrogen, demonstrates the significance of biological nitrogen fixation in agricultural and natural nitrogen cycles.

Azolla

Nitrogen fixing systems offer an economically attractive and ecologically sound means of reducing external inputs and improving internal resources. Symbiotic systems such as that of *Azolla* and *Anabaena* complex and that of leguminous green manures with *Rhizobium* and *Azorhizobium* associations can be of particular value to wet land rice crop and renewable biological source supplementing inorganic **N** for cereals. In **TNAU** significant contributions have been made on *Azolla* biofertilizer particularly on the selection of new strains of *Azolla microphylla* which is highly tolerant to high temperature and also tolerant to salinity with higher biomass producing capacity (25t/ha) (Kannaiyan, 1992). An *Azolla* hybrid was also developed by sexual hybridization and the hybrid is also highly adaptive and has higher N_2 fixing potential. In wetland rice soil ecosystem, the nitrogen fixation by free-living blue green algae or cyanobacteria also contribute significantly to the replenishment of soil **N**. Recently, attempts have been made to improve the associations between rice and nitrogen fixing soil bacteria and possible nitrogen fixing rice endophytes and thus achieve a higher contribution of biological nitrogen fixation to rice crop. In these soils, sustainability depends on symbiotic and non symbiotic **BNF** system.

Blue Green Algae (BGA)

Nitrogen fixing free-living cyanobacteria plays a vital role in the maintenance of soil fertility and sustainability. The large scale production of soil based **BGA** inoculum developed at Tirur by **TNAU** during 1977 is a crude and farmer oriented technology, which is two decades old, and hence has to be updated. Immobilized cyanobacterial inoculum is a recent development in the biofertilizer technology, which was developed by **TNAU**, Coimbatore with the active collaboration of the Division of Life Sciences, King's College, London and the Department of Biochemistry and Molecular Biology, University of Seville, Spain under the **EEC** project. Immobilization of nitrogen fixing cyanobacteria under rice field conditions can be done with solid matrices such as polyurethane foam, polyvinyl foam, sugar cane waste, paper waste, rice husk *etc.*, (Kannaiyan *et al.*, 1997). The immobilized cyanobacteria could continuously excrete the ammonia extracellularly which could be used for continuous supply of nitrogen to rice crop. The immobilized cyanobacterial system in solid matrix when inoculated in rice field is like a mini *in situ* fertilizer factory which produced ammonia continuously. The accumulation of ammoniacal **N** in the flood water as a result of immobilized cyanobacterial inoculation will reflect on the growth and yield of rice.

Rhizobium

Rhizobium is one of the important nitrogen fixing bacteria which helps the legume to maintain the soil fertility by way of their symbiotic association and nitrogen fixation. A legume plant having effective root nodules not only can meet its own nitrogen requirement but can also enrich the soil nitrogen content and thus improving soil fertility and sustainability. Legumes are cultivated widely in different parts of India. The practice of seed inoculation with *Rhizobium* is popular among farmers.

Inoculation of appropriate strains of *rhizobia* to legumes is known to enhance the yield. But both success and failures at field experiments have been documented. Failure to obtain the desired response is due to (i) the presence of native ineffective strains which could not be displaced by the introduced effective strains (ii) the presence of antagonists of *rhizobia* which minimize the numbers of *rhizobia* in the rhizosphere (iii) the soil conditions such as acidity, alkalinity, application of pesticides, herbicides *etc.* and (iv) the presence of effective native *rhizobial* strains in large numbers.

Induction of root nodule structures in cereal crops for N_2 fixation

Non-legumes such as cereal crops, grasses and other non-legume crops do not form the nodules in their root system. Recently it has been investigated at **TNAU** and found that the symbiotic nitrogen fixing bacterium, *Azorhizobium caulinodans* isolated from the stem nodules of *Sesbania rostrata*, could colonise and produce nodular structures in rice root. Among the cereals tested, maize crop responded well for the inoculation of *A. caulinodans* and prominent nodular structures have been observed (Chandrasekar and Kannaiyan, 1995).

The bacterium, *A. caulinodans* is a unique one which fixes nitrogen under free-living condition and also tolerant to combined nitrogen at low concentration. The bacterium could also tolerate flooded condition and *A. caulinodans* also produces the pectic and cellulolytic enzymes with good penetration capacity and in general it fixes nitrogen in rice root system at micro-aerophilic condition (2-3% O_2 level). The nitrogenase enzyme which is responsible for nitrogen fixation was detected from the root nodule structures of rice, maize Sorghum, cumbu and ragi (Amutha and Kannaiyan 1988). The nitrogen fixing rate in the nodular structures of cereal crops is low but the colonization of the bacterium could be induced by biotechnological manipulation of rice plant.

Azospirillum

Field experiments conducted in different places in India have indicated the positive response of rice cultivars to *Azospirillum* inoculation (Rao *et al.*, 1983). Experiments conducted at **IRRI**, Philippines showed that inoculation with *Azospirillum lipoferum* promoted early tillering and also the growth of rice and significantly increased filling rate of grain and the grain weight per plant at harvest. Response of rice to inoculation of *Azospirillum lipoferum* has been reported by several researchers (Rao *et al.*, 1983). However, the response of rice to *Azospirillum* inoculation under low land rice field condition was variable, depending on the initial nitrogen status of the soil and the varieties of the rice. Beneficial effects of *Azospirillum* inoculation have been reported on maize, Sorghum, sunflower, bhendi, cotton and pearl millet.

Acetobacter

More recently, a new species of nitrogen fixing bacterium, *Acetobacter diazotrophicus* was isolated from sugarcane in different parts of the world. *A. diazotrophicus* has been isolated from the root, leaf, bud and stem samples of sugarcane cultivars planted in 10 different locations of Tamil Nadu. Inoculation of the acid tolerant and sucrose loving bacterium to sugarcane enhances the cane yield by 17 to 20 t/ha with a 50 per cent reduction in the recommended chemical nitrogen (Muthukumaraswamy *et al.*, 1994). The results of the field experiments showed that inoculation of *A. diazotrophicus* to sugar cane enhances the cane yield and sugar yield than *Azospirillum* or *Azotobacter* inoculation (Jayakumar, 1996).

Phosphobacteria

Several soil bacteria and fungi notably species of *Pseudomonas, Bacillus, Penicillum* and *Aspergillus etc.*, secrete organic acids and lower the pH in their vicinity to bring about solubilization of bound phosphates in soil. Inoculation of composite culture of phosphate solubilizing microorganism comprising of *Pseudomonas striata* and *Aspergillus awamorii* recorded increase in the yield of cotton than the single strain inoculation. Increase in the yield of wheat and potato were demonstrated due to inoculation of peat based cultures of *Bacillus polymyxa* and *Pseudomonas striata*. Similar results were also obtained in wheat when the mixed cultures of phosphate solubilizing microorganisms were used with *Azotobacter chroococcum*. Saving of 50 per cent of recommended level of P_2O_5 is possible in sugarcane by inoculation with phosphobacteria as the cane yield and sugar yield of 50 per cent P_2O_5 and phosphobacteria treatments are at par with 100 per cent P_2O_5 application. However, it is also observed that the positive influence of phosphobacteria was conspicuous even at 0 and 75 per cent of phosphorus levels in sugarcane. Combined inoculation of *Rhizobium* and phosphobacteria have been tried in different crops and the increase in yield of soybean and groundnut was reported (Gaur, 1990).

Vesicular Arbuscular Mycorrhiza

Mycorrhiza is the mutualistic association between plant roots and fungal mycelia. Vesicular Arbuscular Mycorrhiza (**VAM**) and endotrophic mycorrhizae formed by aseptate phycomycetous fungi are associated as an obligate symbiont with majority of agricultural crops growing under broad ecological range. Many graminaceous plants and legumes are highly susceptible to **VAM** colonization. **VAM** endophytes are not host specific although evidence is showing that certain endophytes may form preferential association with certain host plants. The transfer of nutrients mainly phosphorus and also zinc and sulphur from the soil to the cells of the root cortex is mediated by intracellular obligate fungal endosymbiont of the genera *Glomus, Gigaspora, Acaulospora, Sclerocystis* and *Endogone* which possess vesicles for storage of nutrients and arbuscules for funneling these nutrients into the root system. By far the commonest genus appears to be *Glomus* which has several species distributed in soil.

Experiments with phosphate labelled with ^{32}P indicate that the hyphae of **VAM** fungi obtain their extra phosphate from the labile pool rather than dissolving in soluble phosphates (Raj *et al.*, 1981). The better utilization of sparingly soluble rock phosphate is explained by the hyphae making closer physical contact with the particle surface where the ions are dissociating. These fungi produce phosphatases which allow utilization of organic phosphorus especially under humid tropical conditions where hyphae can be in close contact with finely divided litter where acid soil reaction reduces the sorption of organic phosphates on soil surfaces. Vesicular arbuscular mycorrhiza (**VAM**) has also been reported to improve the uptake of the minor elements zinc, copper and also water (Bagyaraj, 1984).

Soil inoculum is produced using a traditional pot culture technique containing all **VAM** fungal structures which are highly infective. The success of good soil inoculum production depends on the selection of host-plant and the ambient conditions under which a defined **VAM** fungus can be mass multiplied. The dual inoculation with **VAM** fungi and appropriate *Rhizobium* strains on the enhancement of nutrient uptake of legumes like soybean, chickpea, clover, cowpea and groundnut was reported. Several field, laboratory as well as green house experiments have demonstrated that VA-mycorrhizal inoculation can greatly improve growth and nutrition of host plants.

Importance of N_2 fixing green manures in INM

The cultivation of nitrogen fixing legume green manure crop would be an efficient way of improving soil fertility status and sustainability. Many *Sesbania* species are widely used in Asia of which *S. aculeata* and *S. speciosa* are considered to be the most predominant species grown in India. These two are more commonly grown as N_2 fixing green manures because of their vigorous growth habits, higher rates of N_2 fixation and their general ability to withstand a wide range of soil conditions *viz.*, waterlogging, salinity and alkalinity. The role of N_2 fixing green manures and other associated systems which contributes nitrogen in rice field ecosystems and their impact on rice crop is given in Fig. 1. The stem nodulating tropical legume *S. rostrata* has shown better performance as a potential N_2 fixing green manure crop in rice farming systems due to its fast growth rate, N_2 fixing efficiency and tolerance for flooding. The use of stem nodulating *S. rostrata* and root nodulating *S. aculeata* as green manure have not only increased the yield of rice but also improved the soil microbial population and soil enzymes (Kannaiyan and Kalidurai, 1989; Kalidurai and Kannaiyan, 1991 and Kannaiyan, 1993b).

Neem cake and N use efficiency in INM

Nitrification, a strictly aerobic microbial process, occurs in the oxidised surface layer of a flooded soil. Conversion of ammonia to nitrite by *Nitrosomonas* and to nitrate by *Nitrobacter* is nitrification. However, it is difficult to study the nitrification *in situ* in a flooded soil system because as soon an NO_3^- is formed it diffuses down to the reduced layer and is lost from the system by denitrification or reduced to NH_4^+ by dissimilatory NO_3 reduction. Reddy *et al.* (1980) reported that the net nitrification rate in the oxidised surface layer of flooded soil was 2.07 mg NO_3^- N kg^{-1} soil day^{-1}. Among N fertilizer losses, nitrification and denitrification have traditionally been considered the main loss mechanisms. Simpson *et al.* (1984) indicated that 35% of the applied N was lost through denitrification of nitrite and nitrate after nitrification of ammonium near the soil surface. The chemicals that retard the activity of the nitrifying bacteria, *Nitrosomonas* and *Nitrobacter* are called nitrification inhibitors and for effectiveness they must possess the following criteria:

- it should block the conversion of ammonia to nitrite
- it should move with fertilizer or nutrient solution
- it should be stable under oxidised and reduced conditions for longer period
- it should be adaptable to effective and economic use in conjunction with the wide variety of fertilizers
- these materials or compounds should immediately mix with the fertilizer (Hauck, 1972).

Nitrification inhibitors

- synthetic inhibitors or chemical inhibitors such as N serve (2-chloro-6-Trichloromethy-1-pyridine), dicyandiamide **(DCD)**, thiourea *etc.*, (Prasad *et al.*, 1971) and

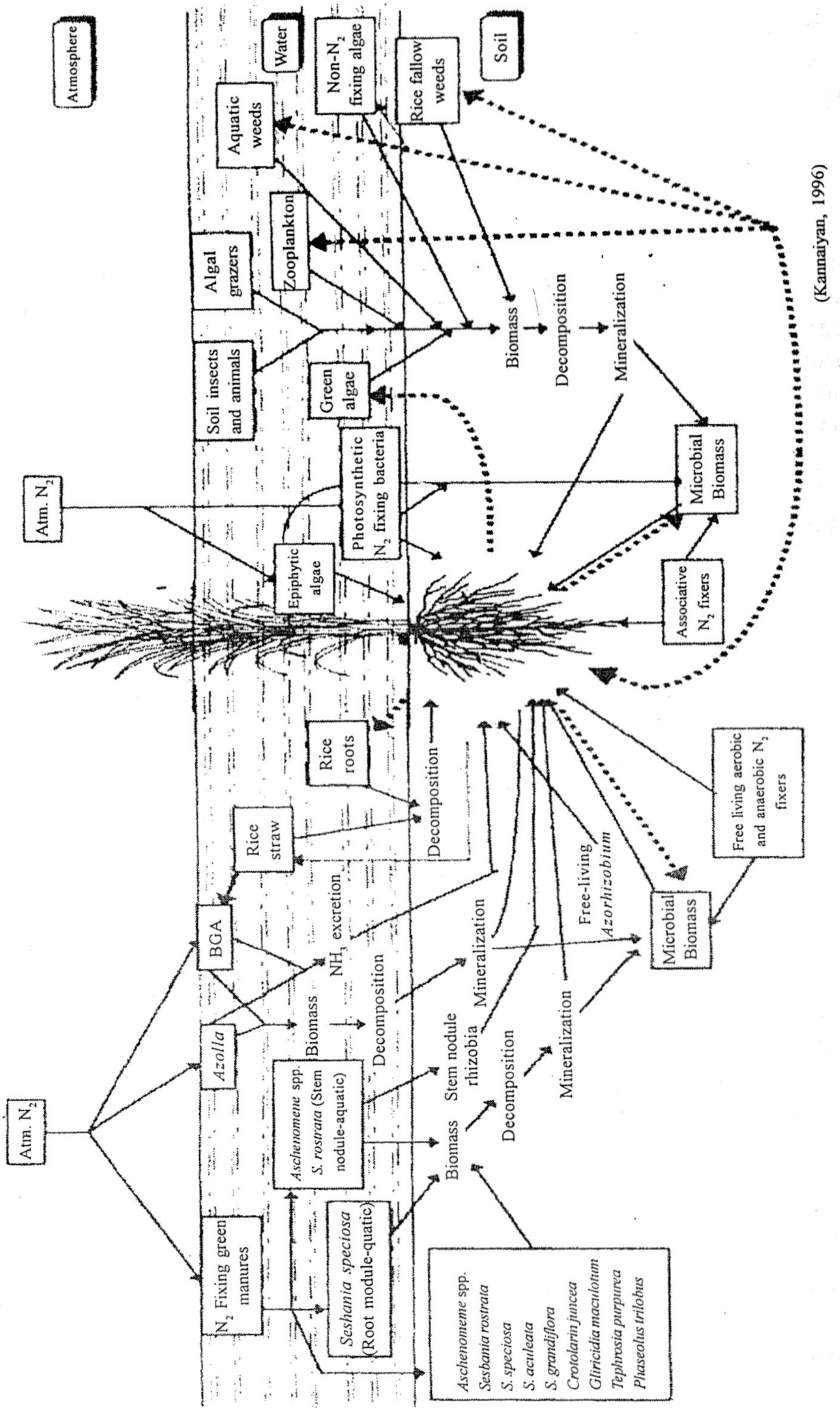

Fig. 1: Natural bioresources input for rice crop nutrition

- indigenous or non-synthetic inhibitors such as neem seed cake *(Azadirachta indica* L.), Mahuva cake (*Bassia latifolia* Roxb) *etc.* (Sahrawat *et al.*, 1974).

Nitrification inhibitor dicyandiamide **(DCD)** blocks transformation of **NH_4^+-N** in mineral or organic **N** fertilizers for a period of 2-3 months, depending on the site conditions. It is decomposed in the soil without leaving any residues. Factors that can influence degradation of **DCD** in soil such as soil moisture and temperature are of primary interest in estimating the length of the nitrification period.

Neem cake is widely used as organic fertilizers for several crops. But it seems to have good nitrification inhibition capacity, due to the presence of alkaloids solanin, meliantrool, azadiractin, nimbin and nimbidin which possess a furan ring system responsible for inhibition. When neem cake is mixed with urea, the rate of inhibition increased with increasing proportion of neem cake. The nitrification inhibition period of neem cake is around 10-25 days. It retains ammonium **N** for longer period and higher **NH_4^+-N** contents are noticed in the soil with neem cake coated urea application. Neem cake at 750 ppm stimulated the growth and nitrogenase activity of *Azolla pinnata*. It was also found that neem cake not only encourages the multiplication and nitrogenase activity of *Azolla* but also helps to combat pests and diseases of *Azolla* (Nandabalan and Kannaiyan, 1998).

The application of nitrification inhibitors along with urea in low land rice could inhibit nitrification and suppress the transformation of ammoniacal **N** and thereby accumulating more ammoniacal **N** for effective utilization by rice crop (Manickam *et al.*, 1976; De Datta, 1986). The increase in rice yields due to the application of **DCD** mixed urea and neem cake mixed urea was reported (Sharma and Sinha, 1977; Simpson *et al.*, 1985). In the experiments conducted at Tamil Nadu Agricultural University, Coimbatore, the grain and straw yields of rice were enhanced significantly due to the application of neem cake and **DCD**. Of the two nitrification inhibitors, neem cake recorded maximum total **N** content in the plant, grain and straw yield when added during the incorporation of **N_2** fixing green manure plants, *viz., Sesbania rostrata* and *Azolla microphylla* compared to **DCD**. Among different treatments with different combinations of green manures, fertilizer **N** and nitrification inhibitors, *S. rostrata* incorporation (20 t ha^{-1}) with fertilizer **N** (75 kg ha^{-1}) and neem cake (250 kg ha^{-1}) registered the highest nitrogen accumulation, grain and straw yield in **IR** 64 rice. It has been apparently shown that both the nitrification inhibitors neem cake and **DCD** have influenced the grain yield of rice considerably. The beneficial effect on **IR** 64 rice, as a result of the addition of nitrification inhibitors during the incorporation of *A. microphylla* and *S. rostrata* was attributed to the significant biological and chemical changes brought by the inhibitors in rice soil which eventually influenced the availability of nitrogen to rice crop. Similar effects of nitrification inhibitors in increasing the yield of lowland rice and reducing the **N** loss have been well documented (Manickam *et al.*, 1976; Kamire and Sohar, 1979; De Datta *et al.*, 1983; Kumar *et al.*, 1992, 1995 and Arumugam, 1996). Kannaiyan (1996) more clearly elucidated the inhibitory action of neem cake on the **N** loss mechanism in rice soil eventually increasing the **N** use efficiency of *Azolla* biofertilizer and inorganic fertilizer **N** by rice crop Fig. 2.

The positive influence of neem cake on the nitrogen use efficiency of *Azolla* biofertilizer and also the growth and yield of rice has been confirmed by the consistent results obtained

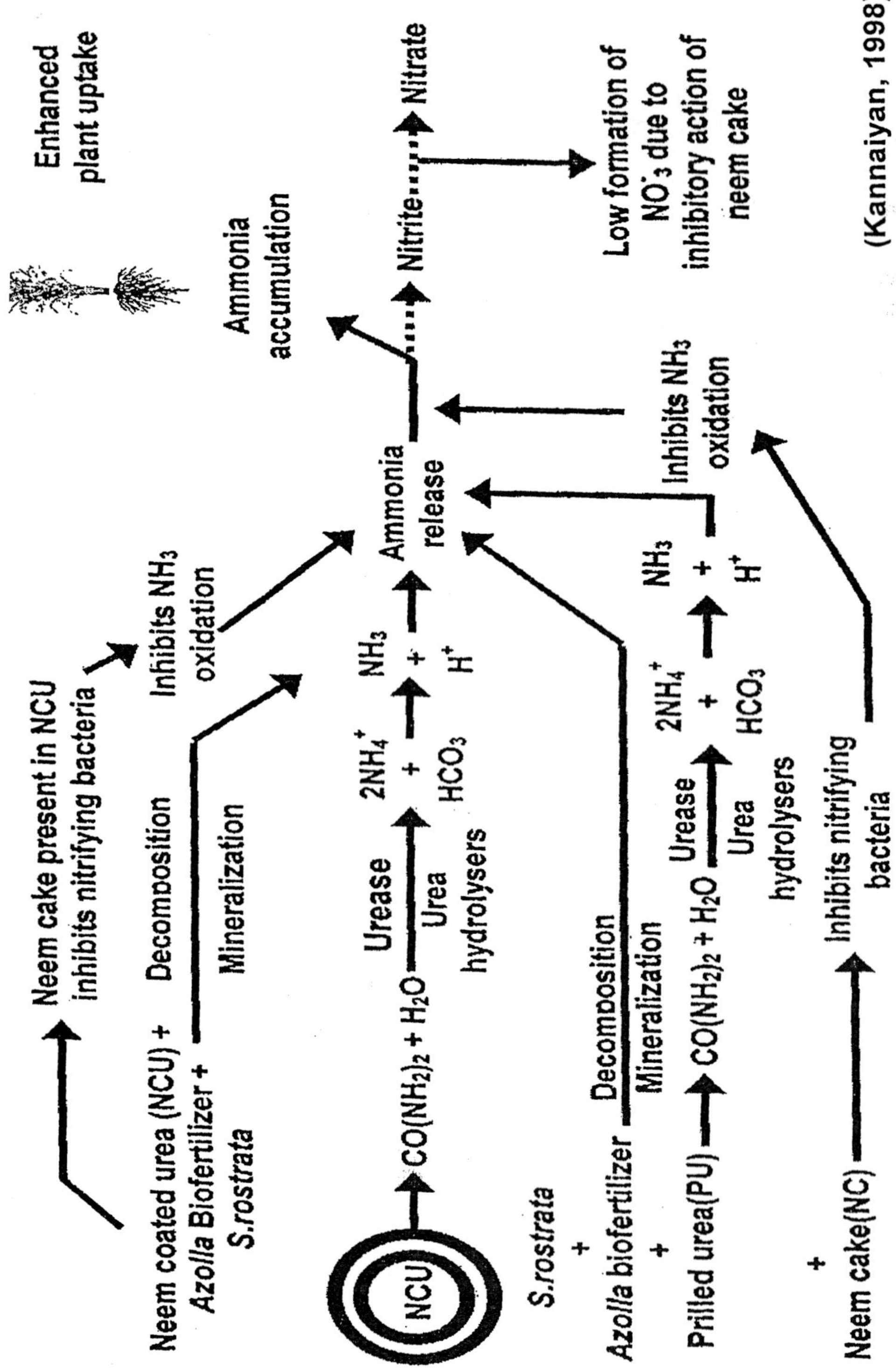

Fig. 2: Role of Neem cake on efficient utilization of Nitrogen in Rice Crop

in the repeated experiments (Tables 1 & 2). Hence the following technology package has been developed for maximizing the nitrogen use efficiency of *Azolla* biofertilizer by rice crop (Kannaiyan and Kumar, 1998).

Table 1: Effect of incorporation of *A. microphylla* as green manure (GM)/Dual cropping (DC) and fertilizer N with neem cake (NC) on total N content and uptake of rice (CO 43).

Treatments	Total N content (%)	N uptake (Kg ha^{-1})	Percent increase over control
Control	0.45	64.6	–
A. microphylla (GM) alone	0.52	85.4	32.2
A. microphylla (GM) + NC	0.54	100.9	56.2
A. microphylla (GM) + PU	0.60	131.9	104.2
A. microphylla (GM) + NC + PU	0.60	138.7	114.2
A. microphylla (DC)	5.51	82.6	27.9
A. microphylla (DC) + NC	0.54	96.1	48.8
A. microphylla (DC) + PU	0.57	112.2	73.7
A. microphylla (DC) + NC + PU	0.59	123.0	90.4
Prilled Urea (PU) alone	0.51	82.1	27.1
SED	0.01	1.779	
CD	0.02	3.658	

PU @ 60 kg N ha^{-1}
NC @ 250 kg ha^{-1}
A.microphylla (GM) 15t ha^{-1}
A.microphylla (DC) 500 kg ha^{-1}

Table 2: Effect of incorporation of *A. microphylla* as green manure (GM)/Dual cropping (DC) and fertilizer N with neem cake (NC) on the yield of rice (CO 43)

Treatments	Grain yield (t ha^{-1})	Percent increase over control	Straw yield (t ha^{-1})
Control	4.72	–	9.64
A. microphylla (GM) alone	5.71	21.0	10.71
A. microphylla (GM) + NC	6.36	34.7	12.32
A. microphylla (GM) + PU	7.41	57.0	14.58
A. microphylla (GM) + NC + PU	7.65	62.1	15.46
A. microphylla (DC)	5.60	18.6	10.60
A. microphylla (DC) + NC	6.10	29.2	11.70
A. microphylla (DC) + PU	6.68	41.5	13.00
A. microphylla (DC) + NC + PU	7.04	49.2	13.80
Prilled Urea (PU) alone	5.57	18.0	10.52
SED	0.106	–	0.239
CD	0.218	–	0.493

PU @ 60 kg N ha^{-1}
NC @ 250 kg ha^{-1}
A. microphylla (GM) 15t ha^{-1}
A. microphylla (DC) 500 kg ha^{-1}

Apply *Azolla* biofertilizer as dual crop at 500 kgs ha^{-1} one week after transplanting

↓

After taking up first weeding 25-30 DAT, drain water from the field

↓

Apply neem cake at 250 kgs ha^{-1} uniformly all over the field

↓

Incorporate the *Azolla* biomass along with neem cake either manually or by the use of Japanese Intercultivator

↓

If a second incorporation is done at the time of second weeding 40-50 **DAT**, repeat the procedure as detailed above.

Economics of the package

The **MRR** (Marginal Rate of Return) for the above package was calculated, which indicated that the maximum return was obtained in the *Azolla* inoculated plots with 60 kg **N** ha^{-1} as prilled urea and neem cake application at 250 kgs ha^{-1} (*i.e.* an income of Rs. 7.90 per one rupee additional investment) (Ramasamy *et al.*, 1995). The **MRR** in the treatment without neem cake application was Rs. 4.98 (Table 3).

Table 3: Economic analysis of the *Azolla* and neem cake package

Treatment	Grain Yield		Straw Yield		Additional Cost Over 100 Kg N ha^{-1}	Additional benefit 100 Kg ha^{-1}	MRR
	(t ha^{-1})	Increase over 100 Kg N ha^{-1}	(t ha^{-1})	Increase over 100 Kg N ha^{-1}			
Azolla (DC) (500Kg ha^{-1}) + PU 60kg N ha^{-1}	6.24	0.17	8.36	0.41	140	697.50	4.98
Azolla (DC) (500kg ha^{-1}) + Neem cake 250kg	6.17	0.12	8.51	0.53	100	552.50	5.53
Azolla (DC) (500kg ha^{-1}) + PU 60 kg N/ha + Neem cake 250kg N/ha	7.93	1.86	10.01	2.03	890	7033.80	7.90
100kg N/ha	6.07	-	7.98	-	-	-	-

Urea super granules (USG) and N use efficiency

Deep placement of fertilizer in reduced soil has been considered the most efficient method to increase fertilizer **N** efficiency in lowland rice and is becoming increasingly relevant since the introduction of modified urea materials like Urea Super Granules **(USG)**. The large size urea granule commonly referred to as urea super granule, can be placed in the soil by hand or with mechanical operator. The advantage of deep placement of urea super granules is that various losses like ammonia, volatilization, denitrification, run off and leaching can be minimized. An additional advantage is that this mode of fertilization does not seem to interfere with dinitrogen fixation in flood water by *Azolla* and blue green algae (Roger *et al.*, 1980).

Several years of agronomic data from the International Network on Soil Fertility and Fertilizer Evaluation for Rice **(INSFFER)** have shown that deep placement is generally more efficient than the traditional split application of urea thereby showing (Craswell and De Datta, 1980; De Datta *et al.*, 1983). Sharma (1995) has documented the positive influence of **USG** on rice soil based on the field experiments conducted at different locations.

Table 4: Effect of incorporation of *S. rostrata* and *A. microphylla* with two forms of urea on the grain and straw yield of CO 43 rice (Rabi).

Treatments	Grain yield (t ha^{-1})	Per cent increase over control	Straw yield (t ha^{-1})	Per cent increase over control
Prilled Urea (PU)**	6.45	40.2	13.47	50.5
Tablet Urea (TU)**	8.93	94.1	18.27	104.1
PU**	5.74	24.8	10.86	21.3
TU**	6.34	37.8	12.92	44.4
S. rostrata alone	6.30	37.0	12.27	37.1
S. rostrata + PU**	7.34	59.6	15.31	71.1'
S. rostrata + TU**	8.32	80.9	17.10	91.1
A. microphylla alone	5.83	26.7	11.30	26.3
A. microphylla + PU*	6.92	50.4	14.50	62.0
A. microphylla + TU*	7.86	.70.9	16.25	81.6
Control	4.60	-	8.95	
SED	0.199		0.371	
CD	0.416		0.774	

*-60 kg N ha^{-1} **-100 kg N ha^{-1}

A.microphylla at 15 t ha^{-1} *S. rostrata* 15 t ha^{-1}

When the nitrogen fixing green manures like *Azolla* and *Sesbania rostrata*, biofertilizer like blue green algae are used, the application of neem cake coated urea and deep placement of **USG** proved better (Kannaiyan, 1993b). The application of different forms of urea along with *Azolla* inoculation has also been evaluated in the field experiments conducted during early and late wet seasons of 1984-85 at Tamil Nadu Agricultural University (Latha *et al.*, 1988). Significant increase in grain yield was obtained over 75 kg **N** ha^{-1} as prilled urea with *Azolla* as dual crop, due to the application of 75 kg **N** ha^{-1} as urea super granule or neem coated urea along with *Azolla* inoculation as dual crop in **IR** 50 rice during early wet season (Thangaraju and Kannaiyan, 1993). The increased yield obtained due to the application of urea super granule and neem coated urea with *Azolla* as dual crop may be attributed to the continuous availability of nitrogen for rice crop. From our results, it is evident that the nitrogen fixing green manure plants *viz., Azolla* and *Sesbania rostrata*, biofertilizer like blue green algae could be effectively utilized by incorporation with different forms of urea *viz.*, neem coated urea, urea super granule *etc.*, to increase the rice production under low land system (Kumar *et al.*, 1995 and Gopalaswamy and Kannaiyan 1997). Incorporation of green manures along with deep placement of **USG** or Tablet Urea **(TU)** has significantly increased the growth and yield of rice than Prilled Urea **(PU)** application (Table 4). Incorporation of *Azolla* hybrids **AH-C2** coupled with **USG** at 75 kg **N** ha^{-1} increased the grain and straw yields of **CO** 43 rice besides increasing the plant **N** content and **N** uptake in sodic soil (Gopalaswamy and Kannaiyan, 2000).

Influence of N_2 fixing green manures on soil enzymes activity

Soil enzymes or soil biocatalysts not only play a major role in the soil chemical and biochemical environment but also affect the rate at which nutrients become available to crop plants as well as to other forms of life in soil. Soil enzyme activity is considered as an index of microbial activity as well as soil fertility (Burns, 1982). Soil enzymes such as urease and dehydrogenase activity were positively correlated with increasing organic carbon, total nitrogen and total phosphorus content (Barush and Mishra, 1984). Bolton *et al.* (1985) reported an increased activity of certain soil enzymes due to the application of green manures. Frankenberger and Tabatabai (1991) observed significant correlation of L-asparaginase activity with organic carbon and total nitrogen.

The incorporation of *Azolla* has resulted in the increased soil urease activity (Subramani and Kannaiyan, 1989; Thangaraju and Kannaiyan, 1989 and Kannaiyan and Kalidurai, 1995). Thanikachalam *et al.* (1984) have reported that the addition of neem cake, carbofuran, rice straw and *Azolla* increased the phosphatase activity in flooded soils. Significant increase in the activity of soil enzymes *viz.*, dehydrogenase, urease, L-asparaginase, phosphatase, cellulase and amylase due to incorporation of N_2 fixing green manures such as *S. rostrata*, *S. speciosa* and *A. microphylla* has been noticed (Kumar and Kannaiyan, 1992; Sundaravarathan and Kannaiyan, 1998).

Constraints and research needs of the green manure technology

Though the beneficial effects of green manures are well known in terms of their effect on soil fertility and productivity of crops, this practice is not widely adopted by farmers because of the following constraints (Palaniappan and Siddheswaran, 1999).

- In a high intensive cropping system, the situation may not induce the farmers to set apart 6 or 7 weeks exclusively for green manure cultivation without any revenue. The fallow season (summer) alone will be left without any crop and during this period the farmers could not think of growing any green manures for want of sufficient irrigation facilities.
- Inadequate availability of quality seeds of preferred green manure types at reduced cost is one of the major problems in the adoption of green manure cultivation.

- Benefits of green manure addition could not be seen very apparently as that of the mineral fertilizer – **N** because; in the farmer's point of view, organic and inorganic source are complementary or synergistic.
- The benefits expected from green manuring may not always be uniform in all the crops, because of the variation in the soils, green manure species, environmental conditions, *etc.*
- The possibility of using green manures as intercrops has got its own limitations since it is believed that these intercrops may compete for the nutrients with the main crops

The following are the agronomic constraints often met by the farmers which need attention of the scientists for wider adoption of the green manure technology:

- Screening and evaluation of green manure plants for higher biomass production and **N** accumulation under field situations.
- The possibility of including non-conventional green manures in the place of already established legume green manures.
- Identifying green manures for the optimum fast **N** supply on mineralisation in matching with crop need so as to minimise the loss of the released **N** through various mechanisms.
- Seed production technology for important green manure crops including plant density and nutrient requirement, appropriate season for green manure seed crop cultivation and seed storage techniques to maintain good germinability and vigour.
- Simple cost effective seed treatment techniques to overcome dormancy and hard seedness.
- Long term economic advantages such as sustainability factors and nutrient contribution in addition to nitrogen for which hitherto green manures were recognized.
- Identification of multipurpose trees and short season legumes, with pests and disease tolerance that can be grown on bunds or wastelands. Legumes that provide both forage and leaf green manure of fuel wood would be more appropriate.
- Research on legume species with efficient bioinoculants for symbiotic **N** fixation even in adverse soil environments.
- Evolving a suitable cropping system for fitting green manures without sacrificing any of the economic crops in the system.

Integration of organics and inorganics for sustainable crop production

Use of mineral fertilizer is the quickest and surest way of increasing the crop production. Major changes in the pricing policies, after the 90's disrupted the balanced fertilizer application. Nitrogen fertilizer dominated for sometimes and thereby affecting the balanced use of major nutrients for crops. Alternatively, therefore, search has been made to exploit the potential alternate sources of plant nutrients harmoniously, by integrating the use of minerals, fertilizers, organics, biological fertilizers and recyclable urban wastes, rural wastes, industrial wastes and farm wastes to increase crop production and at the same time sustain the soil fertility. The basic concept is the supply of the required plant nutrients, for sustaining the desired crop productivity. The (Integrated Plant Nutrient Supply) **IPNS** system, primarily involves combination of mineral fertilizers, organic manures, crop residues, compost or **N**-fixing microbial systems in relation to the system of land use, ecological, soil and economic conditions. The approach focuses mainly, at low to medium external input taking into account, the soil fertility level, crop requirements and management for a targeted yield per unit area (Kannaiyan *et al.*, 1998).

IPNS approach

The present day management practices particularly for **IPNS** are based on the understanding like:

- Organic or biological sources of plant nutrients complement mineral fertilizers in fulfilling the nutrient requirements of crops. The magnitudes of the contributions vary, very widely to sources, combination of sources and agro-ecological conditions.
- In many instances of **IPNS**, where combined application of mineral and organic sources were involved, it resulted in synergistic effects, by increasing significantly the fertilizer use efficiency.
- The humus fraction, which constitutes the residual organic sources substantially, improves the soil physical conditions. Better soil structure, could provide better environment for root development and aeration. Soil aggregation, is quite often improved, which is attributed to the actions of several compounds, particularly the polysaccharides and fulvic acid components of the organic sources. It not only improves the water holding capacity of the soil, but also prevents soil erosions, *etc.* Application of organic resources plays a substantial role, in reducing the plant parasitic nematodes and the soil-borne phytopathogens, besides supplying the nutrients to crop plants. The application of organic resources, also trigger the microbial population dynamics, which directly reflects on the transformation of nutrients in compounds. Long term fertilizer experiments conducted in India and China during the past one century showed that integrated use of organic manures and fertilizers were in addition to maintaining higher productivity, provide stability in crop production. The effect has been more pronounced in acid soils than in normal soils.
- Under high input production system, the productivity curve shows a plateau and it cannot be further increased with any further incremental mineral fertilizer alone. But, addition of organic sources could again increase yields due to increased soil productivity, and fertilizer use efficiency. Again, results of long term experiments, conducted in several agro-ecological location, in our country revealed that responses was higher for farm yard manure along with **NPK** for rice, wheat, maize, soyabean and several other crops.
- The cost of cultivation is held within the reach of the average farmer under **IPNS**, since the crop requirement is partly met from cheaper organic sources like farm yard manure, green manures, legumes and biological fertilizer. The potential N_2 fixing biological systems like *Rhizobium* –legume symbiosis, for increasing the grain legumes production, *Azolla* – *Anabaena* symbiosis, for increasing rice production, *Azospirillum* – Associative symbiosis in cereal crops, *Acetobacter diazotrophicus* in strong

association in sugarcane roots are potential biological agents contributing substantial amounts of nitrogen to the crops. Besides the nitrogen fixing biological systems, phosphorus solubilizing bacteria such as *Bacillus* and *Pseudomonas* and the Vesicular Arbscular Mycorrhizae **(VAM)** plays a triggering role in phosphorus mobilization to crop plants.

IPNS models and packages

Based on the experiences gained in the **IPNS, FAO** has taken the pioneering initiative with the following conceptual programmes like identifying the main organic and biological sources which vary from country to country. For example, in Bangladesh, Pakistan, Sudan and Zambia animal wastes and the biological nitrogen fixing *Rhizobium* are the major natural resources. In Srilanka and India, recycling of rice straw and legume crop residues are the major sources while in Guinea and Zaire it is the crop residues. **FAO** has undertaken development of **IPNS** model systems and field trial networks to generate information for technology transfer programmes in Asia.

It is high time that **IPNS** packages should be developed more in the form of alternative options, and it should be within the overall limits of recommended nutrient applications rates. It also leads sometimes, to a negative trend by loading all available components in a haphazard manner.

Biomass production in fallows or cultivation of leguminous green manure crop like *Sesbania* and Kolingi (*Tephrosia purpurea*) on the field bunds and borders can be attempted, to put it in more appropriate areas where limited farm lands are available.

- Computer models for an **IPNS** calculations programmes of plant nutrients should be developed based on general agronomic assumption.
- Site specific packages based on farming systems, farm size, availability of inputs and socio-economic factors should be developed.
- **IPNS** should be developed for cropping system, rather than for individual crops.
- **IPNS** practices for major agro-ecological zones and for various categories of farms should be developed.
- Forests and pastures are to be encouraged, in non-cultivable areas, for transport of biomass to the nearby farm lands.

Soil health management through IPNS

The basic concept of integrated plant nutrient supply is the promotion and maintenance of soil fertility for sustaining crop productivity through optimizing all possible sources like organic, inorganic and biological in an integrated manner appropriate to each farming situation in its ecological, soil and economic possibilities. The principal aim of this **IPNS** is efficient and judicious use of all the major sources of plant nutrients in an integrated manner, so as to get maximum economic yield without any deleterious effect on physico-chemical and biological properties of the soil. The **IPNS** system mainly concentrates on the following objectives.

- To achieve potential yield of crops
- To optimize and to sustain the resource base

- To minimize the cost of cultivation
- To demonstrate the need for soil health management
- To inculcate the principle of integrated nutrient management

The major components of integrated plant nutrient supply system are fertilizers, **FYM/** compost, green manures, crop residues/recyclable wastes and biofertilizers. These components possess great diversity in terms of chemical and physical properties, nutrient release efficiencies, positional availability, crop specificity and farmer's acceptability. Therefore, the combination of different components to ensure optimum nutrient supply to a prediction system may depend on land use, ecological, social and economic conditions.

Besides these components, **IPNS** considers some factors for increasing the available nutrients in the soil. These factors are,

- Management of physical condition of soil,
- Reclamation of problem soils,
- Water management, *etc.*

Soil health management is fundamental to sustainable intensification of agriculture. The **IPNS** approach affords the opportunity to include stem-nodulating legumes such as *Sesbania rostrata* in the farming system and to incorporate *Azolla* or blue-green algae and other sources of symbiotic and non symbiotic nitrogen fixation. The nitrogen fixing biological systems such as blue green algae, *Azolla*, *Azospirillum*, *Acetobacter*, and *Azorhizobium* are compatible to different cropping systems and could fix considerable amounts of nitrogen. The biomass of these biological systems decomposes rapidly in soil and supply the nitrogen to various crops grown in the field. Besides nitrogen addition, these biofertilizers also contribute significant amounts of phosphorus, potassium, sulfur, zinc, iron, molybdenum and other micronutrients. The biological fertilizers were known to influence the total soil microbial population and soil enzymes activity which directly reflect on soil fertility index. The use of biological fertilizers would reduce the cost of chemical fertilizers involved in crop production. The effective utilization of these biological fertilizers for different cropping systems will not only provide economic benefits to rice farmers, but also improve and maintain the soil fertility and sustainability of the soil ecosystem.

Conclusion

The management of native soil **N**, which is the principal **N** source for rice and biological N_2 fixation, may play a role in sustaining the soil **N** under intensive rice cropping. Biological nitrogen fixation is an important process in rice farming systems because it is an inexpensive source of nitrogen for increasing the productivity of crops. The nitrogen fixing bio systems such as *Azolla,* cyanobacteria and legume green manures were able to adopt well under wetland rice field ecological conditions and fixes considerable amounts of nitrogen. The biomass of these biosystems decomposes rapidly in rice soil and supply nitrogen to rice crop. Besides nitrogen addition, these biofertilizers or biomanures also contribute significant amount of phosphorus, potassium, sulphur, zinc, iron, molybdenum and other micronutrients. The organic acids released during the mineralization process of the biomass of biomanures would accelerate the phosphorus availability in rice soil. These biological fertilizers or biomanures were known to influence the total soil microbial population such as bacteria, fungi, actinomycetes and also substantially activates the population dynamics of nitrogen fixing free-living bacteria like *Azotobacter, Azospirillum* and also the symbiotic bacterium such as *Rhizobium*. It is evidently clear now that the application of biological fertilizers were greatly involved in the accumulation of soil enzymes which directly reflects on soil fertility index. The use of biofertilizers would reduce the cost of chemical fertilizers involved in crop production. The effective utilization of biofertilizers for crops will not only provide economic benefits to the farmers but also improve and maintain the soil fertility and sustainability in natural soil ecosystem. Since these biosources represent a great diversity in chemical, physical and biological characteristics, their efficient use depends upon the particular agro-ecological environment and local availability. Therefore, it is very much essential to develop a strong, workable and compatiable package of nutrient management through organic sources for various crops based on scientific facts, local conditions and economic viability.

References

Anonymous. 2000. RTDP review. INMnet Bullettin No. 8. IRRI.

Amutha. K. and S. Kannaiyan. 1998b. Effect of **NAA** and inoculation of *A. caulinodans* and induction of paranodules and nitrogen fixation in cereal crops. Paper presented in short term workshop on "Association of *A. caulinodans* in rice roots and its role in nitrogen fixation", sponsored by British Council. Chennai. India. 2 & 3rd April. 1988.

Aruna Geetha, S. and Thiyagarajan, T.M. 1998. Rice grain type and nitrogen management. Paper presented in the National Seminar on Developments in Soil Science held at Hisar. Haryana, India. p. 159.

Bagyaraj, D.J. 1984. Biological interactions with VA mycorrhizal fungi. In: VA Mycorrhiza. (Eds.) C.L. Powell and D.J. Bagyaraj. CRC Press. USA. pp. 131-152.

Baier, W. 1990. Organic Sources. Proc. Inter. Symp. Sustainable Agriculture. New Delhi, India.

Balasubramanian, V. 1999. Managing soil quality for sustainable rice production: How does **CREMNET** help? In: **Vistas of Rice** (ed.) M. Subramanian, Tamil Nadu Rice Research Institute, Aduthurai, Tamil Nadu, India. pp. 11-21.

Barush, M. and Mishra, R.R. 1984. Dehydrogenase and urease activities in rice field soils. *Soil Biol. Biochem.*. **16**: 423-424.

Bolton, H. Jr.. L.F. Elliot, R.I. Papendick and D.F. Bezdicek. 1985. Soil microbial biomass and selected soil enzyme activites. Effect of fertilization and cropping practices. *Soil Biol. Biochem*.. **17**: 297-302.

Broadbent. F.E. 1978. Transformation of soil nitrogen. In: *Nitrogen and Rice. IRRI. Los Banos, Philippines*, pp. 543-559.

Bradey, N.C. and Weil. R.K. 1998. The Nature and Properties of Soils. 12th edition. McMillan Publishing Company, New York.

Burns, R.C. 1982. Enzyme activity in soil. Location and possible role in microbial ecology. *Soil Biol. Biochem*.. **14**: 107-108.

Chandrasekar. M.V. and Kannaiyan. S. 1995. Induction of nodular structures by *Azorhizobium caulinodans. Tamil Nadu Agric. Univ. Newslett.*. **24**(12): 3.

Craswell, E.T. and De Datta. S.K. 1980. Recent developments in research on nitrogen fertilizers for rice. IRRI Research paper Series 49, pp. 11.

De Datta, S.K.. Fillery. I.R.P. and Craswell. E.I. 1983. Results from recent studies on nitrogen fertilizer effiency in wetland rice. Outlook on Agriculture. **12**: 125-134.

De Datta. S.K. 1986. Improving nitrogen fertilizer efficiency in lowland rice in tropical Asia. *Fert. Res.*, **9**: 171-186.

De Wit, C.T. 1990. Understanding and managing changes in agriculture. In: System Theory Applied to Agriculture and the Food Chain (ed.) J.G.W. Jones. and P.R. Street. Elsevier. New York. pp. 235-249.

Dobermann, A. and P.F. White. 1999. Strategies for nutrient management in irrigated and lowland rice systems. In: Resource Management in Rice Systems: Nutrients. (ed.) V. Balasubramanian. J.K. Ladha and G.L. Denning. Kluwer Academic Publishers. the Netherlands. pp. 1-27.

FAO. 1989. Sustainable agriculture production: For International Agricultural Research **(CGIAR)**. Pergamon. New York. USA.

Frankenberger, W.T. Jr. and M.A. Tabatabai. 1991. L. asparaginase activity in soils. *Biol. Fertil. Soils*, **11:** 6-12.

Gaur, A.C. 1990. Phosphate solubilising microorganisms as biofertilizers. Orega Scientific Publishers, New Delhi.

Gopalaswamy, G. and Vidhyasekaran, P. 1989. Response of different rice cultivars to *Azospirillum inoculatin Curr. Sci.*, **58**(13): 752-754.

Gopalaswamy, G. and S. Kannaiyan. 1997. Performance of *Azolla* hybrids in lowland rice culture. Inter. Rice Res. Newsletter. **23 (3):** 28-29.

Gopalaswamy, G. and S. Kannaiyan. 2000. Bioremediation of sodic soil using *Azolla* hybrids. Paper presented in the National Symposium on Microbes in Bioremediation for an Eco-friendly Environment in the New Millennium 6th -7th January. 2000. University of Madras. Chennai (Abstr.) p. 13.

Goswami, N.N. and Rattan, R.K. 2000. Ecofriendly and efficient integrated nutrient management in sustainable agriculture. In: Souvenir on International conference on managing natural resources for sustainable agricultural production in the 21st Century. Indian Society of Soil Science, New Delhi. pp. 226-258.

Hauck, R.D 1972. Synthetic release fertilizer and fertilizer amendments. In: Organic chemicals in the soil environment (eds.). Goring C.A.I. and Hamaker U.W. Marcel Dekker. Inc., New York. Vol. 2: 633-690.

Jotwani, M.G. and Srivastava, K.P. 1981. Neem insecticide of the future: Chemistry, toxicology and future strategy. *Pesticide*. **15**(2): 12-14.

Kalidurai, M. and Kannaiyan, S. 1989. Effect of *Sesbania rostrata* on nitrogen uptake and yield of low land rice. *J. Agron. Crop Sci.*, **163**: 284-289.

Kalidurai, M. and Kannaiyan, S. 1991. *Sesbania* as a biofertilizer for rice. *Bioresource Technol.*, **36**: 141-145.

Kamire, K.K. and Sonar, K.R. 1979. Mineralization of urea-N with 'Neem' (*Azadirachta indica*) and 'Karanj' (*Pongamia glabrarent*) cakes. *J. Maharashtra Agric. Univ.*, **4**: 10-13.

Kannaiyan, S. 1989a. *Azolla* biofertilizer for rice. In: Proc. Natl. Seminar on Biotech. Mol. Biol., C.M.S. College, Kottayam, Kerala, India. p. 15.

Kannaiyan, S. 1992. *Azolla* biofertilizer technology for rice. Tech. Bull., Tamil Nadu Agric. Univ., Coimbatore, p. 56.

Kannaiyan, S. 1993a. Nitrogen fixing green manures. INSURF Progress Report. Tamil Nadu Agri. Univ., Tamil Nadu. pp. 1-56.

Kannaiyan, S. 1993b. Biofertilizers for rice. Tamil Nadu Agric. Univ., Coimbatore, Tamil Nadu, India. p. 60.

Kannaiyan, S. 1995. Biofertilizers for rice status and scope. In: Rice Management Biotechnology (ed.) S. Kannaiyan, Associated Publishing Co., New Delhi, pp. 237-265.

Kannaiyan, S. 1996. Immobilization of cyanobacteria in solid matrix for ammonia production and development of immobilized substrate based cyanobacterial inoculants for rice crop. Paper presented in X Southern Regional Conference on Microbiol Inoculants. A.V.V.M. Sri Pushpam College, Poondi, India. p. 1.

Kannaiyan, S. 1998. Integrated Plant Nutrient Supply System. In: Integrated Plant Nutrient Supply System for Sustainable Productivity. Indian Institute of Soil Science. Bhopal, M.P. India.

Kannaiyan, S. and M. Kalidurai. 1989. Nitrogen fixing potential of stem nodulating *Sesbania rostrata*. In: *In vitro* approaches to Plant Breeding. Proc. Natl. Seminar on Molecular Biol. and Biotechnol (ed.) Joseph Varghese. Kottayam, Kerala, India. pp. 75-84. (abst).

Kannaiyan, S. and M. Kalidurai. 1995. Nitrogen fixing potential of stem nodulating *Sesbania rostrata* and its effect on rice yield. In: Rice Management Biotechnology (ed.) S. Kannaiyan. Associated Publishing Co., New Delhi, pp. 173-195.

Kannaiyan, S. and K. Kumar. 1998. Use of neem cake for enhancing the nitrogen use efficiency of *Azolla* biofertilizer. Nandini Chemical Journal. July. 1998. 31-34.

Kannaiyan, S., S.J. Aruna, S. Merina Prem Kumari and D.O. Hall. 1997. Immobilized cyanobacteria as a biofertilizer for rice crops. *J. Appl. Phyco.*. **9**: 167-174.

Kannaiyan. S., S. Sundaravarathan and S.P. Palaniappan. 1998. Nitrogen fixing green manures for sustainable rice production. Tamil Nadu Agricultural University. Coimbatore. p. 289.

Krishna. K.R. and D.J. Bagyaraj. 1984. Growth and nutrient uptake of peanut inoculated with mycorrhizal fungus *Glomus fasciculatum* compared to non inoculated ones. *Plant Soil*. **77**: 405-408.

Kumar. K. and S. Kannaiyan. 1992. Changes in the activity of soil enzymes during decomposition of N fixing green manures. Paper presented in the 33rd Annual Conference of the Association of Microbiologists of India. Goa University. Goa. (Abstr.) p. 57.

Kumar, K. and A. Balasubramanian. 1986. Field response of rice to Azospirillum biofertilizer *Curr. Res.*. **15**: 74-76.

Kumar. K., G. Gopalaswamy and S. Kannaiyan. 1992. Effect of incorporation of nitrogen fixing plants *Azolla microphylla* and *Sesbania rostrata* with nitrification inhibitors neem cake and dicyandiamide. In: Biological Nitrogen Fixation and Biogas Technology (eds.) S. Kannaiyan. K. Ramasamy. K. Ilamurugu and K. Kumar, Tamil Nadu Agric. Univ., Coimbatore. Tamil Nadu. India. pp. 49-53.

Kumar, K., M. Thangaraju and S. Kannaiyan. 1995. Use of nitrification inhibitors and nitrogen use efficiency in low land rice soil. In: Rice Management Biotechnology (ed.) S. Kannaiyan. Associated Publishing Co., New Delhi, pp. 101-100.

Latha. K.R., S. Kannaiyan and S. Subramanian. 1988. Uptake of nitrogen in different growth phases of rice through *Azolla* biofertilizer. *Acta Agron. Hungarica*. **37**(3-4): 239-243.

Ladha. J.K., R.P. Pareek and M. Becker. 1992. Stem nodulating legume—*Rhizobium* symbiosis and its agronomic use in lowland rice. *Adv. Soil Sci.*. **20**: 147-192.

Lynam. J.K. and R.W. Herdt. 1988. Sense and sustainability. Sustainability as an objective in international agricultural research. CIP. Rockfeller Conference. "Farmers and Food Systems". Lima. Peru.

Manickam. T.S., A. Selvaseelan and K.K. Krishnamurthy. 1976. Nitrification of urea treated with nitrification inhibitors, coated fertilizers and slow acting fertilizers. *Fert. News*. **21**(9): 64-65.

Mikkelsen, D.S. and S.K. De Dutta and W.N. Obcemae. 1978. Ammonia volatilization losses from flooded rice soils. *Soil Sci. Soc. Am. J.*, **442**: 725-730.

Mosse. B. 1977. Plant growth responses to vesicular arbuscular mycorrhizae. Responses of *Stylosanthes* and maize to inoculation in unsterilized soils. *New Phytol.*. **78**: 277-288.

Palaniappan. S.P. and K. Siddeswaran. 1999. Green manuring for crop production. In: Bioresources Technology for Sustainable Agriculture (Ed.) S. Kannaiyan. Associated Publishing Co.. New Delhi. pp. 41-55.

Powell. C.L. and D.J. Bagyaraj. 1984. VA Mycorrhiza. CRC Press. Florida. USA.

Prasad. R., G.B. Rajale and B.A. Lakhdive. 1971. Nitrification retardants and slow release nitrogen fertilizer. *Adv. Agron.*. **23**: 337-383.

Ramasamy, C., S. Kannaiyan, D. Suresh Kumar and K. Kumar (1995): *Azolla* biofertilizer use and its productivity impact on rice-An economic analysis. In: Rice Management Biotechnology (ed.) S. Kannaiyan. Associated Publishing Co., New Delhi. pp. 287-292.

Raj. J.. D.J. Bagyaraj and A. Manjunath. 1981. Influence of soil inoculation with Vesicular arbuscular mycorrhiza and a phosphate dissolving bacterium no plant growth and ^{32}P uptake. *Soil Biol. Biochem.*. **13**: 105-108.

Rao. V.R., D.N. Nayak. P.B.D. Charyulu and T.K. Adhya. 1983. Yield response of rice root inoculation with *Azospirillum*. *J. Agric. Sci.*. **100**: 689-691.

Reddy, K.R., W.H. Patrick Jr. and R.E. Phillips. 1980. Evaluation of selected processes controlling nitrogen loss in flooded soil. *Soil. Sci. Soc. Am. J.*, **44:** 1241-1246.

Roger, P.A., S.A. Kulasooriya, A.C. Third and E.T. Craswell. 1980. Deep placement of nitrogen fertilizer application, compatible with algal nitrogen fixation in wetland rice soils. *Plant and Soil*, **57:** 137-142.

Roy, R.N. and H. Braun. 1983. Fertilizer and Plant Nutrition Service Land and Water Development division Fao, Rome. Paper presented at the Joint FAO/ECE (Economic Commission for Europe) Fertilizer Symposium held in Geneva, 17-21, January, 1983.

Sahrawat, K.L., B.S. Parmar and K. Mukerjee. 1974. Note on the nitrification inhibitors in the seeds, bark and leaves of Pongam vent. *Indian J. Agric. Sci.*, **44**: 415-418.

Savant, N.K., S.K. De Datta. 1982. Nitrogen transformations in wetland rice soils. *Adv. Agron.*, **35**: 241-302.

Savant, N.K., E.T. Craswell and R.E. Kiamond. 1983. Use of urea supergranules for wetland rice: A Review. *Fert. News*, **28**(8): 27-35.

Sharma, N.N. and U.P. Sinha. 1977. Neem cake blending of urea for raising efficiency of applied nitrogen to wheat. *Inian J. Agron.*, **22**: 106-107.

Sharma, S.K. 1995. Urea Super Granules—Its performance and prospects for rice production. In: Rice Management Biotechnology (ed.) S. Kannaiyan. Associated Publishing Co., New Delhi, pp. 117-134.

Simpson, J.R., J.R. Freney, W.A. Muirhead and R. Leuning. 1984. Effect of phenyl phosphoro diamidate and dicyandiamide on nitrogen loss from flooded rice. *Soil Sci. Soc. Am. J.*, **49**: 1426-1431.

Simpson, J.R., J.R. Freney, W.A. Muirhead and R. Leuning. 1985. Effect of phenyl phosphorodiamidate and dicyandiamide on nitrogen loss from flooded rice. *Soil. Sci..Soc. Am. J.*, **49**: 1426-1431.

Subbiah, S. and G.V. Kothandaraman. 1980. Effect of neem cake blending of urea on ammonium and nitrate N content in soil under field conditions. *J. Indian. Soc. Soil. Sci.*, **28**: 136-137.

Subramani, S. and S. Kannaiyan. 1989. Effect of the water fern *Azolla* on enzyme activity in soil under rice cultivation. *Mikrobiologiya*, **58(1):** 118-121.

Summer, M.E. and M.P.W. Farina. 1986. Phosphorus interactions with other nutrients and lime in field cropping system. *Adv. Soil Sci.*, **5:** 201-236.

Subba Rao, N.S., K.V.B.R. Tilak, C.S. Singh and P.V. Nair. 1986. Yield and nitrogen gains of *Sorghum* as influenced by *Azospirillum brasilense*. In: Cereal nitrogen fixation (ed.) S.P. Wani, Inte. Crop Res. Inst. for Semi Arid Tropics, Patancheru, India, pp. 69-74.

TAC, 1988. Sustainable Agricultural Production. Implications for international agricultural research Report of the Technical Advisory Committee Consultative Group on International Agricultural Research (CGIAR), Washington DC.

Thangaraju, M. and S. Kannaiyan. 1989. Studies on the urease activity and fertility status of soil by incorporating certain *Azolla* cultures. 29th Ann. Conf. Assoc. Microbiologists of India, HAU, Hissar (Abstr.) p. 87.

Thangaraju, M. and S. Kannaiyan. 1993. Effect of nitrogen fixing water fern *Azolla* and different forms of urea application on growth of nitrogen uptake and grain yield of rice crop. *Acta. Agro. Hungarica*, **42 (1-2):** 69-76.

Thanikachalam, A., K. Rajakannu and S. Kannaiyan. 1984. Effect of neem cake, carbofuran and *Azolla* application on phosphatase activity in soil. 25th Ann. Conf. Assoc. Microbiologists of India, Pant Nagar, (Abstr.). p. 7.

Soil Microbiology and Soil Biotechnology : Vol. 2
Editor : S. Kannaiyan
Associated Publishing Co., New Delhi, 2010, pp. 27-69

2

Biological Fertilizers

S. Kannaiyan and K. Kumar***

Introduction

Nitrogen inputs through Biological Nitrogen Fixation (BNF) support sustainable and environmentally sound agricultural production. BNF conserves soil nitrogen resources. Soil fertility and soil tilth could be improved by practices such as green manuring with dinitrogen (N_2) fixing legumes, and companion crops or subsequent non-legume crops can also be benefited from BNF (Zahran, 2005). The selection of crop sequences and the integration of farm inputs for maximal returns are all components of effective soil management. Further, the beneficial interactions associations between plants and microorganisms (Table 1) are important in maintaining soil fertility.

Table 1: Major types of beneficial interactions/associations between plants and soil micro-organisms (Subba Rao, 1993)

Nature of interactions/ associations	Examples of higher plants involved	Examples of microorganisms involved
Rhizosphere, Rhizoplane and Phyllosphere microflora	All plants with roots and leaves	Bacteria, fungi and *Actinomycetes*
Ectomycorrhizae	Forest trees - *Pinus*	Mostly basidiomycetous fungi *Boletus, Lactarius, Armillaria*
Endomycorrhizae (Arbuscular Mycorrhiza fungal association)	Certain orchids cereals, grasses and legumes	*Rhizoctonia, Endogone* and *Glomus*
Root nodules of nodulating legumes	Soybean, gram, Lucerne *etc.*	*Rhizobium* spp. and *Bradyrhizobium* sp.
Root nodules of plants other than legumes	*Alnus, Myrica, Casuarina*	Actinomycetous endophytes (Frankia)
Leaf nodules	*Psychotria, Pavetta*	*Klebsiella*
Algal associations with higher plants	*Cycas, Zamia, Heterozamia, Gunnera scabra, Azolla*	*Anabaena, Nostoc*
Associative symbiosis	Grasses, Sorghum and millets	*Azospirillum*

* National Biodiversity Authority, Chennai – 600 041
** Department of Agricultural Microbiology, TNAU, Coimbatore – 641 003

To guarantee food for all, the chemical fertilizer consumption has increased from 113 (mt) in 1980 to 307 (mt) in 2000, most of the deficit noticeably occurring in developing countries where infrastructure facilities for chemical fertilizer production are poor. In contrast, biofertilizers/organic fertilizers are cost-effective, eco-friendly and renewable source of plant nutrients to supplement chemical fertilizers in sustainable agricultural system (Emmerling, 2005). Despite sizable increase in the use of chemical fertilizers over the years the gap between nutrient removal and replenishment is significantly high in many developing countries including India. The role of biofertilizers in sustainable agriculture assumes special significance particularly in the present context of very high cost of chemical fertilizers and environmental hazards associated with its excessive use. The production and application of biofertilizers for leguminous crops, oilseeds, rice, millets and other important crops besides forest nursery plants are very common in India (Kannaiyan, 2003).

In this volume, application and biotechnology of rhizobial biofertilizers have been discussed by Teamroong and Boonkerd in Chapter 2 and phosphorus (P)-solubilizing biofertilizer by Mikanova and Kubat in Chapter 5. In this chapter, we are limiting our discussion of various biofertilizers commonly applied to rice agro-ecosystem and the underlying biotechnological principles are discussed.

Azospirilium biofertilizer

The concept of "**associative symbiosis**" has widened our knowledge of free-living N_2 fixing microorganisms. Dobereiner and Day (1976) reported for the first time, the occurrence of a N_2 fixing bacterium *Spirillum lipoferum* in the roots of digit grass in Brazil. The bacterium was found to fix nitrogen in association with roots of various plants. This discovery opened new vistas for world's most staple food crops like rice, maize, sorghum, wheat and millets, which were previously believed to have no potential for biological N_2 fixation.

Azospirillum spp. is of ubiquitous distribution in many parts of the world consisting of tropical, sub-tropical and temperate climatic conditions and also with various standing crops grown in a variety of soil types. Lakshmikumari *et al.* (1976) reported the occurrence of N_2 fixing *Spirillum* in the roots of rice, sorghum and maize. Reynders and Vlassak (1976) from Belgium isolated this organism from more than 50 per cent of soil samples representing a wide range of soil types. Amer *et al.* (1977) reported that various geographical regions of the world were rich in *Azospirillum*. Neyra and Dobereiner (1977) reported that nearly 10 per cent of the soils and roots from temperate region have *Azospirillum* while more than 50 per cent tropical soils were found positive for *Azospirillum.*

In India, Rao *et al.* (1978) reported the presence of *Azospirillum* in different rice soil types (alluvial, laterite, pokkali and kari). The occurrence of *Azospirillum* in the roots of cotton, maize, ragi, kudiraivali, guinea grass, hariyali grass (Purushothaman *et al.,* 1980), plantation orchard plants (Subba Rao, 1983), pearl millet, arecanut, banana, cashew, coconut, coca, rubber and cardamom (Purushothaman and Govindarajan, 1984; Govindan and Purushothaman, 1985) and upland rice (Purushothaman *et al.,* 1987) was well documented.

A new acid tolerant sp. *i.e., Azospirillum amazonense* was reported by Magalhaes *et al.* (1983) from mason region in Brazil. Charyulu and Rao (1980) and Palaneeswar *et al.* (1992) observed the occurrence of *Azospirillum* with appreciable N_2 fixing ability in highly acidic soils. Reinhold-Hurek *et al.* (1980) reported *Azospirillum halopmeferens* characterized by its high salt tolerance. Salt tolerant *Azospirillum* isolates have also been reported by Sundaram *et al.* (2001). Thiruchenthilvasan and Purushothaman (1999) observed better colonization of sorghum rhizosphere by genetically engineered *Azospirillum* cultures.

So far taxonomists have identified five species in the genus *Azospirillum i.e., A. brasilense, A. lipoferum, A. amazonense, A. halopraeferens* and *A. irakense*. They are Gram-negative, vibrioid in shape, motile with a long polar flagellum and occasionally with peritrichous flagella. Organic acids such as malate and succinate are the preferred carbon sources for *Azospirillum.* However, the different species vary in their ability to use different carbon sources (Okon, 1985). Among the free living N_2 fixing bacteria *Azospirillum* is considered to be more efficient with nitrogenase properties comparatively better than other N_2 fixers. It has been well demonstrated that *Azospirillum* inoculated plants were able to absorb nutrients from solutions at faster rates than uninoculated plants resulting in accumulation of more dry matter, N,P and K in the stem and leaves (Okon, 1985).

Table 2: Effect of *Azospirillum* inoculation on the grain yield of rice (var. ADT 36) under graded levels of nitrogen

Treatment	Grain yield ($t.ha^{-1}$)			
	1985	1986	1987	Mean
0 kg $N.ha^{-1}$	3.1 d*	3.0 d	3.1 c	3.1 d
0 kg $N.ha^{-1}$ + *Azosprillum*	3.1 d	3.4 d	3.3 d	3.5 d
25 kg $N.ha^{-1}$	3.6 d	3.4 d	3.5 d	3.5d
0 kg $N.ha^{-1}$ + *Azosprillum*	3.8 bc	4.0 c	3.9 cd	3.9 c
50 kg $N.ha^{-1}$	4.1 ab	4.3 bc	3.7 cd	4.0 bc
50 kg $N.ha^{-1}$ + *Azosprillum*	4.4 a	4.5 ab	4.2 bc	4.4 b
75 kg $N.ha^{-1}$	3.6 cd	4.5 ab	3.9 cd	4.0 d
75 kg $N.ha^{-1}$ + *Azosprillum*	4.4 a	4.9 a	5.0 a	4.7 a
100 kg $N.ha^{-1}$	4.3 ab	4.7 ab	4.2 bc	4.3 b
100 kg $N.ha^{-1}$ + *Azosprillum*	4.0 ab	4.7 ab	4.5 ab	4.4 ab
CD (P= 0.05)	0.5	0.4	0.5	0.4

* In a column means followed by the same letter are not significantly different at 5% level by DMRT (Gopalaswamy *et al.*, 1989).

Azospirillum inoculation has been reported to increase the growth, nitrogen uptake and yield parameters in a number of crops. Field trials conducted in many places with different crops have revealed that biological N_2 fixation by root associated *Azospirillum* contribute significant amounts of N to the plants thereby saving valuable N-fertilizers. The results also revealed that the *Azospirillum* inoculation benefits plant growth and increases yield of crops by improving root development, mineral uptake and plant-water relationship. Several experiments have been conducted at different locations in India and all over the world to find out the response of crop plants to *Azospirillum* inoculation under field conditions. A three year field experiment conducted to assess the response of rice to *Azospirillum* inoculation under graded levels of inorganic nitrogen (0,25,50,75 and 100 kg N ha^{-1}) revealed that *Azospirillum* inoculation at 75 kg N ha^{-1} significantly increased the grain yield (Gopalaswamy *et al.*, 1989). At lower and higher levels of inorganic nitrogen the inoculation effect was not significant (Table 2).

Okon and Labandera-Gonzalez (1994) evaluated worldwide data accumulated over the past 20 years on field inoculation with *Azospirillum* and concluded that these bacteria are capable of promoting the yield of agriculturally important crops in different soils and climatic regions. The review of the extensive data collected revealed 60-70 per cent success with statistically significant increases in yield in 5-30 per cent of the trials.

Azolla biofertilizer

Azolla is a free-floating water fern, which in symbiotic association with the cyanobacterium—*Anabaena azollae* contributes substantial amount of biologically fixed nitrogen to the rice crop. The N_2 fixing water fern, *Azolla* has been used as a biofertilizer for rice crop in India, China, Vietnam, Thailand, Philippines, Korea, Sri Lanka, Bangladesh, Pakistan, Nepal, Burma, Indonesia, Brazil and West Africa (Kannaiyan, 1987). The technology of utilizing the N_2 fixing water fern *Azolla* has been attempted by these countries but the full potential of this biological system has not been exploited. The *Azolla – Anabaena* association is a "**living floating nitrogen manufacturing factory**" which is utilizing the energy from photosynthesis to fix atmospheric nitrogen amounting to 150 to 200 kg N ha^{-1} year $^{-1}$. *Azolla* is used as biofertilizer for rice crop and it could contribute 40-60 kg N ha^{-1} per crop (Kannaiyan, 1989). *Azolla* is also used as biofertilizer in agriculture/fish culture (Garg and Bhatnagaer, 2005).

The important factor in using *Azolla* as a biofertilizer for rice crop is its quick decomposition in soil and efficient availability of its nitrogen to rice plants (Kannaiyan, 1990a). In tropical rice soils, the applied *Azolla* mineralizes rapidly and its nitrogen is made available to the rice crop in very short period.

Growth

The growth rate of *Azolla* is very rapid so much so that it doubles its weight in a mere 2-3 days. Due to fast relative growth rate the biomass production is heavy in a given time and hence the biomass is exploited as bio-fertilizer for rice. *Azolla* has the potential of maintaining an exponential growth rate under optimum conditions. The growth of *Azolla* is initially slow in the growth medium as well as in the rice field followed by fast growth (Kannaiyan, 1993a). The growth of *Azolla* is subjected to various environmental factors such as temperature, light intensity, season, humidity, wind, pH and salinity (Kannaiyan and Sompom, 1987; Latha *et al.*, 1988; Rajagopal *et al.*, 1994). The response of *Azolla* to phosphorus fertilization is necessary for its growth and N_2 fixation (Kannaiyan *et al.*, 1981; Singh and Singh, 1989). *Azolla* grown in phosphorus deficient solution had a reddish brown discolouration that spread from the centre of the frond to the tip of the body with reduction in size of the frond. Addition of phosphorus is most effective in stimulating the growth of *Azolla* (Singh, 1979, Kannaiyan 1985a) reported higher biomass of *Azolla filiculoides* at 20 mg. g^{-1} of phosphorus but found that 5-10, 20 mg. g^{-1} level was adequate for growth and multiplication. Singh (1977a) recommended 4-6 kg P_2O_5 ha^{-1} per week for large scale multiplication of *A. pinnata* under field conditions. Kannaiyan *et al.* (1982) have demonstrated the effectiveness of split application of phosphorus for *Azolla* growth and multiplication.

Mass production of *Azolla* biomass

A simple *Azolla* nursery method for large-scale multiplication of *Azolla microphylla* in the fields has been evolved for easy adoption by the farmers (Kannaiyan, 1982; 1989). The field selected for *Azolla* nursery must be thoroughly prepared and leveled uniformly. The field is divided into one cent plots (20m ′ 2m) by providing suitable bunds and irrigation channels. Water is maintained at a depth of 10 cm. Ten kilos of fresh cattle dung mixed in 20 litres of water is sprinkled in each plot and *A. microphylla* SK -TNAU inoculum at 8 kg is inoculated into each plot. Super phosphate (100 g) is applied in three split doses at four days interval as top dressing fertilizer for *A. microphylla* SK - TNAU. For insect pest control, Furadan^R [(carbofuran) (2, 3-dihydro-2, 2-dimethybenzofuran-7-yl methyl carbamate; Rallis India Ltd.)] granules at 100 g. $plot^{-1}$ (40 m^2 area) are applied seven days after inoculation. Fifteen days after inoculation *A. microphylla* SK - TNAU is harvested and introduced into the main field as a source of primary inoculum. From one harvest, 100-150 kg fresh *Azolla* is obtained from each plot. Again *Azolla* is inoculated in the same plot and cattle dung, super phosphate, furadan granules are applied and harvested. Two cents (80 m^2) nursery plots are required for raising *A. microphylla* SK-TNAU for inoculation to one hectare main field.

Azolla spore production technology

Sporulation in different *Azolla* species are not well understood and the factors responsible for sporulation and spore germination are not well defined. *Azolla microphylla* SK-TNAU a sporulating culture was selected for spore inoculum development. Fresh young fronds of *A. microphylla* SK-TNAU were inoculated in puddle and well prepared field under flooded condition. A basal nutrition of 10 kg fresh cattle dung was mixed in 20 litres of water and applied in 40 m^2. A week after inoculation of *A. microphylla* SK-TNAU, the growth regulator Gibb relic acid (GA_3) at 100 mg. ml^{-1} was applied as foliar spray. Foliar spray with GA_3 induced the growth and stimulated the production of sporocarps. These sporulated fronds were harvested after four weeks and formed a heap with flat surface. One hundred kilogram of sporulated fresh fronds of *A. microphylla* SK-TNAU formed a heap. The heap was then covered with a thin layer of clay soil slurry (20%) and allowed to undergo partial decomposition for a period of 21 days. Subsequently the heaps were removed and dried well under shade for a week period. The decomposed and well dried frond materials are called "**Frond based spore inoculum**" of *Azolla* (Kannaiyan, 1990b, Shanmugasundaram and Kannaiyan, 1992a).

Mahesh and Kannaiyan (1994) have investigated the influence of growth regulators *i.e.*, GA_3 naphthyl acetic acid (NAA), 2, 4-dichlorophenoxy acetic acid (2, 4-0) and benzyl amino purine (BAP) at various concentrations on the germination and fertilization of the megasporocarps of *A. microphylla* and found that GA_3 and BAP at 20 to 80 mg. g^{-1}. Stimulated the germination and fertilization of megasporocarps. The growth regulators also enhanced the survival of the spore-lings and the heterocyst frequency of the algal symbiont in the spore-lings (Shanmugasundaram and Kannaiyan, 1992b).

Inoculation of fresh biomass of *Azolla*

Fresh biomass of *Azolla* is broadcasted in the main field 7-10 days after transplanting rice. Inoculation of the fresh biomass of *A. microphylla* at 200 kg ha^{-1} could multiply faster and cover the rice fields as a green mat in 2-3 weeks periods with 15-25 tonnes biomass accumulation. However, inoculation of even 50-100 kg biomass of various *Azolla* spp., *i.e.*, *A. microphylla, A. pinnata, A. caroliniana* and *A. filiculoides* per hectare will multiply and establish in rice fields but it takes 4-5 weeks for full coverage in the transplanted rice. Under optimum conditions with cloudy days and low temperature in day/night cycle of 25±1°C/ 18±1°C, the inoculated *Azolla* spreads so fast in transplanted rice. *Azolla* technology is very efficient in terms of N_2 fixation and biomass accumulation during wet season due to the better environmental conditions prevailing during the second rice season in India particularly cloudy days coupled with low temperature favors its regetative multiplication. The best suitable season for *Azolla* utilization as dual crop with rice is wet season. It can also be used for late dry season crop. Season is very critical for its multiplication and biomass accumulation. *A. microphylla* is known to survive even in summer season but its full potential as biofertilizer could be exploited only during second wet rice season crop in tropical conditions. Inoculation of *Azolla* in transplanted rice crop supports the growth and multiplication. Growing *Azolla* along with rice crop is called dual culture and it is not affecting the growth of rice crop in any way but it suppresses the weed growth in wetland rice (Kannaiyan *et al.*, 1984).

Application of *Azolla* bio-fertilizer for rice crop

Srinivasan (1980) reported that *Azolla* as green manure could reduce the nitrogen input by 25-30% in lowland rice. *Azolla* incorporation as green manure increased the yield of rice by 9-38% (Singh, 1977a), 35.8% (Kannaiyan, 1981) and 44% (Baratharkur and Talukdar, 1983). Subramani and Kannaiyan (1987) reported that application of gypsum at 5 tonnes ha^{-1} with 60 kg N ha^{-1} as urea and *Azolla* inoculation as dual crop recorded the highest grain yield of rice. Jayanthi *et al.* (1994) have reported that incorporation of 10 tonnes ha^{-1} fresh *Azolla* biomass along with 50 kg N ha^{-1} resulted in increased N uptake, available nitrogen in soil and rice yield. Gopalaswamy *et al.* (1994) have investigated the potential of *Azolla* hybrids and reported that inoculation of *Azolla* hybrids significantly improved the soil organic carbon status besides increasing rice yield.

In wetland rice ecosystem the productivity of rice could be increased by raising the efficiency of applied nitrogen either by natural means or by chemical fertilization. Latha *et al.* (1988) have established that the inoculation of *Azolla* coupled with Urea Super Granule (USG) application have recorded significantly higher grain yield of rice than Prilled Urea (PU) application. Thangaraju and Kannaiyan (1990) reported that *A. microphylla* and *A. filiculoides* coupled with USG application significantly increased the grain yield of rice. Kannaiyan (1992 a) reported increased rice yield by the combined application of USG and

A. microphylla. Shanmugasundaram and Kannaiyan (1994) have reported increased rice yield by the combined application of USG and *A. microphylla.* Shanmugasundaram and Kannaiyan (1994) have reported the inoculation of *Azolla* hybrid RS-SK-TNAU-1 with PU as well as USG recorded higher grain yield and N uptake in rice. The N_2 fixing green manure plants *i.e. Azolla* and *Sesbania rostrata* could be effectively utilized by incorporation with USG to increase the rice production in lowland rice ecosystem (Kumar *et al.,* 1995). The nitrogen use efficiency in lowland rice system could be increased substantially by adopting the integrated nutrient management system using N_2 fixing green manures or *Azolla* biofertilizer.

Azolla inoculation has considerably increased the soil microbial as well as enzymes activity (Kumar, 1994; Kannaiyan, 1994). Zhang Zhuang *et al.* (1987) have reported that incorporation of 22.5.-37.5 fresh *Azolla* ha^{-1} biomass before planting increased the rice yield by 9.6 -13.0 per cent. Patil and Mandhare (1991) reported that incorporation of 5 t *Azolla* ha^{-1} with 100 per cent recommended nitrogen level yielded 6.7 t ha^{-1} which is more or less equal to the incorporation of 5 t *Azolla* ha^{-1} with 75 per cent of the recommended dose of nitrogen. Alexander *et al.* (1992) observed maximum rice yield increase with an output ratio recovery of 2.07 when 7.5 t fresh *Azolla* ha^{-1} biomass incorporated along with 90 kg N ha^{-1}. An increase in the grain yield of rice by the application of *Azolla* as green manure, dual crop or both with rice was equivalent to 30-40 kg N ha^{-1} and this has been well documented by several investigators in various countries such as China (liu Chung Chu, 1987), The Philippines (Mabbayad, 1987; Watanabe, 1987), India (Singh, 1977b; Kannaiyan, 1981), Thailand (Loudhapasitiporn and Kanareugsa, 1987), Sri Lanka (Kulasooriya *et al.,* 1987), USA (Talley *et al.,* 1977) and West Africa (Diara *et al.,* 1987). The use of *Azolla* in rice field has considerably increased the soil microbial as well as enzymes activity (Kumar, 1994 Kannaiyan, 1994), which in turn maintained soil fertility. It is important low cost input, which plays a vital role in improving soil quality management in sustainable rice farming (Kannaiyan, 1992b, 1993b, 1996). Dual cropping of *Azolla* is a most effective, eco-friendly practice, which plays a major role in sustainable rice farming (Kannaiyan, 1992b 1993b, 1996). Though *Azolla-Anabaena* symbiosis is the only plant-cyanobacterial symbiosis used as a rice biofertilizer for maintaining soil health and productivity, this system could still be improved physiologically by using the growth regulator GA_3 (Gopalaswamy and Kannaiyan, 1996). *Azolla* hybrids were found to undergo rapid decomposition and to increase the floodwater ammoniacal N content, which is very vital for the rice crop. By carefully screening of different *Azolla* cultures, *Azolla* hybrid AH-C2 was found to be saline tolerant. Rice production could be significantly increased with *Azolla* hybrids in combination with 75 kg N ha^{-1} USG (Gopalaswamy and Kannaiyan, 2000).

Kumar and Kannaiyan (2000) developed a mutant, *A. microphylla* – KK-SK-E20 that exhibited higher nitrogenease and ammonia assimilating enzyme activity than the wild parent. Variation among N_2 fixing *Anabaena azollae* strains isolated from four different *Azolla* cultures were characterized based on their total protein profile and RAPD (random amplified polymorphic DNA) profile. Considerable variation in protein banding at 40-70 kDa (Dalton) protein subunits was shown by different isolates and specific polymorphic bands were generated for different isolates with different primers (Subhashini *et al.,* 2000).

Blue Green Algae (BGA)

The agronomic potential of cyanobacterial N_2 fixation in rice fields was first recognized in India during 1939 by P.K. De, who attributed the natural fertility of tropical rice fields to N_2 fixing blue green algae. The rice field ecosystem provides an environment favourable for the growth of blue green algae with respect to their requirements for light, water, high temperature and nutrient availability.

Algal bio-fertilizer constitutes a perpetual source of nutrients and they do not contaminate ground water and deplete the resources. In addition to contributing 25-30 kg N ha^{-1} of biologically fixed nitrogen, they also add organic matter to the soil, excrete growth promoting substances, solubilize insoluble phosphates and amend the physical and chemical properties of the soil (Venkataraman, 1972; Kannaiyan, 1978; Goyal, 1982).

Distribution of BGA in Indian soils

Under submerged soil in tropical conditions in rice fields of India, BGA form the most promising biological system in addition to contributing nitrogen; benefit the crop in many other ways (Goyal, 1982). The occurrence and distribution of BGA in Indian rice fields have been reported by several workers (Mitra, 1951; Pandey, 1965; Venkataraman, 1972; Bendra and Kumar, 1975; Sinha and Mukherjee, 1975; Prasad *et al*, 1978; Kannaiyan, 1978; Bongale and Bharati, 1980; Sardeshpandey and Goyal, 1981a; Kannaiyan, 1993a). The available information shows that while some forms like *Nostoc, Anabaena, Calothrix, Aulosira* and *Plectonema* are ubiquitous in Indian soils while others like *Haplosiphon, Scytonema* and *Cylindrospermum* have localized distribution (Venkataraman, 1972; Kannaiyan, 1985b).

The beneficial effect of algal inoculation on the grain yield of rice has been well documented at a number of locations (Venkataraman, 1972; Kannaiyan, 1978). Algal inoculation to rice crop was effective in different agro climatic condition and soil types (Goyal and Venkataraman, 1971; Kannaiyan, 1983). Blue green algal inoculation with composite cultures was found to be more effective than single culture inoculation (Kannaiyan 1978). A technology for mass scale production of composite culture of BGA under rice field condition was developed by Kannaiyan (1978) and the soil based BGA inoculum could survive for more than two years (Kannaiyan, 1990c). At many sites where algal inoculation was used for three to four consecutive cropping seasons, the inoculated algae established well and the effect persisted over subsequent rice crop (Kannaiyan, 1990b). Technologies for utilizing N_2 fixing organisms in lowland rice were reported (Venkataraman, 1972; Roger and Watanabe, 1986) and the beneficial role of blue green algal inoculation in rice soils of Tamil Nadu province of India was reported by Kannaiyan (1985b).

N_2 Fixing BGA

The species of BGA which are known to fix atmospheric nitrogen are classified into three groups: (i) heterocystous – aerobic forms (ii) aerobic unicellular forms and (iii) non-heterocystous, filamentous; microaerophilic forms. The blue green algal culture's composite inoculum consisting of *Nostoc, Anabaena, Calothrix, Tolypothrix, Plectonema, Aphanothece, Gloeocapasa, Oscillatoria, Cylindrospermum, Aulosira* and *Scytonema* have been used for inoculation in rice (Kannaiyan, 1985b). These N_2 fixing cultures have better adaptability in wetland rice.

Mass scale multiplication under field conditions

Mass production of dried soil based algal flakes under rice field condition was developed by Kannaiyan (1978). The field selected for mass multiplication of BGA is prepared well and leveled uniformly. The field is divided into small plots size of 20m ′ 2m (400m^2, by providing suitable bunds and irrigation channels. Water is maintained into a height of 10 cm. Super phosphate at 2 kg is added to the plot and inoculated with 5 kg of starter culture of BGA. Carbofuran (furadan) at 200g is also added to prevent the activities of soil insects. Water is let in at periodical intervals, so that the height of water level is always maintained at 10cm. The growth of algae is rapid and in about 10 days, a thick mat is formed which floats on the surface of the water. After 15 days of inoculation, the plots are allowed to dry up in the sun and the algal flakes are collected and stored. From each harvest, 30-40 kg of dry algal flakes are obtained per plot.

Algal inoculation to rice crop

Inoculation of BGA is highly suitable during dry season. Application of dried soil based blue-green algal flakes at the rate of 10 kg ha^{-1} is recommended for rice crop. At the time of inoculation of BGA, 500 g of single super phosphate may be mixed with 10 kg of soil based BGA which encourages better growth and establishment in rice field. A composite starter culture consisting of more than one species or genera is more preferable. Fast growing species which can tolerate certain amount of extremities are more suitable than highly susceptible ones. The algal species should have a fair amount of tolerance to agricultural chemicals such as pesticides and weedicides. The algal inoculant is to be applied 10 days after transplantation of rice crop. Continuous application of blue green algae to the rice fields for at least three or four consecutive seasons sustains the high crop yield at reduced levels of commercial nitrogen fertilizers input in ensuing cropping seasons (Kannaiyan, 1993a).

Residual effect of BGA inoculation

Blue green algae are living bio-systems and once they establish their biological activity is a continuous process. Inoculation of blue green algae for three or four cropping seasons continuously results in an appreciable population build up and consequently their beneficial effect can be seen in the subsequent years without any further inoculation unless some unfavourable soil ecological conditions operates. The result of several field trials conducted in Tamil Nadu province of India have shown the positive residual effect of algal inoculation on rice yield (Kannaiyan, 1978).

Induction of native BGA

Phosphorus nutrition is known to induce the native BGA (Kannaiyan, 1985b). Top dressing (5 kg/ha) of single super phosphate (SSP) on 10 -15 days after transplanting induces the native N_2-fixing algal flora. Top dressing of 5 kg of SSP not only stimulate the growth of native BGA but also encourages the growth of inoculated BGA. Environmental conditions prevailing during dry season are more favourable for multiplication and establishment of BGA. It has been shown that the incorporation of green manures in rice fields have induced the N_2-fixing algal flora and certain N_2 fixing algae exhibits heterotrophic types in rice fields (Kannaiyan, 1992a). Algal populations were able to colonize well in rice straw. Addition of 2 kg of rice straw bits per 40 m^2 plot induces the populations of BGA in algal multiplication plots.

Blue-green algae are ideal biological system under tropical situation for better N_2-fixation and conservation in rice soil. Blue-green algae are renewable biofertilizer and they represent a self-supporting biological agent because they can photo synthetically provide energy for N_2 fixation. Rice cropping system provides favourable aquatic environment for the growth, multiplication and N_2 fixation by utilizing the solar energy. Thus, these potential groups of N_2 fixating algae may be better exploited for harnessing atmospheric nitrogen so as to increase the productivity of rice.

Stress tolerant *Cyanobacteria*

In India, out of a total of 157 million ha of cultivable area, 49 million ha of land are acidic in nature (Das, 1996). Moore (1963) reported the occurrence of many cyanophyceae in acid soils. Many cyanobacterial cultures were isolated from acid soils of Kerala (India) having a pH of 3.8 (Aiyer, 1965; Madhusoodanan and Dominic, 1995). The Survival of cyanobacteria under acid soil rice field condition was studied by Jurgensen and Davey (1968). Later, Sardeshpandey and Goyal (1981b) observed that the native isolates survived better under acid soils than the introduced cultures. Singh *et al.* (1997) isolated many acid tolerant strains of cyanobacteria from acid soils of Nagaland (India). The effect of low pH (4-6) on growth and biomass production of different cyanobacterial cultures was examined and compared with a normal non-acid soil cyanobacterium *Anabaena*–NS-TGK-A1, *Westiellopsis* - AT-TGK-W1 showed maximum *Anabaena* TGK A1 (Gopalaswamy *et al.*, 2002).

Cyanobacteria react to salt stress either through osmotic adjustment or by secretion of excess polysaccharide (Flowers *et al.* 1977; Saxena and Kaushik, 1991). Salt tolerance in *cyanobacteria* mediated by an osmo–regulatory mechanism, which involved the accumulation of carbohydrates, amino acids and quaternary ammonium compounds (Reed *et al.* 1986). Salt adaptation strategy of cyanobacteria includes two main processes: (i) maintenance of low internal contents of inorganic ions by export mechanism and (ii) synthesis and accumulation of osmoprotective compounds corresponding to the osmotic potential of the surrounding medium.

The first process is achieved by Na^+/H^+ anti-porters that are coupled to cytochrome oxidase (Militor *et al.*, 1986). Several compounds like sucrose, trehalose, glucosyl glycerol and glycine acting in the second process were identified in salt loaded cyanobacteria (Reed and Stewart, 1988). The physiological basis of salt tolerance may be due to extrusion of Na^+ and maintenance of low intracellular Na^+, accumulation of K^+ or ionic balance between K^+ and Na^+ (Miller *et al.*, 1976; Apte and Thomas, 1986; Jha and Kaushik, 1988).

There is increasing evidence that respiration plays a primary role in the adaptation of cyanobacteria to higher concentration of salt (Erber *et al.*, 1986 Fry *et al.*, 1986; Wastynt

1987). In *Anacystis nidulans* increasing salt concentration led to decreased photosynthesis and respiration (Vonshak *et al.,* 1988). When cyanobacterial cells were exposed to salt stress a term increase in the cellular sodium concentration was noticed which was due to a transient increase in the permeability of plasma membrane during the first few seconds of exposure to high salt and that led to the immediate inhibition of photosynthetic and respiratory system (Reed *et al.,* 1986). Erdmann *et al.* (1992) have reported a decreased photosynthesis and chlorophyll- a contents in *Microcystis firma* and *Synechocystis* sp. PCC 6803 during the first days after a salt shock.

With increased sodium-chloride (NaCl) concentration the filaments of *Anabaena doliolum,* a salt sensitive fresh water cyanobacterium was shorter with less number of heterocysts which was an indirect evidence of reduced N_2 fixation, decrease in relative protein content and chlorophyll-a content (Rai and Abraham, 1993). Schubert *et al.* (1993) investigated that *Synechocystis* sp. PCC 6803 when grown in different salt concentration, the optimum growth was found to be around 34.2 mM NaCl and at this salinity, photosynthesis and chlorophyll and phycocyanin contents showed the highest values. Selva Kumar *et al.* (2002) analyzed pigment content in butachlor (N-butoxymethyl-2 chloro-2, 6 diethyl acetanilide) tolerant cyanobacteria and showed that the C-phycoerythrin content of *Anabaena* HT-SGK-A_2, *Westiellopsis* –HT –SGK-W_2 was not affected even at 12 mg. ml^{-1} level of butachlor. Interestingly butachlor was found to be stimulatory to *Anabaena* –HT-SGK-A_2 at 9 g. ml^{-1} concentration.

Anand *et al.* (1994) have reported that the salt tolerance level of different cyanobacteria *Chroococcus minor* and *Oscillatoria saline* grew upto 3% salinity, *Nostoc piscinale, Tolypothrix tenuis* and *Gloeocapsa polydesmatica* failed to grow beyond 1.5% salinity. The cells of *Anabaena* strain L-31 when exposed to 350 mM sucrose and 10mM KNO_3, osmotic stress induced proteins (OSPS) were stimulated which resulted in significant osmo-protection (Iyer *et al.,* 1994).

Immobilization of Cyanobacteria

Cyanobacteria could be immobilized into the pores of polyurethane foam (PU). It has been well documented that N_2-fixing cyanobacteria are known to colonize outside PU foams as well as immobilized into pores of the foam (Kannaiyan *et al.,* 1994). In polyurethane foam, colonization on the foam surface followed by penetration into the pores by the cyanobacteria was reported earlier by Brouers *et al.* (1988). *Anabaena azollae* and *A. variabilis* when immobilized in polyurethane and polyvinyl foams and calcium alginate beads have been shown to release ammonia extracellularly (Kerby *et al.,* 1986; Brouers *et al.,* 1988). The algal symbiont *A. azollae* is known to colonize and immobilize well in polyvinyl foam (Kannaiyan *et al.,* 1994). Kannaiyan *et al.* (1994) have shown ammonia excretion by *A. azollae* (AS-DS) and *Nostoc muscorum* immobilized in polyvinyl foam at different foam density. Significantly higher quantities of ammonia were excreted by *Anabaena* sp. immobilized in polyurethane foam in the presence of the glutamine analogue MSX (L-methionine DL-Sulphoximine) (Brouers and Hall 1986; Shi *et al.,* 1987a; Kannaiyan *et al.,* 1994).

Boussiba (1997) reported that the N_2- fixing activity of the algal symbionts was 6-10 times higher in immobilized condition than the activity of free-living cyanobacteria and that the independent isolate of *Anabaena azollae* from *Azolla filiculoides* registered higher heterocyst frequency and nitrogenase activity. An *in vitro* study with *A. azollae*-AS-DS has recorded higher nitrogenase activity followed by *variabilis*- SAo. Interestingly the immobilized cyanobacterial cultures treated with the systemic fungicide Bavistin (methyl benzimidazol -2-yl carbamate; BASF (India) Ltd.,) t 5 mg. ml^{-1} level stimulated the nitrogenase activity (Table 3).

Table 3: Effect of bavistin on nitrogenase activity of cyanobacteria immobilized in PU foam (Kannaiyan *et al.*, 1997).

Cyanobacterial cultures	Nitrogen activity(n moles ethylene produced. H^{-1} g^{-1} dry wt.)	
	Control	Bavistin (5 mg. ml^{-1})
Anabaena variabilis SAo	2678	3086
A. variabilis SA1	1904	2226
A. azollae AS-DS	3571	3874
N. muscorum SK6	1309	1454
N. muscorum SK	1492	1651
N. muscorum DOH	2237	2417
SEd	9.55	
CD	19.8	

SEd: Standard Error of difference; **CD:** Critical Difference

Kannaiyan *et al.* (1994) have shown that the systemic fungicide Benlate (methyl-1-butyl- carbamoyl-2 carbamoyl-2 benzimidazole carbonate) stimulated ammonia production by *A. azollae* in the immobilized state in a photo bio-reactor system. The N_2- fixing cyanobacterial cultures were used for immobilization in polyurethane foam (PUF) and investigated for their performance on the ammonia production under laboratory condition. Hall *et al.* (1997) immobilized two cyanobacterial cultures *i.e., A. azollae* –AS-DS-SK and *A. variabilis*- SAo in polyurethane foam and inoculated in transplanted rice. They assessed the ammonia production in the floodwater of rice and found positive influence on rice crop.

N_2-fixing BGA both symbiotic and free-living types are known to contribute nitrogen to rice plants in terms of ammonia under natural ecological conditions. The algal symbionts *A. azollae*-MPK-SK-AF-38, *A. azollae*-MPK-SK-AM-24 and *A. azollae*-MPK-SK-AM-27 have excreted significantly higher ammonia level into the medium and the same cyanobacterial cultures when immobilized in polyurethane foam further increased the ammonia production (Table 4 and 5). Higher ammonia excretion by *A. azollae* immobilized in sodium alginate and polyurethane foam was reported (Kannaiyan, 1994).

Table 4: Ammonia excretion by the algal symbionts under free-living condition at different intervals (Kannaiyan *et al.*, 1997).

Algal symbionts	Ammonia excretion (n moles.ml-[1])			
	7th day	14th day	21st day	28th day
Anabaena azollae-MPK-SK-AM-4	510.5	516.8	562.8	489.2
A. azollae-MPK-SK-AM-7	455.6	475.5	518.5	535.7
A. azollae-MPK-SK-AM-24	625.7	657.9	735.6	677.4
A. azollae-MPK-SK-AM-27	556.3	610.8	712.9	646.8
A. azollae-MPK-SK-AF-38	599.2	571.1	687.1	605.3

	SEd	CD
Algal symbionts	6.26	12.66
Days	5.60	11.32
Algal symbionts ´ Days	12.5	25.29

Table 5: Ammonia excretion by the algal symbionts under PU (polyurethane) foam immobilized condition at different intervals (Kannaiyan *et al.*, 1997).

Algal symbionts	Ammonia excretion (n moles. Ml-[1])			
	7th day	14th day	21st day	28th day
Anabaena azollae-MPK-SK-AM-4	622.8	598.3	718.9	671.4
Anabaena azollae-MPK-SK-AM-7	575.5	656.1	635.4	602.1
Anabaena azollae-MPK-SK-AM-24	732.9	766.1	843.1	771.8
Anabaena azollae-MPK-SK-AM-27	672.4	691.7	761.7	723.6
Anabaena azollae-MPK-SK-AM-38	705.1	745.4	809.2	756.3

	SEd	CD
Algal symbionts	8.36	16.89
Days	7.47	15.09
Algal symbionts × days	16.70	33.75

Systemic fungicides on ammonia production

The treatment of the PU foam immobilized cyanobacterial symbionts with the systemic fungicides Bavistin and Vitavax (5, 6 dihydro-2methyl-N-Phenyl-1, 4-oxathin-3-carboxamide) induced ammonia production at significantly higher rates. The cyanobacterial culture, *A. azollae*-MPK-SK-AM-24 immobilized in PU foam excreted maximum amount of ammonia on 21st day (Kannaiyan *et al.,* 1997). They have also noticed that the immobilization of the algal symbionts in PU foam and treatment with the systemic fungicides Bavistin and Vitavax greatly decreased the activity of glutamine synthetase. Of the algal symbionts, *A. azollae*-MPK-SK-AM-24 registered very low glutamine synthetase (GS) activity and resulted in higher ammonia production. Paul and Kannaiyan (1994) have developed mutants of *A. azollae*-AS-DS using colchicine and these mutants have recorded higher excretion of ammonia when treated with the herbicide butachlor. Uma and Kannaiyan (1994) have shown that incorporation of the granular insecticide carbofuran (2, 3-dihydro -2, 2- dimethybenzofuran-7-yl methyl carbamate; Rallis India Ltd.,) at 5 ppm into N-free medium stimulated ammonia excretion by the cyanobacterial cultures immobilized in PU foam. This might be due to the membrane permeability of the cells by the immobilization process (Shi *et al.*, 1987b) which would allow a greater accessibility of substrates like inhibitors of ammonia assimilating enzymes, which would facilitate excretion of ammonia. Kannaiyan *et al.* (1994) found that Benlate (methyl-1-(butylcarbamoyl) benzimidazole-2-yl carbamate; (Dupont) due to its systemic action is likely to diffuse into the cyanobacterial cells where it might, like methyl sulphoxamine (MSX) inhibit ammonia assimilating enzyme, inhibit glutamine synthetase and thereby increasing ammonia production. It is apparently clear that the systemic fungicides Bavistin and Vitavax could induce significant amounts of ammonia production by the algal symbionts under immobilized condition (Kannaiyan *et al.*, 1997).

Nitrogen regulation in rice by immobilized Cyanobacteria

Fertilizer nitrogen usually from an inorganic source as well as nitrogen fixed by biological agents derived from the mineralization of organic matter is subject to a series of physical, chemical and biological changes which ultimately determines its fate in flooded rice soil and subsequently available for rice crop. Integrated use of organics and fertilizer nitrogen apparently increases the nitrogen utilization in rice crop. In low land rice soil ecosystem, nitrogen could be efficiently supplied by the use of free-living cyanobacterial and symbiotic *Azolla* bio-fertilizers. The nitrogen excreted as ammoniacal form by the immobilized cyanobacteria or due to mineralization of both symbiotic as well as free-living cyanobacterial biomass is utilized by the rice plants at all its critical stages of growth like tillering, panicle initiation and grain filling stage. Hall *et al.* (1997) have shown that the application of immobilized N_2 fixing cyanobacteria increased the growth as well as grain and straw yield of rice. Kannaiyan (1990a) has shown significant increase in grain yield of rice when *Azolla* bio-fertilizer was used along with USG. Recently, Kannaiyan (1996) has established the use of Neem cake (*Azadirachta indica*) at 200 kg. ha^{-1} along with *Azolla* or neem coated urea (NCU) *i.e.*, 20% Neem cake powder coated urea with *Azolla* bio-fertilizer significantly increased the nitrogen use efficiency in rice. The alkaloid "**azadirachtin**" present in neem cake (NC) inhibits nitrifying bacteria, which in turn inhibits the ammonia oxidation in rice field and thereby the accumulation of ammonia occurs. The inhibitory action of NC on efficient utilization of nitrogen in rice crop is well documented by Kannaiyan (1996) by developing a pathway, which reflects in increasing the grain and straw yield of rice.

Carrier based immobilized Cyanobacterial inoculum

The symbiotic N_2 fixing cyanobacterium *i.e., Anabaena azollae*—AS-DS and the free-living N_2 fixing cyanobacterium *i.e., Anabaena variabilis*—SAo were immobilized in solid matrices such as polyurethane foam (PUF) and polyvinyl foam (PVF) and it was found that the ammonia excretion was stimulated as a result of the immobilization of cyanobacterial cells. The effect of inoculation of immobilized cyanobacterial inoculum in the transplanted rice was quite significant. Similarly inoculation of sugarcane waste immobilized strains of *A.*

azollae—AS-DS and *A. variabilis*—SAo in transplanted rice field significantly increased the flood water ammonia content and also increased the grain and straw yield of rice. Among the different treatments tested, the inoculation of *A. azollae* immobilized in sugarcane waste along with 60 kg N ha^{-1} recorded maximum grain and straw yield of rice (var. ADT 36). Hence, a carrier based immobilized cyanobacterial inoculant was developed to maintain better quality. The polyurethane foam, sugarcane waste and paper waste immobilized cyanobacteria were the components of the inoculant. Rice husk and sterilized soil were used in equal proportion as a carrier material for preparing this inoculant. The carrier based immobilized cyanobacterial inoculum may be applied for rice crop at 2 kg. ha^{-1} one week after transplanting.

A field experiment was conducted to study the effect of inoculation of immobilized cyanobacteria along with fertilizer N on the floodwater ammonia of rice. High ammonia level was recorded in the floodwater of the experimental plots due to inoculation with immobilized cyanobacteria than un-inoculated control (Table 6). It is seen from the data that the PU foam and sugarcane waste immobilized cyanobacteria could produce ammonia in the floodwater of rice crops. Both *A. azollae*—AS-DS-SK and *A. variabilis*—SAo immobilized in PU foam and sugarcane waste have excreted significant amounts of ammonia in experimental plots and the results have clearly indicated that the immobilized cyanobacterial inoculum in rice field have produced ammonia in flood water.

Table 6: Effect of inoculation of immobilized cyanobacteria on ammonia production of rice flood water.

Treatments	Ammonia production (n moles. Ml^{-1})				
	2nd day	8th day	11th day	14th day	28th day
Control	318	341	428	727	311
Anabaena azollae (PUF)	890	1137	1027	942	341
A. variabilis (PUF)	850	1559	1267	1195	350
A. azollae (SCW)	1013	1234	1169	1199	331
A. variabilis (SCW)	1072	1364	1445	1120	360
A. azollae (PUF) + 60 Kg N ha^{-1}	1653	1624	1462	1608	389
A. variabilis (PUF) + 60 Kg N ha^{-1}	1234	1786	1364	1624	370
A. azollae (SCW) + 60 Kg N ha^{-1}	1150	1786	1559	1494	380
A. variabilis + Kg + 60 N ha^{-1}	1267	1591	1656	1786	389
60 Kg N ha^{-1}	357	357	584	890	321
SEd	5.950	2.654	1.431	3.446	0.995
CD	11.890	5.307	3.005	6.893	1.909

PUF: Polyurethane foam; **SCW**: sugarcane waste

Sesbania rostrata

Legumes that nodulate on above ground stems are referred to as stem nodulating legumes. The term "**Stem Nodule**" was first reported in histological studies on *Aeshynomene* *aspera* by Hagerup (1928). Since then, several workers reported stem nodulation in *Aeschynomene paniculata* (Suessenguth and Beyerle, 1936), *Neptunia oleracea* (Schaede, 1940), *Aeschynomene indica* (Schaede, 1940), *Aeschynomene indica* (Arora, 1954), *A. elaphroxylon* (Jenik and Kubikova, 1969), *A. evenia* and *A. villosa* (Barrios and Gonzales, 1971). However, the potential importance of stem nodulation was recognized only when Dreyfus and Dommergus (1981) reported profuse stem nodulation on fast growing leguminous plant *Sesbania rostrata*. Until now, 26 different stem-nodulating legume species from different genera *i.e., Aeschynomene, Neptunia* and *Sesbania* have been reported. However, fast growing, extensively nodulating, erect leguminous plants such as *S. rostrata, Aeschynomene afraspera* and *A. nilotica* have been reported as most promising N_2 fixing green manures for lowland rice (Rinaudo *et al.,* 1983; Alazard and Becker, 1987; Becker *et al.,* 1990 and Kannaiyan, 1993a).

The most distinctive characteristic of stem nodulating legumes is the presence of predetermined nodulation sites on the stems, which are independent of infection with *rhizobia* (Dreyfus *et al.,* 1984). They comprise primordia of adventitious roots that are able to grow out under waterlogged conditions (Tsien *et al.,* 1983) and they are arranged in straight vertical rows (*Sesbania*) or in spiral shaped rows winding around the stem (*Aeschynomene*). Ladha *et al.* (1992) defined "**stem nodules**" as those formed on aerial portions of the plant, whose location prevents contact with soil and whose formation is initiated on a primordium and not on an outgrowing adventitious root.

Among the leguminous green manure crops, *S. rostrata* was found to be highly sensitive to photoperiod (Visperas *et al.,* 1987). Low biomass production and low nitrogen content of *S. rostrata* in The Philippines between October and February 1998 have been ascribed to the day length of less than 12 hour (Becker *et al.,* 1990). *S. rostrata* flowered after 37 and 125 days when planted during December and April respectively, while *A. afraspera* took 54 and 85 days to flower respectively in the same season in The Philippines (Becker *et al.,* 1990).

The specific acetylene reduction activities in stem nodules of *S. rostrata* and *Aeschynomene scabra* were 1.5 and 9 times higher respectively than in root nodules. Stem nodules have higher enzyme activities and metabolic contents than root nodules, suggesting that they fix N_2 with greater energy efficiency, which was correlated to photosynthesis in the cortex of stem nodules (Hungria *et al.,* 1992). Sadasivam *et al.* (1993) found close relationship between nodule phosphoenolpyruvate carboxylase and nitrogenase activities and suggested that nodule CO_2 fixation contributed directly to N assimilation in stem nodules of *S. rostrata.* Moreover, nitrogenase component II was localized in bacteriods within the outermost layers of infected cells, suggesting that a low CO_2 was maintained, despite the chloroplasts being nearby (James *et al.,* 1996).

Characteristics of the symbiont *Azorhizobium caulinodans*

Rhizobia isolated from root and stem nodules of *Sesbania* and *Aeschynomene* were able to produce nodules in both stem and root Yatazawa and Yoshida, 1979, Legocki *et al.,* 1983; Alazard, 1985; Subba Rao, 1988; Dreyfus *et al.,* 1988. Dreyfus *et al.* (1984) isolated two types of strains; one that fixes atmospheric N_2 under free-living and nodulates both root and stem of *S. rostrata*, while the other, do not fix N_2 in culture and nodulates only roots of *S. rostrata, Rhizobial* strains from stem nodules of *A. indica, A. aspera* and *S. rostrata* were characterized and the strains were found to be intermediate type, sharing characteristics of both fast and slow growers (Stowers and Eaglesham, 1983, Chakrabarti *et al.*, 1986).

Dreyfus *et al.* (1983) observed free-living N_2 fixation and growth of *Azorhizobium* with N_2 as sole N source. Gebhardt *et al.* (1984) investigated the important characteristics of free-living N_2 fixation by *Azorhizobium* and noticed higher tolerance to oxygen. Urban *et al.* (1986) reported that *Azorhizobium* strain ORS 571 fixes nitrogen at a relatively high oxygen concentration (3%) and temperature (37°C). Free-living growth and N_2 fixation have also been shown for isolates of *A. indica, A. sensitiva, A. protensis* and *A. afraspera* (Ladha *et al.*, 1992).

Kalidurai and Kannaiyan (1988) have shown that *S. rostrata* stem isolates recorded relatively more *in vitro* nitrogenase activity in arabinose and mannitol followed by arabinose and sucrose in semi-solid agar medium. The role of *azorhizobia* as free-living nitrogen fixer in rice soil was demonstrated by Ladha *et al.* (1998). Free living N_2-fixation by *A. caulinodans* strain ORS 571 was ten times higher than that fixed by *Bradyrhizobium* strain CB 756 (Alazard, 1990).

Spermospheric presence of *Azorhizobium* in seeds of *S. rostrata* was noticed by Ladha *et al.* (1988) and Balasubramani and Kannaiyan (1991a) and when released from stem nodules can survive and grow in flooded soils (Ladha *et al.*, 1998). The epiphytic occurrence of *A. caulinodans* in leaves and flowers of *S. rostrata* was reported (Adebayo *et al.*, 1989; Balasubramani and Kannaiyan, 1991b). The blister beetle (*Mylabnis pustulata*) generally seen during the flowering season of *S. rostrata* was found to transmit *A. caulinodans* from one plant to another (Balasubramani and Kannaiyan, 1991a). The isolates obtained from *S. rostrata* leaves, seeds, flowers and from the insect-blister beetle were found to nodulate *S. rostrata* (Table 7). The internal movement of *A. caulinodans* in *S. rostrata* stems was also reported (Table 8) (Chitra, 1992). Robertson and Alexander (1994) suggested that rain and wind blown soil are important means of transmitting *Azorhizobium* for stem nodulation in *S. rostrata*. The occurrence and distribution of *Azorhizobium* in four vegetation zones of Senegal was determined by Robertson *et al.* (1995).

Table 7: Studies on the population in roots and stem by the isolation of *Azorhizobium* in *S. rostrata*

Inoculation	Per cent recovery			
	Root bits		Stem bits	
	IRRI (+) NO_3	IRRI (-) NO_3	IRRI (+) NO_3	IRRI (-) NO_3
Inoculated	20.0	100.0	40.0	100.0
Un-inoculated	-	30.0	30.0	20.0

IRRI: International Rice Research Institute

Table 8: Internal movement of *A. caulinodans* in seedlings of *S. rostrata*

Azorhizobium isolates	Stem nodule number/plant		
	30 DAS	40 DAS	60 DAS
Insect isolate	33	42	50
Seed isolate	42	53	67
Leaf isolate	34	47	60
Flower isolate	24	31	43

DAS: Days after Sowing

Based on whole cell protein analysis, DNA-DNA and DNA-rRNA hybridization, (Dreyfus *et al.*, 1988) proposed a new genus and species *Azorhizobium caulinodans* for the stem nodulating isolates of *S. rostrata*. Rinaudo *et al.* (1991) studied DNA homology among members of the genus *Azorhizobium* and other stem and root nodulating *rhizobia* isolated from *S. rostrata* and found that most isolates belonged to the genus *Azorhizobium* with G+C content ranging from 66 to 68%. The sequence surrounding the *caulinodans* regulatory *nod D* gene was analyzed by Geele *et al.* (1995). Tomekpe *et al.* (1996) reported that the greater infectivity of *caulinodans* strain ORS 571 was due to more efficient production of *nod* factors at infection sites located on the stem.

Seed/foliar application of *Sesbania rostrata*

The primary objective of seed inoculation is to introduce *rhizobia* in soil in such a way that the inoculum will remain viable until the host seedlings would be able to accept an infection and compete with native *rhizobia* for optimum nodulation and higher N_2-fixation. In general, the seed coat of different leguminous green manure seeds is very thick which affects the seed germination. Pre-soaking the seeds of *Sesbania* spp. and subsequent inoculation with *Rhizobium/ Azorhizobium* have substantially increased the colonization and survival of *Rhizobium/ Azorhizobium* and higher population load in pre-soaked seeds than un-soaked seeds.

Though the nodules on the stem appear spontaneously on the preformed infection sites, foliar inoculation by foliar spraying with *A. caulinodans* has been found to activate the nodule initiation. During summer season when the temperature is high (37±1°C), the infection by *A. caulinodans* on *S. rostrata* stem is very low. But the external supply of *A. caulinodans* inoculation by foliar spraying triggers the infection and nodule formation. Eaglesham and Szalay (1983) showed that a single inoculation of the stem of *Aeschynomene* plants with *rhizobial* strain BTAI-1 induced the formation of stem nodules within 8 days.

Rinaudo *et al.* (1982) reported that stem inoculation of *S. rostrata* in Northern Senegal resulted in increased N_2 fixing activity. The positive effect of stem inoculation on the growth and nitrogen fixing activity of *S. rostrata* was reported by Ladha *et al.* (1998). Chitra and Kannaiyan (1995) found that foliar inoculation of *A. caulinodans* significantly increased the fresh weight, nodule number and nutrient content of *S. rostrata* (Table 9). Kalidurai and Kannaiyan 1991 have also shown that foliar inoculation with *A. caulinodans* favoured stem nodulation in *S. rostrata*.

Table 9: Effect of foliar inoculation of *A. caulinodans* on nodulation, biomass production and nutrient content of *S. rostrata*

Foliar Inoculation	Nodule numbers.Plant^{-1}		Fresh weight (g)	Nutrient content (%)		
	Root	Stem		N	P	K
Sprayed	34.5	1173.5	972.59	3.31	0.42	2.00
Unsprayed	29.2	87.7	510.00	2.06	0.40	1.90
SEd	1.02	95.94	92.84	0.05	0.01	0.01
CD	2.06	201.56	195.72	0.12	0.02	0.02

SEd: Standard Error of difference; **CD:** Critical difference.

Stem inoculation seems to be appropriate particularly in acid soil where root infection of *rhizobia* is not possible (Ladha *et al.,* 1998). Foliar inoculation of *A. caulinodans* by foliar spraying increases the population load considerably on the stem surface of *S. rostrata.* It is also presumed, that the inoculated *A. caulinodans* might survive in the plant parts such as leaves and flowers and later might spread to the stem portion of *S. rostrata* for increasing the colonization, infection and nodulation.

Ladha *et al.* (1998) have demonstrated that both seed and stem inoculation have improved the growth and N_2 fixation in *S. rostrata.* Seed inoculation coupled with foliar spraying has substantially increased the stem nodule numbers per plant (Chitra, 1992). Seed inoculation and foliar spraying have also increased the nitrogen, phosphorus and potassium content of *S. rostrata.*

Foliar spraying with appropriate strains of *A. caulinodans* on the stems of 2-7 weeks old plants have been shown to produce more number of stem nodules (Ladha *et al.,* 1992). Chitra (1992) has reported that foliar inoculation of 20-25 days old plants of *S. rostrata* recorded maximum number of stem nodules per plant. Foliar inoculation both in the morning and evening recorded maximum number of nodules per plant. However, foliar inoculation in the evening resulted in more number of stem nodules than the morning spray.

Stem nodulation and N_2 fixation in *S. rostrata*

The tolerance of leguminous green manures to flooding is an important trait for species used in wetland rice. Eaglesham and Szalay (1983) studied the effect of flooding on root and stem nodulation of *Aeschynomene scabra* and found that the number of root nodules decreased to almost zero compared with 1000 stem nodules per plant in non-flooded condition. *Sesbania rostrata* was found to tolerate water logging and produce about 25 t ha^{-1} biomass around 45 days growth (Ndoye and Dreyfus, 1988 and Kalidurai and Kannaiyan, 1988). Becker *et al.* (1990) reported higher biomass production and N_2 fixation in 40-50 days old *S. rostrata afraspera.* Stem nodulation in *S. rostrata* was twice in flooded condition (Table 10) compared to water saturated condition (Balasubramani and Kannaiyan, 1993). Becker and George (1995) found maximum N accumulation of 275 mg/plant by *S. rostrata* which was 20 per cent more than *S. cannabina,* in flooded soils. Flooding the soil increased the relative humidity of the crop micro-environment by 4 to 11% and induced early appearance of stem nodules in *S. rostrata* (Patel *et al.,* 1996).

Table 10: Influence of water-logging condition on stem nodule formation in *S. rostrata*

Azorhizobium isolates	Stem nodule number plant^{-1}					
	Water logging			Saturation		
	35 DAS	45 DAS	60 DAS	35 DAS	45 DAS	60 DAS
SRS2	104	118	134	48	55	62
SRR1	75	87	98	35	43	54

DAS: Days after Sowing

Nutritional factors are responsible for the wide variation in the amount of N_2 fixed by legume-*Rhizobium* symbiosis (O'Hara *et al.,* 1988). The growth of *S. rostrata* and *A. afraspera* was vigorous on both fertile and relatively poor sandy soils (Becker *et al.*, 1990).

The requirement of phosphorus for efficient nodulation and maximum nodule activity was much greater in nodules than the host plant growth (DeMooy *et al.,* 1973). Balasubramani and Kannaiyan 1991a) have shown enhanced nodule induction and biomass production in *S. rostrata* by P nutrition. Becker *et al.* (1991) showed that the application of mineral P and K (potassium) fertilizers stimulated nodulation and acetylene reduction activity (ARA) in *S. rostrata* and *A. afraspera,* which led to a 40 per cent increase in N content. Potassium alone mainly enhanced plant growth, whereas P activated. A positive effect of P and K fertilizers on biomass production of *S. rostrata* has also been reported by Meelu and Morris (1988). Becker *et al.* (1991) recommended the application of phosphorus and potassium fertilizers to green manure crops for higher biomass production.

The type and concentration of mineral N influenced both nodulation and N_2 fixation in root nodulating legumes (Eaglesham and Szalay 1983). However, the N_2-fixation by stem nodulating legumes is reported to be unaffected to some extent by mineral N (Dreyfus and Dommergues, 1981 Eaglesham and Szalay, 1983; Becker *et al.,* 1986; 1989; 1991 and Kannaiyan, 1993a, Rinaudo *et al.,* 1983) have reported complete inhibition of root and stem nodule ARA in hydroponically grown *S. rostrata* where the ammonium nitrate concentrations were constantly maintained at 5-3.0 mM and 6 mM, respectively.

Kwon and Beevers (1993) found that 0-9.0 mM NO_3 had no adverse effect on stem nodulation in *S. rostrata*. However, when the NO_3 concentration was raised above 20 mM, the growth of nodules and ARA were strongly inhibited. Balasubramani *et al.* (1992) have reported that the dry weight of stem nodule and nitrogenase activity of root and stem nodules of *S. rostrata,* were not affected adversely upto 90 kg N ha^{-1} but interestingly 30 to 90 kg N ha^{-1} increased the fresh and dry weight of nodules. Reduction in the amount of BNF due to applied mineral N was greater in aerobic soil than in flooded soil. The N_2 fixing activity of stem nodules was less sensitive to NH_4^+ concentration in flooded soils than root nodules and resulted in more biologically fixed N in *S. rostrata* (Becker and George, 1995).

Ghai *et al.* (1988) evaluated 36 *Sesbania* spp. in the wet season and recorded an average contribution of 63 kg N ha in 30-45 days. Under flooded conditions, the N accumulation potential ranged from 41 kg N ha^{-1} in *A. indica* (Crozat and Sangchyosawat, 1985) to more than 200 kg N ha^{-1} in *S. rostrata* (Morris *et al.,* 1987; Becker *et al.,* 1990). Ladha *et al.*

(1992) found that 50 day old *S. rostrata* could accumulate 190-267 kg N ha^{-1}. Pareek *et al.* (1990) have estimated N_2 fixation of *S. rostrata* by ^{15}N technique and found *S. rostrata* to be the fastest N_2 fixing plant known, fixing 70-95 per cent of the total plant N within 45-55 days. They also found that *S. rostrata* could accumulate more nitrogen than *S. cannabina* during dry season.

Stem nodulation has been reported to be more efficient N_2 fixing system than root nodulation. Eaglesham and Szalay (1983) reported activities of over fixing system than root 160 mM C_2H_4 h^{-1} plant of above ground portions of *A. scabra*. Dreyfus *et al.* (1985) reported nitrogenase activity of about 600 mM C_2H_4 h^{-1} plant in a three month old *S. rostrata* and most of the nitrogenase activity was attributed to stem nodules. Negligible ARA has also been recorded in *S. rostrata* root nodules (Rinaudo *et al.,* 1982; Becker *et al.,* 1990). The nitrogenase activity and the nodule dry weight were higher in stem nodules than in root nodules (Balasubramani *et al.,* 1992 and Saraswati *et al.,* 1992).

Xylem colonization by *A. caulinodans* in *S. rostrata*

Rhizobia are bacteria that invade legumes after activating cell division in the plant root cortex, initiating a new organ, the N_2- fixing nodule (Long, 1996). Since the nodule is generally considered the only endophytic destination of invading *rhizobia*, most previous studies have focused on the *rhizobial* invasion pathway into and within the cortex (Kijne, 1992). Many *non-rhizobia* endophytic bacteria colonize the vascular system of plants, sometimes, but not always leading to plant disease (Bell *et al.,* 1995) or example, the xylem of healthy *alfalfa*, a legume, is colonized by *non-rhizobia* endophytes (Gagne *et al.,* 1987). Agro-bacteria and plant pathogens closely related taxonomically to *rhizobia*, invade the xylem of several species including, interestingly the topical legume *Sesbania rostrata* (Vlachova *et al.,* 1987). Inoculation of *S. rostrata* with *Azorhizobium caulinodans* ORS 57 (Dreyfus *et al.,* 1988) results in production of nodules on roots (Ndoye *et al.,* 1994) and stems (Tsien *et al.,* 1983). Interestingly, strain ORS 57 is able to establish itself endophytically in the roots of rice (Christiansen-Weniger 1996) and wheat (Sabry *et al.,* 1997). The presence of *azorhizobia* in the xylem was confirmed by light and electron microscopy. The recent availability of various genetic tools and reporter genes based on fusions with *lac Z* and *gus* A is having a major impact on studying the xylem colonization (Vande Broek *et al.,* 1993; Arsene *et al.,* 1994 Gough *et al.,* 1996). Bacteria re-isolated from inoculated plants and plated onto selective media containing X-Gal, produced blue staining colonies, confirming that some *azorhizobia* still retained pXLGD4. Re-isolated bacteria also exhibited typical *azorhizobial* colony morphology on semi-solidified TY (Tryptone yeast extract) medium and failed to take up Congo red dye (a diagnostic test for *rhizobia;* Somasegaran and Hoben, 1994). Recently it was found that ORS 571 colonizes xylem elements, in addition to inducing and invading nodules in the root cortex of *S. rostrata*. However xylem colonization is not regulated in the same way as nodulation. Thus, for the first time, a species of legume nodule bacteria has been found to colonize, reproducibly, regions of the host plant other than nodules. This novel endophytic interaction will be of interest to phytopathologists, to researchers investigating legume—*rhizobia* interactions, and to workers attempting to extend endophytic *rhizobial* N_2 fixation to non-legumes. (O'Callaghan *et al.,* 1997).

Azorhizobium caulinodans

N_2 fixing nodules are usually found on the roots of the host leguminous plants. Only some legume species belonging to the genera *Neptunia*, *Aeschynomene* and *Sesbania* bear nodules on both roots and stem. Dreyfus *et al.* (1988) found that the tropical legume *Sesbania rostrata* was symbiotically associated with *Azorhizobium caulinodans*.

The cells of *A. caulinodans* are Gram-negative, small rods (0.5-0.6 mm by 0.5-2.5 mm) and motile. The cells have peritrichous flagella on solid medium and one lateral flagellum in liquid medium (Dreyfus *et al.*, 1988). On agar medium, the colonies appear circular, viscid and creamy colour and fix N_2 in microaerophilic condition. Bacteria growing within the Para-nodule displayed variable morphology, differing in size and cell contents. Electron transparent regions within some bacterial cells of these sections may have accumulations of poly-b-hydroxybutyrate. Considerable amounts of indeterminate material were found together with these bacteria. This was surprising, since the bacteria were extracellular in the middle lamellar space or in rare cases, present in dead cells. *Azorhizobium caulinodans* ORS 571 induces the formation of root and stem nodules on the tropical legume, *Sesbania rostrata*. *Azorhizobium* (can use nitrogen while free living) and organic acids such as lactate and succinate are the favourite carbon substrates for both NH_4^+ and N_2 dependent growth.

This bacterium contains essential nodulation genes as that of *nod ABC* of *rhizobia* (Goethals *et al.*, 1989). Hence it is possible to elucidate the *nod* genes in the biosynthesis of the *nod* factor induced by host plant exudate. In view of the reports of associative N_2 fixation by *Azorhizobium* in rice fields, it will be interesting to study this behaviour and identify functions that contribute to the interactions between *Azorhizobium* and rice plants. The induction of Para-nodules and the invasion of rice by *rhizobia*, either spontaneously or after the treatment of the roots with cell wall degrading enzymes have been reported (Cocking *et al.*, 1990). The nodule like structures induced by 2, 4-D can be inhabited by diazotrophs (Nie, 1983). It is possible that 2, 4-D has at least two effects:

(1) An auxin-like activity, controlling development of nodular structure, and
(2) An effect enabling increased frequency of infection of plant root cells and colonization of nodular structures with certain bacteria.

The choice of the bacterial strain *A. caulinodans* ORS 57 was based on three characteristics (de Bruijn *et al.*, 1995). Firstly, as it is unlikely that there will be bacteroid differentiation in non-legume plants, ORS 571 has the unusual ability among the *rhizobia* to fix nitrogen in free-living conditions without differentiation into bacteroids. Secondly, as it is also unlikely that non-legumes will be able to reproduce the sophisticated oxygen barriers which exist in legume nodules to protect the nitrogenase enzyme complex from inhibitory oxygen levels, ORS 571 strain can tolerate up to 12 mM dissolved oxygen while fixing nitrogen in culture. Thirdly, ORS 571 forms nodules on its host plant following crack entry infection *i.e.* inter-cellularly between adjacent cells. It has been proved that the *Azorhizobium caulinodans* strain ORS 571 is able to enter the hosts (rice, wheat, sorghum, oil seed rape and *S. rostrata*) through crack entry at the point of emergence of bacterial roots, their colonization and invasion were significantly stimulated by the flavonoid naringenin. Though in rice, *Azorhizobium caulinodans* was found to stimulate the root-hairs formation and no root-hair curling was observed. However abnormal lateral root development was observed within 2 weeks after inoculation. More than 80% of the rice roots exhibited thick short lateral root.

Flavonoid as signal molecule for endophytic colonization by *A. caulinodans*

Flavonoids are a class of phenolics secreted by plant roots as part of the regulatory system for colonization of *A. caulinodans*. Four families of flavonoids such as flavones, flavonols, flavonones and iso-flavonones have been shown to have *nod*-gene inducing activity. Naringenin- a specific flavonoid is known to stimulate the colonization of *A. caulinodans* in rice roots. It is a signal molecule for attracting the population of *A. caulinodans*. But rice root did not produce the flavonoid naringenin. The flavonoid naringenin is considered as a potential signal molecule to activate the colonization of *A. caulinodans* in rice roots and, therefore, a specific gene could be introduced into rice plants for synthesis of Naringenin. Flavonoids also enhance the growth rate of certain *rhizobia* and phenolic compounds in general could induce expression of a variety of genes in plant associated bacteria. The flavonoids naringenin and dizdzein, at low concentration were found to significantly stimulate the frequency of lateral root colonization (LRC) and inter-cellular colonization by *A. caulinodans* in non-legumes by acting as chemical stimulants (Gough *et al.*, 1996). Flavonoids have been shown to accumulate in many plants under infection. However, there are few reports in cereal crops showing such accumulation. Naringenin was detected in rice leaves exposed to UV (ultra violet) irradiation. Kumar *et al.* (1996) while studying the effect of flavonoid naringenin on the endophytic colonization by *A. caulinodans* in rice, reported favourable stimulation of lateral root formation by naringenin in rice, (var. ADT 36). They also reported that the main mode of entry of *A. caulinodans* in rice roots is by crack entry through the emerging lateral roots. The effect of naringenin (3.0 mg. ml^{-1}), enzyme mixture and *A. caulinodans* on induction of Para-nodules in rice and maize was reported (Senthil Kumar, 2000 and Buvana and Kannaiyan, 2002). It was found that addition of naringenin enhanced the short thickened lateral root formation. The seedlings treated with naringenin, enzyme mixture and *A. caulinodans* recorded maximum number of lateral rootlets rhizoplane population, nodule like structures and nitrogenase activity. Lateral root development of rice and colonization of lateral root cracks by bacteria were shown to be stimulated by the flavonoid naringenin (Table 11). The effect of flavonoid on root morphology of different rice varieties was studied and found that flavonoid naringenin at 5 ′ 10^{-5} M concentration significantly stimulated the lateral root formation in rice varieties (Gopalaswamy *et al.*, 2000). The symbiont *A. caulinodans* ORS 571 upon inoculation stimulated the lateral root development as well as formation of nodular like structures in the roots of rice and found to be stimulated by the flavonoid naringenin (Kannaiyan *et al.*, 2001).

Table 11: Effect of naringenin on LRC (lateral roots colonization) colonization of four varieties of *Oryza sativa* inoculated with *Azorhizobium caulinodans* ORS 571 (pXLGD4)

Variety	Treatment	Percentage of LRCs colonized per plant (± S.D)	
		Seminal root	Adventitious root
Lemont	-	9.4 ± 0.7	5.7 ± 1.2
	+ naringenin	20.9 ± 1.5	10.7 ± 0.6
CO 43	-	8.3 ± 0.6	11.0 ± 0.4
	+ naringenin	10.2 ± 1.3	16.1 ± 1.0
ADT 36	-	6.7 ± 1.1	14.5 ± 1.5
	+ naringenin	9.3 ± 0.6	37.6 ± 3.1
CR 1009	-	5.3 ± 0.6	11.3 ± 1.1
	+ naringenin	8.8 ± 0.9	16.5 ± 1.2

- Without naringenin

Endophytic colonization by *A. caulinodans*

"Endophytic N_2 fixation" is a process where the N_2 fixing bacteria enter the plant roots through cracks or fissures in the lateral roots, colonize the inter-cellular spaces (cortex, within few cells of epidermis, inner cortex to epidermis, vascular tissues and xylem), where without differentiating into bacteroids fix atmospheric nitrogen micro-aerobically, with the help of the enzyme "**nitrogenase**" (Kennedy *et al.*, 1997). In the earlier days, it was thought that only by inducing nodular structures to provide a place for the organism specifically BNF could be extended to cereals (Arora, 1954). But now recent researches have vividly brought down the difficulties and increased the scope for N_2- fixation in cereals by means of endophytic nitrogen fixation (Stoltzfus *et al.*, 1997; Kannaiyan, 2000).

The extension of biological N_2 fixation to non-legume crops such as rice, maize *etc.*, would be of enormous economical and environmental impact. The discovery that bacteria of the genera *Rhizobium* and *Bradyrhizobium* can form effective N_2 fixing nodules on the non-legume *Parasponia* (Davey *et al.*, 1997) provided added impetus to this approach. Many of these nodular structures have to be artificially induced by adding 2, 4-D (Kennedy *et al.*, 1997; Tchan *et al.*) or cell wall degrading enzymes (Al-Mallah *et al.*, 1989; Tchan *et al.*, 1991; Cocking *et al.*, 1990). Cocking *et al.* (1993) reported that *Rhizobium* of *Parasponia* and *Aeschynomene* produce nodule like structure on the emerging lateral roots of rice and maize without any enzyme treatment. Kennedy and Tchan (1992) studied various approaches for the induction of nodule like structures on the roots of rice and other cereals.

Zeman *et al.* (1992) reported that the exogenous application of the synthetic auxin 2, 4-D to rice induces modified root outgrowths (MROs), which result from the induction of meristem. Reports have been made that certain *rhizobia* can enter rice at low frequency (Cocking *et al.*, 1994). Christiansen-Weniger (1996) reported nodule like structure formation in rice roots treated with 2, 4-D and inoculated with *A. caulinodans*. It is suggested to have a thorough knowledge about the plant genes involved in nodulation and symbiotic N_2 -fixation and the function of their gene products before the development of true nodulation of cereals (de Bruijn *et al.,* 1995; Kennedy *et al.,* 1997; Webster *et al.,* 1997). Now, opinion has consequently changed and the emphasis is to try to establish stable endophytic associations between diazotrophic bacteria and non-legume crops.

Two major approaches are currently being undertaken to try to extend biological N_2 fixation to non-legume crops. The first is to look for natural endophytic diazotrophs of non-legume crops and to assess such endophytic N_2 fixation. This is being exploited with sugarcane (Boddey *et al.,* 1995) and other crops such as rice and wheat (Sabry *et al.,* 1997). The second approach is to determine whether the *rhizobia* can internally colonize and ultimately fix nitrogen in non-legumes. The first step is to study the possibility of inter-cellular colonization of diazotrophs with non-legume including entry spread and the mechanism involved. A sensitive and quantifiable system is needed for this. Microscopic observation of colonization is greatly facilitated by using a genetically modified bacterial strain tagged with a reporter gene (Webster *et al.,* 1997). The efficacy of endophytic colonization by diazotrophs *Azospirillum* and *Azorhizobium* tagged with *lac Z* marker genes in rice was compared. It was observed that the colonization by diazotrophs was mainly through crack entry. In the case of *Azospirillum* the bacteria entered the lateral cracks and their invasion was not so deep and very much limited to endodermis and cortex, whereas, *Azorhizobium* exhibited deep colonization. They first got entry through lateral cracks and traverse inter-cellularly and colonized xylem of rice var. ADT 36 (Gopalaswamy *et al.,* 2000).

Pre-requisites for an endophytic relationship

There are several pre-requisites for the establishment of an endophytic situation. (Quispel, 1974)

(1) Hostile reactions from the host, like hypersensitive reactions and production of phytoalexins have to be prevented.

(2) The bacteria must find an inters or intra-cellular niche, where sufficiently reducing conditions are provided and the pO_2 is regulated by diffusion barriers and/or leghaemoglobins

(3) Membrane systems around the endosymbiont of microbial and host origin must be adapted in such a way that a bi-directional transport of substances is possible.

Recently Chaintreuil *et al.* (2000) have investigated the presence of endophytic *rhizobia* within the roots of the wetland wild rice *Oryza breviligulata*, which is the ancestor of the African cultivated rice *Oryza glaberrima*. This primitive rice species grows in the same wetland sites as *Aeshymomene sensitiva*, an aquatic stem-nodulated legume associated with photosynthetic strains of *Bradyrhizobium.* Twenty endophytic and aquatic isolates were obtained at three different sites in West Africa (Senegal and Guinea) from noda! roots of *O. breviligulata* and surrounding water by using *A. sensitiva* as a trap legume. Most endophytic and aquatic isolates were photosynthetic and belonged to the same phylogenetic *Bradyrhizobium, Blastobacter* subgroup as the typical photosynthetic *Bradyrhizobium* strains previously isolated from *Aeschynomene* stem nodules. N_2- fixing activity, measured by acetylene reduction, was detected in the rice plants inoculated with endophytic isolates. A 20% increase in the shoot growth and grain yield of *O. breviligulata* grown in a green house was also observed upon inoculation with one endophytic strain and one *Aeschynomene* photosynthetic strain. The photosynthetic *Bradyrhizobium* sp. strain ORS 278 extensively colonized the roots surface, followed by inter-cellular and rarely intracellular, bacterial invasion of the rice root, which was determined with a *lac Z*-tagged mutant of ORS 278. The discovery that photosynthetic *Bradyrhizobium* strains, which are usually known to induce N_2 -fixing nodules on stem of the legume *Aeschynomene*, are also natural true endophytes of the primitive rice *O. breviligulata,* could significantly enhance cultivated rice production.

Para nodulation by *A. caulinodans*

Recently, de Bruijn *et al.* (1995) have reviewed the potential and pitfalls of extending symbiotic interactions between N_2 fixing organisms and cereals such as rice. They considered chemical signaling between plant and microbe, nodulation and the prospects for symbiotic N_2 fixation with such systems. The results of experiments carried out in China on the induction of "**nodule-like structures**" on rice roots by *rhizobia* were highlighted. They concluded that the formation of hypertrophies on rice roots infected and colonized with microbes had been confirmed, but that little evidence supported their designation as "**nodules**" or even "**nodule-like**". The frequency of formation of the root structures was low but non-saprophytic colonization of the rice-root endorhizosphere had clearly been observed and deserved further study in this important area of research.

During the past four years the status of biological N_2- fixation in non-leguminous field crops were reviewed (Kattupitiya *et al.,* 1995a, 1995b; Kennedy, 1994, 1997; Yu and Kennedy, 1995; Kennedy *et al.,* 1997). The approach is based on the initial observation of Nie (1983) at Shandong University (China) that 2, 4-D acting as an auxin could stimulate colonization by *rhizobia* in '**nodular structures**' modified from the lateral roots of many plants species.

Although the use of synthetic auxin such as 2, 4-D presents difficulties for field application, we consider that the use of this procedure is justified as a laboratory model, providing improved colonization and N_2 fixation that allow the controlling factors to be studied. It may be possible to provide similar outcomes to the use of 2, 4-D biologically, such as by using mutants producing higher amounts of indole acetic acid (IAA). However, it should be recognized that many systems of agricultural monoculture already apply very stringent chemical treatments, sometimes with dramatic effects on plant development (*e.g.* cotton production using hormones and chemical defoliants).

Para-nodular bacteria: mode of entry and colonization

Para-nodules are nodular structures produced by modification from the lateral roots of many plant species due to the stimulating effect of auxins such as 2, 4-D (Tchan and Kennedy, 1989). Theses structures inhabited by diazotrophs may be quite dissimilar to legume nodules (Kennedy and Tchan, 1992).

The invasion of root cells by endophytic bacteria and their colonization has attracted the interest of many workers. The bacteria are attracted by certain root exudates. In cereals like rice and maize, since the exopolysaccharide production and the production of flavonoids like naringenin, luteoline and maiden are low, by supplying them artificially at very low concentrations invasion can be enhanced (Kennedy *et al.,* 1997; Sabry *et al.,* 1997). The bacteria searches for cracks or fissures through which it can enter the roots. The bacteria enter through cracks in the lateral roots. It moves through the intercellular spaces and finds a site of low redox potential, possibly better suited for N_2-fixation. A large number of bacteria were also found associated with the Para-nodules (Kennedy *et al.,* 1997).

Structure of para-nodules

Para-nodular structures were formed when seedlings were treated with 2, 4-D. The Para-nodules were spheroid in shape, and less than a millimeter in diameter. The frequency of Para nodulation on the young roots was 1-4 nodules per cm of root length. Sections of Para-nodules show the presence of central vascular tissue, connected to the root stele. Several layers of cortical cells, usually without recognizable cytoplasmic contents, surrounded this rudimentary. This is in contrast to the organization of vascular tissue in legume nodules, in which stele surrounds the infected cell. In early stages, root-cap like structures were found on the Para-nodules. In later stages, outer cells in the region showed swelling and sloughed off frequently. Occasionally bacteria were observed inside the outer '**empty**' cortical cells but never in large numbers. In contrast, extensive bacterial infection was confined to the inter-cellular spaces. Serial sectioning of Para-nodules showed that the bacteria and associated material were present in large, irregular channels open to the exterior. Examination of roots by phase-contract and light microscopy indicated that extensive colonization of the root surface and root hair by the endophytic bacteria. Electron microscopical observation of the Para-nodules revealed the presence of large numbers of bacteria in inter-cellular spaces. These irregular spaces or channels were developed by splitting of the middle lamella, during bacterial colonization. These channels surrounded by shrunken and almost empty cortical cells. Only cytoplasmic debris was observed in these empty cells.

Effect of growth regulators and cell wall degrading enzymes on induction of Para-nodules in rice

The inoculation of the bacterium *A. caulinodans* with growth regulators produces lateral rootlets formation in rice. During the development of later root emergence a minute way is formed on both sides of the lateral root by which the bacterium *A. caulinodans* could colonize well and invade into the roots through crack entry and gradually reach to the inner region of the rice root tissues. Seven days old seedlings of rice (var. ADT 36), when inoculated with the symbiotic N_2 - fixing bacterium, *A. caulinodans* along with the growth regulators such as 2, 4-D, NAA, kinetin could induce the formation of nodular structures. Buvana and Kannaiyan (2001) studied the effect of growth regulator NAA and cell wall degrading enzymes (cellulase and pectinase) on induction of Para nodules in rice by *A. caulinodans*. Inoculation of *A. caulinodans* with NAA and cell wall degrading enzyme mixture (cellulase + pectinase) decreased the root growth of seedlings compared to inoculation of *A. caulinodans* alone. Inoculation of *A. caulinodans* and cell wall degrading enzyme mixture developed lateral rootlets which were short and thickened. The number of nodule like structures and total nitrogen content was maximum with inoculation of *A. caulinodans* strains in combination with cell wall degrading enzyme mixture and NAA. Inoculation of *A. caulinodans* was found to increase the root and shoot growth, total biomass and lateral rootlets of rice seedlings over un-inoculated control. The growth regulator kinetin was used at 3.0, 5.0 and 7.0 μg. ml^{-1} concentration and the levels above 3.0 mg. ml^{-1} concentration was found to be inhibitory to the seedlings by recording decreased root and shoot growth, total biomass and lateral rootlets. The inoculation of *A. caulinodans* could nullify the adverse effect of kinetin to some extent compared to the incorporation of growth regulator treatment alone. Increased concentrations of kinetin with inoculation of *A. caulinodans* increased the number of nodule like structures in the roots of young rice seedling. The number of nodule like structure were relatively more in rice variety TKM-9 compared to the rice variety ADT-36. It is interesting to note that the increased concentrations of kinetin with inoculation of *A. caulinodans* increased the number of nodule like structures in rice seedlings. However, there is no nodule like structure formation in rice roots when *A. caulinodans* and growth regulator kinetin were treated in the seedlings separately (Amutha and Kannaiyan, 1999, 2000).

Xylem colonization by *A. caulinodans* in rice

The diazotroph, *Azorhizobium caulinodans* ORS-571, colonizes xylem of roots of the legume *Sesbania rostrata* (O' Callaghan *et al.*, 1997) in addition to colonizing nodules in the legume led us to investigate whether this *rhizobial* strain, might nevertheless be able to colonize the xylem of rice roots. Colonization of the xylem of rice by *A. caulinodans* ORS 571 is of interest because it might provide a suitable nutritional and environmental niche for N_2 fixation, comparable to the situation in sugar cane in which endophytic diazotrophs internally colonize the plant, including the xylem, and are probably major contributors to the observed high levels of N_2 fixation (Boddey *et al.*, 1995).

A large number of plant species exhibit xylem colonization by bacteria without disease symptoms and xylem colonization is increasingly being seen as a common aspect of plant microbe interactions. Moreover, it is becoming increasingly realized that the xylem may be more robust structurally and physiologically than previously envisaged and that it should no longer be regarded as a vulnerable pipeline on the edge of disaster (Canny, 1998). The *non-rhizobial* diazotrophic *Acetobacter diazotrophicus* has been shown to penetrate sugarcane roots inter-cellularly at the root tip and at cracks is lateral roots junctions and to colonize xylem vessels following the inoculation of aseptically grown plants (James *et al.*, 1994). In the presence of the flavanoid naringenin, *A. caulinodans* ORS 571 entered the rice through lateral root cracks and present in large numbers in cortex (Webster *et al.*, 1997). Reddy *et al.* (1997) have also observed that *azorhizobia* invade inter-cellularly the outer cracks. Little is known however, as to how exactly bacteria reach and invade the xylem. It has been suggested that xylem elements are possible sites of N_2 fixation by diazotrophs, since the xylem elements could provide the low pO_2 (partial pressure O_2) and a site for exchange of metabolites necessary for nitrogen fixation (James *et al.*, 1994; Rolfe *et al.*, 1998). Recently Gopalaswamy *et al.* (2000) performed microscopic analyses of sections of the junction regions of lateral and primary roots of rice inoculated in the presence of naringenin with ORS 571 ($pXLGD_4$). Carrying the *lac Z* reporter gene and demonstrated for the first time the xylem colonization of rice roots by *A. caulinodans*. It fact, the number of plant species found to exhibit xylem colonization by bacteria without plant symptoms has increased dramatically in recent years, suggesting the xylem colonization is a common aspect of plant

–microbe interactions (Kloepper *et al.*, 1992). Since *A. caulinodans* ORS 571 is able to fix nitrogen in the free-living state without differentiation into bacteriods up to 3% oxygen (Kitts and Ludwing 1994), this first report of the colonization of xylem elements of rice by *azorhizobia* may be of significance in this respect. The partial pressure of O_2 (pO_2) in rice roots is likely to be influenced by water-logging of the rice plant and more extensive invasion will probably be required to have an impact on the nitrogen balance.

Ecological and economic aspects

Here it has to be considered whether under field conditions, these non-legumes show the properties that would be expected on the basis of green house experiments. As it happens in the case of legumes, it is now being realized that these endophytes also aid in improving soil properties (Dobereiner *et al.*, 1995). The crop residues and stubbles is a major route for the transfer of fixed nitrogen to the soil. The decomposition of dead roots and nodules, and a possible excretion of nitrogen have also to be considered (Rinaudo *et al.*, 1983). It is thought that these organisms have an ability to act as pioneers in nitrogen poor soils and subsequently raise the level of soil nitrogen (Jain and Patriquinn, 1984).

In summary, there is reasonable evidence that this endophytic N_2 fixation can attain a magnitude, which is of definite ecological and economic significance. Additionally, it should be noted that these endophytes constitute a group of species, which can tolerate adverse conditions of growth such as low temperature, or water-logged and acid soils. Thus the group is in a position to fulfill a role of nitrogen providers under very varied circumstances.

Phosphobacterial biofertilizer

Phosphate solubilizing bacteria otherwise called phosphobacteria play a major role in the solubilization and uptake of native and applied soil phosphorus. The P requirement of crops is often met by the addition of phosphatic fertilizers, but the use efficiency of applied phosphorus use is very low due to its fixation either in the form of aluminium or iron phosphates in acidic soils or in the form of calcium phosphate in neutral or alkaline soils. The introduction of efficient P solubilizers in the rhizosphere of crops and in soils increases the availability of P. Phosphobacteria bring about dissolution of bound forms of phosphates in soil and hence advocated as bio-fertilizer to crop plants. Several soil bacteria and fungi and rhizosphere micro-organisms notably species of *Bacillus, Pseudomonas, Aspergillus* and *Penicillium* secrete organic acids and lower the pH in their vicinity to bring about solubilization of bound phosphates in soils (Sundara Rao and Sinha, 1963; Rodriguez and Fraga, 1991).

Beneficial effects due to inoculation with phosphate solubilizing organisms had been reported for different crops under diverse agro climatic conditions (Gaur, 1990; Mikanova and Kubat (2005). Phosphobacteria inoculation with rock phosphate has shown increased benefits in terms of crop yield (Rodriguez and Fraga, 1999). Significant increase in grain yield of wheat and paddy was obtained in field inoculations with *Pseudomonas striata* and rock phosphate application to soil (Table 12).

Table 12: Effect of phosphate solubilizing micro-organisms in yield of Wheat and paddy

Treatment	Yield (Kg ha^{-1})	
	Wheat	Paddy
Control	4975	4866
Bacillus polymixa – no P	5000	4966
B. polymyxa with MRP*	5000	5300
Pseudomonas striata – no P	5066	5083
P. striata with MRP	5466	5200
Aspergillus awamori – no P	5166	4783
A. awamori with MRP	4966	5000
Super phosphate at 50 kg P_2O_5 ha^{-1}	5500	5083
Super phosphate at 100 P_2O_5 ha^{-1}	5633	5600
MRP at 100 kg P_2O_5 ha^{-1}	5250	5166

*MRP –Mussoorie rock phosphate (Gaur, 1990)

Potential for future research

Extending endophytic N_2 fixation to all cereal crops of economic significance may be tried by the following two methods.

Genetic modification of host plant

- To avoid hypersensitive reactions of the host plant.
- To avoid production of phyto-alexins
- To increase host compatibility to accommodate endophytic N_2 fixers.
- To provide areas of low redox potential for the establishment of endophytes by means of diffusion barriers or leghaemoglobins.
- To make the membrane systems around the endosymbiont facilitate bi-directional transport of substances.
- To study *nif* gene transfer.

Genetic modification of the micro-organisms

- To induce production of auxins such as 2, 4-D, NAA, *etc.*
- To produce cellulases and pectinases, which can degrade the cell walls.
- To make the endophytes more tolerant to oxygen concentrations.
- To increase the ability of endophytes to colonize both intra- and inter-cellularly.
- To increase the span of microbial life in vegetative phase.
- To make the micro-organisms use the host materials as energy source but without any phytopathogenic effect.
- To make micro-organisms produce sufficient amount of oxygen dismutase.
- Selection of promising strains possessing high nitrogenase activity.

- To screen germplasm for diazotrophic colonization.
- To investigate root morphogenesis in relation to invasion.
- To study in detail the interactive responses between flavonoids and *nod*-factor.
- To investigate the metabolic programming of cereals for attachment of micro-organism protection of nitrogenase from oxygen inactivation.

Conclusion

The management of native soil N, which is the principal N source for rice and biological N_2- fixation, may play a role in sustaining the soil N under intensive rice cropping. Biological N_2 fixation is an important process in rice farming systems because it is an inexpensive source of nitrogen for increasing the productivity of crops. The N_2 fixing bio systems such as *Azolla, Cyanobacteria* and legume green manures are able to adopt well under wetland rice field ecological conditions and fixes considerable amount of nitrogen. The biomass of these bio systems decomposes rapidly in rice soil and supply the nitrogen addition, these bio-fertilizers or bio-manures also contribute significant amount of phosphorus, potassium, sulphur, zinc, iron, molybdenum and other micro-nutrients. The organic acids released during the mineralization process of the biomass of bio-manures would accelerate the phosphorus availability in rice soil. These biological fertilizers or bio-manures are known to influence the total soil microbial population such as bacteria, fungi, and actinomycetes and also substantially activate the population dynamics of N_2 fixing free living bacteria like *Azotobacter, Azospirillum* and also the symbiotic bacterium such as *Rhizobium*. It is evidently clear now that the application of biological fertilizers is greatly involved in the accumulation of soil enzymes, which directly reflects on soil fertility index (Fioretto and Fuggio, 1995). The use of bio-fertilizers would reduce the cost of chemical fertilizers involved in crop production. The effective utilization of bio-fertilizers for crops will not only provide economic benefits to the farmers but also improve and maintain the soil fertility and sustainability on natural soil ecosystem. Since these bio-sources represent a great diversity in chemical, physical and biological characteristics their efficient use depends upon the particular agro-ecological environment and local availability. Therefore, it is very much essential to develop a strong, workable and compatible package of nutrient management through organic sources for various crops based on scientific facts, local conditions and economic viability.

References

Adebayo, A., Watanabe, and J.K, Ladha, 1989. Epiphytic occurrence of *Azorhizobium caulinodans* and other *rhizobia* on the host and non-host legumes. *Appl. Environ. Microbiol,* **55(9)**: 2407-2409.

Aiyer, R.S. 1965. Comparative algological studies in rice fields of Kerela state. *Kerela.* **3(1)**: 100-104.

Alazard, D. 1985. Stem and root nodulation in *Aeschynomene* spp. *Appl. Environ. Microbiol,* **50**: 732-734.

Alazard, D. 1990. Nitrogen fixation in pure culture by *rhizobia* isolated from stem nodules of tropical *Aeschynomene* species. *EMS Microbial, Lett.,* **68(1-2)**: 77-182.

Alazard, D and M. Becker, 1987. *Aeschynomene* as green manure for rice. *Plant Soil,* **101:** 141-143.

Alexander, D., N. Sadanandan and K. Karunakaran, 1992. Evaluation of *Azolla* as an organic matter for rice production. In: New Trends in Biotechnology (eds.) N.S. Subba Rao, C. Balagopalan and S.V. Ramakrishnan, Oxford and IBH, Publishing Co., New Delhi. p. 255.

Al-Mallah, M.K, M.R. Davey and E.C Cocking, 1981. A new approach to the nodulation of non-legumes by *rhizobia* and the transformation of cereals by agrobacteria using enzymatic treatment of root hairs. *Int. J. Gen/ Manip. Plants,* **5**: 1-7.

Amer H., N.A. Hegazi and M. Nonib, 1977. A survey on the occurrence of nitrogen fixing *Spirillum* in Egypt. Non-symbiotic Nitrogen fixation. *Newsletter*, **5**:101-102.

Amutha, K and S Kannaiyan, 2000. Effect of kinetin and inoculation of *Azorhizobium caulinodans* on induction of lateral rootlets and Para nodules in rice. *J. Microb. World.* **2(1)**: 1-7.

Amutha, K and S. Kannaiyan, 1999. Effect of 2, 4-D, NAA and inoculation with *Azorhizobium caulinodans* on induction of lateral rootlets and Para nodules in rice. *Ind. J. Microbiol.* **39**: 125-128.

Anand, N., R.S.S. Hopper and G. Jagatheswari, 1994. Response of certain blue- green algae (*cyanobacteria*) in salinity. In: Recent Advances in Phycology (eds.) A.K. Kashyap and H.D. Kumar, Rastogi Publications, Varanasi, India. pp. 125-129.

Apte, S.K and J. Thomas, 1986. Membrane electro-genesis and sodium transport in filamentous nitrogen fixing *cyanobacteria. Eur. J. Biochem.*, **154**: 395-401.

Arora, N. 1954. Morphological development of the root and stem nodules of *Aeschynomene indica,* L. *Phytomorphology,* **4**: 211-216.

Arsene, F., S. Katupitiya, L.R. Kennedy and C. Elmerich, 1994. Use of *lac Z* fusions to study the expression of *nif* genes of *Azospirillum brasilense* in association with plants. *Molecular Plant Microbe Interactions,* **7(6)**: 748-757.

Balasubramani, G and S. Kannaiyan. 1991a. Effect of phosphorus on nodulation and biomass production in *Sesbania rostrata. INSURF Netwatch,* **1**: 2-3.

Balasubramani, G and S. Kannaiyan, 1991b. Epiphytic occurrence of *Azorhizobium caulinodans* in leaves and flowers of *Sesbania rostrata.* INSURF Netwatch 1: 4-5. Nodule formation in *Sesbania. Madras Agric. J. India,* **80**: 657-658.

Balasubramani, G., K. Kumar and S. Kannaiyan, 1992. Studies on nitrogen fixation activity in stem nodulating *Sesbania rostrata.* In: Biological Nitrogen fixation and Biogas Technology. (eds.) S. Kannaiyan, K. Ramasamy, K. Ilamurugu and K. Kumar, Tamil Nadu Agricultural University, Coimbatore, India, pp. 84-89.

Baratharkur, H.B. and H. Talukdar, 1983. Use of *Azolla* and commercial nitrogen fertilizer in Jorhat, India. *Int. Rice Res. Newslett.*, **8(2)**: 20.

Barrios, S and V. Gonzales, 1971. *Rhizobial* symbiosis on Venezuelan savannas. *Plant Soil*, **34**: 707-719.

Becker, M., D. Alazard and J.C.G. Ottow, 1986. Mineral nitrogen effect on nodulation and nitrogen fixation of the stem nodulating legume *Aeschynomene afraspera. Z. pflazenemaehr Boden.*, **149**: 485-491.

Becker, M. and T. George, 1995. Nitrogen fixation response of stem and root nodulating *Sesbania* species to flooding and mineral nitrogen. *Plant Soil,* **175(2)**: 189-196.

Becker, M.., J.K. Ladha and J.C.G. Ottow, 1990. Growth and N_2 fixation of two stem nodulating legumes and their effect as green manure on lowland rice. *Soil Biol. Biochem.*, **22**: 1109-1119.

Becker, M., K.H. Diekmann, J.K. Ladha, S.K. De Datta and J.C.G. Ottow, 1991. Effect of N, P, K on growth and nitrogen fixation of *Sesbania rostrata* as green manure for lowland ice *(Oryza sativa). Plant Soil*, **132**: 149-158.

Becker, M., R.P. Pareek, J.K. Ladha and J.C.G. Ottow, 1989. Bio-fertilizer production of stem cut planted and seeded *Sesbania rostrata. Int. Rice Res. Newslett.*, **14(2)**: 30-31.

Bell, C.R., G.A. Dickle, W.L.G. Harvey and J.W.Y. Chan. 1995. Endophytic bacteria in grapevine. *Can. J. Microbiol.*, **41**: 46-53.

Bendra, A.M. and S. Kumar, 1975. Cyanophyceae of Merits. *Phykos*, **14(1-2)**: 1-7.

Boddey, R.M., O.C. de Oliveira, S. Urquiaga, V.M. Reis, F.L, de Olivares, V.L.D. Baldani and J. Dobereiner, 1995. Biological nitrogen fixation associated with sugarcane and rice: contributions and prospects for improvement. *Pl. Soil*, **174**:195-209.

Bongale, U.D and S.G. Bharathi, 1980. On the algal flora of cultivated soils of Karnataka State, India. *Phykos*, **19(1)**: 95-113.

Boussiba, S. 1997. Ammonia assimilation and its bio-technological aspects in *Cyanobacteria*. In: Cyanobacterial Nitrogen Metabolism and Environmental Biotechnology (ed.) A. K. Rai, Narosa Publishing House, New Delhi, India, pp. 36-62.

Brouers, M. and D.O. Hall, 1986. Ammonia and hydrogen production by immobilized cyanobacteria. *J. Biotechnol*, **3**: 307-321.

Brouers, M., H.D. Jong, D.I. Shi and D.O. Hall. 1988. Immobilized cyanobacteria for sustained ammonia production. In: Algal Biotechnology (eds.) T. Stadler, J. Mollion, M.C. Verdus, Y. Karamanos, H. Morran and D. Christian, Elsevier Appl. Science Publishers, New York. U.S.A, pp. 265-275.

Buvana, R and S. Kannaiyan, 2001. Influence of cell wall degrading enzymes on colonization of N_2 fixing bacterium *Azorhizobium caulinodans* in rice. *Indian. J. Exp. Biol.*, **40**: 369- 372.

Buvana, R and S. Kannaiyan 2002. Effect of 2, 4-D and naringenin on the induction of Para-nodules by *Azorhizobium caulinodans* in cereals. *Indian J. Microbiol.*, **42**: 111-115.

Canny, M.J., 1998. Applications of the compensating pressure theory of water transport. *Am. J. Bot.*, **85**: 897-909.

Chaintreuil, C., J. Prin, A.B. Lourquin, M. Gillis, P. de Lajudie and B.L. Dreyfus, 2000. Photosynthetic *bradyrhizobia* are natural endophytes of the African wild rice *Oryza breviligulata. Appl. Environ. Microbiol*, **66**: 5437- 5447.

Chakrabarti, S.K., A.K. Mishra and P.K. Chakrabarty, 1986. A note on physiological characteristics and genetic relatedness of a fast growing *Rhizobium* from stem-nodules of *Aeschynomene aspera. L.J. Appl. Bacteriol.*, **60**: 463-468.

Charyulu, P.B.B.N and V.R. Rao. 1980. Influence of various soil factors on nitrogen fixation by *Azospirillum* spp. *Soil Biol. Biochem.*, **12**: 343-346.

Chitra, C. 1992. Studies on seed coating with certain carbon sources on the survival of *Azorhizobium caulinodans* in seeds and foliar inoculation on nodulation nitrogen fixation and biomass production in *Sesbania rostrata.* M.Sc. (Ag.) thesis, Tamil Nadu Agric. Univ., Coimbatore, Tamil Nadu., India, pp. 230.

Chitra, C. and S. Kannaiyan. 1995. Application of stem nodulating *Sesbania rostrata* for rice production. In: Rice Management Biotechnology (ed.) S. Kannaiyan. Associated Publishing Co., New Delhi, India, pp. 217-236.

Christiansen-Weniger, G. 1996. Endophytic establishment of *Azorhizobium Caulinodans* through auxin-induced root tumors of rice *(Oryza sativa* L). *Biol. Fertil. Soils*, **21**: 293- 302.

Cocking, E.C., M.K. Al Mallah, E. Benson and M.R. Davey, 1990. Nodulation of non legumes by *rhizobia*. In: Nitrogen fixation: Achievements and Objectives (eds.) P M. Gresshoff, E.C. Roth G. Stacey and W.E. Newton, Chapman and Hall, New York. U.S., pp. 813-823.

Cocking, E.C, M.R. Davey, S.L. Kothari, J.S. Srivastava, Y.X. Jing, R.W. Ridge and B.G. Rolfe, 1993. Altering the specificity control of the interaction between *rhizobia* and plants. *Symbiosis*. **14**:123-130.

Cocking, E.C., G. Webster, S.L. Kothari, K. Pawlowski, J. Jones. J.P. Batchelor, P. Jotham, S. Jain and M.R. Davey, 1994. Nodulation and nitrogen fixation in non-legume *Parasponia* species: Dues for non-legume Crops. In: Proceedings of the 1st European Nitrogen Fixation Conference (eds.) G.B. Kiss and G. Endre, Szeged. Hungary, Officina Press, pp. 215-219.

Crozat, I and C .Sangehyosawat, 1985. Evaluation of different green manures on rice yield in Sonkla Lab Basin, Sonklanadarin. *J. Sci. Technol*, **7**: 391-397.

Das, D.K 1996. Soil acidity. In: Introductory Soil Science. Kalyani Publishers. Ludhiana, India, p.498.

Davey, M.R., G. Webster, G. Manders, F.L. Ringrose, J.B. Power and E.C. Cocking, 1997. Effective nodulation of micro-propagated shoots of the non- legume *Parasponia andersonii* by *Bradyrhizobium. J. Exp. Bot.*, **44**: 863-867.

de Bruijn, F.J., Y. Ling and F.B. Dazzo, 1995. Potential and pitfalls of trying to extend symbiotic interactions of nitrogen-fixing organisms to presently non-nodulated plants, such as rice. *Plant Soil*, **174**: 225-240.

de Mooy, J.C., J. Pesek and E. Spaldon, 1973. Mineral nutrition. In: Soybean Improvement, Production and Uses, (ed.) H.E. Caldwell, Am. Soc. Agron., Madison, USA, pp. 267-352.

Diara, H.F., H. van Brandt, A.M. Diop and C. van Hove, 1987. *Azolla* and its use in rice culture in West Africa. In: *Azolla* Utilization, Proc. Workship on *Azolla* use, Fuzhou, Fujian, China, Int. Rice Res. Inst., Los Banos, Philippines, pp.147-152.

Dobereiner, J and J.M. Day, 1976. Associative symbiosis in tropical grasses: Characterization of micro-organisms and nitrogen fixing sites. In: Proceedings of the First International Symposium on Nitrogen Fixation (ed.) W.E Newton, Washington State University Press, Pullman, U.s., pp. 515-536.

Dobereiner. J., V.L.D. Baldani and V.M. Reis, 1995. Endophytic occurrence of *diazotrophic* bacteria in non-leguminous crops. In: NATO ASI Series, Vol. G37 *(Azospirillum* and related micro-organisms) (eds.) Fedrik, M. del Gallo, J. Vanderleyden and Zamoroczy, Springer Verlag, Berlin, Heidelberg, Germany, pp. 3-14.

Dreyfus, B and Y.R. Dommergues. 1981. Stem nodules on the tropical legume *Sesbania rostrata.* In: Current Perspective in Nitrogen Fixation. (ed.) Sci., Canberra, Australia, p. 615 A.H. Gibson, Aust.

Dreyfus, B., 0. Alazard and Y.R. Dommergues. 1984. Stem nodulating *rhizobia.* In: Current Perspectives of Microbial Ecology (eds.) M. Klug and Reddy American Society of Microbiologists, Washington D.C., pp. 161-169

Dreyfus, B.L., G. Rinando and Dommergues, 1985. Observations on the use of *Sesbania rostrata* as: green manure in paddy fields. *Mircen. J. Microbiol*, **1:** 111-121.

Dreyfus, B.L. and C. Rinaudo and Y.R. Dommergues, 1983. Use of *Sesbania rostrata* as a green manure in paddy fields. ORSTOM, Dakar, Senegal, pp. 18-19

Dreyfus, B.L., J.L. Garcia and M. Gillis, 1988. Characterization of *Azorhizobium caulinodans* gen. novo sp. nov., a stem nodulating nitrogen fixing bacterium isolated from *Sesbania rostrata. Int. J. Syst. Bacteriol*, **38**: 89-98.

Eaglesham, A.R.J. and A.A. Szalay, 1983. Aerial stem nodules on *Aeschynomene* sp. *Plant Sci. Lett.*, **29**: 265- 272.

Emmerling, C. 2005. Soil microbial biomass, and activity and the lasting impact on agricultural deintensification. In: Microbial Biotechnology in Agriculture and Aquaculture (ed.) R.C. Ray, Science Publishers, New Hampshire, in press.

Erber, W.W.A., W.H. Nitschmann, R. Munchl and G.A. Peschek, 1986. Endogenous energy supply to the plasma membrane of dark aerobic *cyanobacterium*, *Anacystis nidulans* ATPase-independent efflux of and Na^+ from respiring cells. *Arch. Biochem. Biophys.*, **247**: 28-39.

Erdmann, N., S. Fulda and M. Hagemann, 1992. Glucosylglycerol accumulation during salt accumulation of two unicellular cyanobacteria. *J. Gen. Microbiol.*, **138**: 363-368.

Fioretto, A. and A. Fuggio, 2005. Biotechnology of soil enzymes. In: Microbial Biotechnology in Agriculture and Aquaculture (ed.) R.C. Ray, Science Publishers, New Hampshire, in press.

Flowers, T.J., P.K. Troke and A.R. Yeo, 1977. The mechanism of salt tolerance in halophytes. *Annu. Rev. Plant Physiol.*, **28**: 89-121.

Fry, J.V., M.A. Hufleji, W.W. Erber, G.A. Peschek and L. Packer, 1986. The role of respiration during adaptation of the fresh water *cyanobacterial Synechococcus-6311* to salinity. *Arch. Biochem. Biophys.*, **244**: 686-691.

Gagne, S., C. Richard, H. Rosseau and H. Antoun, 1987. Xylem -residing bacteria in *alfalfa* roots. *Can. J. Microbiol.*, **33:** 996-1000.

Garg, S.K. and A. Bhatnagar, 2005. Use of microbial bio-fertilizers for sustainable aquaculture/ fish culture. In: Microbial Biotechnology in Agriculture andAquaculture (ed.) R.C. Ray Science Publishers, New Hampshire, in press.

Gaur, A.C. 1990. Phosphate solubilizing micro-organisms as bio-fertilizers. Omega Scientific Publishers, New Delhi, India, pp.

Gebhardt, C., C. Turner, A.H. Gibson, B.L Dreyfus and F.J. Bergerson, 1984. Nitrogen fixing growth in continuous culture of a strain of *Rhizobium* sp. isolated from stem nodule of *S. rostrata. J. Gen. Microbiol.*, **130**: 843-848.

Geelen, D., K. Goethals, M.V. Montagu, and M. Holsters, 1995. The *nod* D locus from *Azorhizobium caulinodans* is flanked by two repetitive elements. *Gene*, **164(1)**: 107-

Ghai, S.K., D.N. Rao and L. Batra. 1988. Nitrogen contribution to wetland rice by green manuring with *Sesbania* sp. in an alkaline soil. *Biol. Fertil. Soils*, **6**: 22-25.

Goethals, K., M. Rao, K. Tomekpe, *nod ABC* genes in *Nod* locus M. van Montagu and M. Holsters, 1989. Common of *Azorhizobium Caulinodans:* nucleotide sequence and plant-inducible expression. *Mol. Gen. Genet.*, **219**: 289-298.

Gopalaswamy Anthoni Raj, S and A. Abdul Kareem, 1994. Interaction of herbicid with *Azolla* and soil microbes. *Indian. J. Weed Sci.*, **26(3&4)**: 28-34.

Gopalaswamy, G, S. Kannaiyan, K.J. 0' Callagham, M.R. Davey and C. Cocking 2000. The xylem of rice *(Oryza sativa)* is colonized by *Azorhizobium caulinodans. Proc. Royal Soc. Lond.*, **267**: 103 -107.

Gopalaswamy, G and S. Kannaiyan, 2000. Changes in soil microbial and enzyme activities during decomposition of biomass of *Azolla* hybrids. *J. Microb. World*, **2(2)**: 1-8

Gopalaswamy, G. and S. Kannaiyan, 1996. Effect of growth regulator *Phykos*, **35(1&2)**: 183-188.

Gopalaswamy, G., B. Tamil Selvam and S. Kannaiyan, 2002. Biomass production and biochemical composition of acid tolerant cyanobacteria. *Indian J. Microbiol*, **42**:121-124.

Gopalaswamy, G., P. Vidhyasekaran and S. Chelliah, 1989. Effect of *Azospirillum lipoferum* inoculation and inorganic nitrogen on wetland rice. *Oryza*, **26**: 378-380.

Gough, C., G. Webster, J. Vasse, C. Galera, C. Batchelor, K.J. 0' Callaghan and M.R. Davey, 1996. Specific flavonoids stimulate intercellular colonization of non-legumes by *Azorhizobium caulinodans.* In: Biology of Plant Microbe Interactions (eds.) G. Stacey, B. Mulin and P.M Gresshoff, Proc. 8th Int. Congo Molec. Plant Microbe Interactions, 14-19 July 1996, Knoxville, USA, pp. 409-415.

Govindan, M and D. Purushothaman, D. 1985. Association of nitrogen fixing bacteria with certain plantation crops. *Natl. Acad. Sci, Lett.*, India, **8**: 1631-1665.

Goyal S.K and G.S. Venkataraman, 1971. Response of high yielding rice varieties to algalization. II. Interaction of soil types with algal inoculation. *Phykos*, **10 (1- 2)**: 32-33

Goyal, S.K. 1982. Blue-green algae and rice cultivation. Proc. Natl. Sym. Biol. Nitrogen Fixation, Indian Agric. Res. Inst., New Delhi, India, pp. 346-376.

Hagerup, O. 1928. En hygrotil Bael-g-plante *(Achynomene aspera* L.) med bakterieknolde paa staengelen. *Dan Bot.*, **5**: 19.

Hall, D.O., S. Kannaiyan and M. Vander Leij, 1997. Ammonia bio-fertilizer production in rice paddies using immobilized cyanobacteria. In: Asian Paddy Fields (eds.) Oshirra, Y.E. Spratt and J.W.B. Stewart, Univ. Saskatchewan, Canada, pp. 218.

Hallman, J., A.M. Quadt, W.E. and J.W. Klopper, 1997. Bacterial endophytic association in agricultural crops. *Can. J. Microbiol.*, **43**: 895-914.

Hungria, M., A.R.J. Eaglesham and R.W.F. Hardy, 1992. Physiological comparisons of root and stem nodules of *Aeschynomene scabra* and *Sesbania rostrata. Plant Soil*, **139(1)**: 7-13.

Iyer, V., T. Fernandes and S.K. Apte, 1994. A role for osmotic stress induced proteins in the osmotolerance of nitrogen fixing cyanobacterium *Anabaena sr. strain L-31. J. Bacteriol.*, **176(18)**: 5868-5870.

Jain, D. and D. Patriquin, 1984. Root hair deformation, bacterial attachment and plant growth in wheat - *Azospirillum* interaction. *Appl. Environ. Microbiol.*, **48**: 1208-1213.

James, E.K., P.P.M. Fannetta, P. Nixon, A.I. Whiston, L. Peat, R.M.M. Crawford, J.I. Sprent and N.J. Brewin, 1996. Photosystem II and oxygen regulation in *Sesbania rostrata* stem nodules. *Plant Cell Environ.*, **19(8)**: 895-910.

James, E.K., V.M. Reis, F.L. Olivares, J.I. Baldani and J. Dobereiner, 1994. Infection of sugarcane by the nitrogen fixing bacterium *Acetobacter diazotrophicus. J. Exp. Bot.*, **45**: 757-766.

Jayanthi, M., V. Rajagopal and M.V. Shantaram, 1994. Studies on conjunctive use of *Azolla* and fertilizer in rice cultivation. Natl. Seminar on *Azolla* and Algal Bio-fertilizers for Rice, Tamil Nadu Agricultural University, Coimbatore, Tamil Nadu, India (abst.), p. 23.

Jegocki, R.P., A.R.J. Eaglesham and A.A. Szalay, 1983. Stem nodulation in *Aeschynomene*. A model system for bacterium-plant interaction. In: Molecular Genetics of the Bacteria- Plant Interaction (ed.) A. Puhler, Springer Verlag, Berlin, Heidelberg, pp. 210-219.

Jenik, J and Kubikova, J. 1969. Root systems of tropical trees. *Peslia, Prague, Czech. Rep.*, **14:** 59-65.

Jha, M.N and Kaushik, B.D. 1988. Response of *Westiellopsis prolifca* and *Anabaena* sp. to salt stress II. Uptake of Na^+ in the presence of K^+ as chloride, nitrate and phosphate. *Curr. Sci.*, **57:** 667-668.

Jurgensen, M.F and C.B. Davey, 1968. Nitrogen fixing blue green algae in acid forest and nursery soils. *Can. J. Microbiol.*, **14:** 1179-1183.

Kalidurai, M. and S. Kannaiyan, 1998. Studies on biomass production, nodulation and nitrogen fixation in stem nodulating *Sesbania rostrata.* In: Int. Symp. Biological Nitrogen Fixation in Rice Soil. Central Rice Res. Inst., Cuttack, India. (Abstr.), p. 18.

Kannaiyan, S. 1978. A note on *Azolla and* blue-green algae. Special Bull., Dy. Director Agric., Madras, Tamil Nadu, India, pp. 18.

Kannaiyan, S. 1981. *Azolla* bio-fertilizer for rice. International Network for Soil Fertility and Fertilizer Evaluation for Rice (INSFFER) Training Seminar, Int. Rice Res. Inst., Los banos, Manila, Philippines, pp. I-II.

Kannaiyan, S. 1982. *Azolla* and rice. In: Multiplication and use of *Azolla* Bio-fertilizer For Rice Production. Tamil Nadu Agric. Univ., Coimbatore, India, pp. 1-56.

Kannaiyan, S. 1983. Studies on the effect of *algal* bio-fertilizer inoculation of rice crop. Natl. Conf., Soc. Basic Appl. Microbiologist, Dungar College, Bikaner, India, p. 26.

Kannaiyan, S. 1985a. Potentiality of *Azolla* bio-fertilizer for rice. In: Soil Biology (eds.) M.M. Mishra and K.K. Kapoor, Haryana Agric. Univ., Hisar, India, pp. 255-259.

Kannaiyan, S. 1985b. Studies on the Algal Application for Lowland Rice Crop. Technical Bulletin, Tamil Nadu Agric. Univ., Coimbatore, Tamil Nadu, India, p. 21.

Kannaiyan, S. 1987. Sheath blight disease of rice. In: Advances in Rice Pathology (ed.) S. Kannaiyan, Tamil Nadu Agri. Univ. Coimbatore, India, pp. 21 0-233.

Kannaiyan, S. 1989. *Azolla* bio-fertilizer for rice. In: Proc Natl. Seminar Biotech. Mol Biol, C.M.S. College, Kottayam, Kerala, India, p. 15.

Kannaiyan, S. (1990a). Bio-technology of Bio-fertilizer for Rice Crop. Tamil Nadu Agric. Coimbatore, Tamil Nadu, India, pp. 212.

Kannaiyan, S. 1990b. *Azolla-Anabaena* complex as bio-fertilizer for rice. Proc. Current Trends in Biotechnol., Cochin Univ. Sci. Technol., Cochin, Kerala, India, pp. 16-27.

Kannaiyan, S. 1990c. Blue green algal bio-fertilizers. In: Biotechnology of Bio-fertilizers for Rice Crop. (ed.) S. Kannaiyan, Tamil Nadu Agric. Univ. Publication, Coimbatore, Tamil Nadu, India, and p. 225.

Kannaiyan, S. 1992a. Potentiality of *Azolla* as bio-fertilizer in sustainable rice farming. In: Biodiversity Implications for Global Food Security (eds.) M.S. Swaminathan and S. Jana, M.S. Swaminathan Res. Foundation, Chennai, India, pp. 221-223.

Kannaiyan, S. 1992b. *Azolla* Bio-fertilizer Technology for Rice Crop. Tamil Nadu Agric. Univ., Coimbatore, Tamil Nadu, India, pp. 54.

Kannaiyan, S. 1993a. Bio-fertilizers for Rice. Nadu, India. Tamil Nadu Agric. Univ. Coimbatore, pp. 60.

Kannaiyan, S. 1994. Ammonia production by immobilized cyanobacteria for its use as bio-fertilizer for rice crop. Final Technical Report, Tamil Nadu Agric. Univ., Coimbatore, Tamil Nadu. India, p. 58.

Kannaiyan, S. 1996. Technology Development and Demonstration of Bio-fertilizer-*Azolla* and BGA. Technical Report. Tamil Nadu Agricultural University, Coimbatore, Tamil Nadu, India, pp. 65.

Kannaiyan, S. 2000. Bio-fertilizer Technology and Quality Control. Publication Directorate, Tamil Nadu Agric. Univ., Coimbatore, India. pp. 256.

Kannaiyan, S. 2003. Inoculant production in developing countries: problems, potentials and success. In: Maximizing the use of Biological Nitrogen Fixation in Agriculture (eds.) G. Hardarson and W. J Broughton, Kluwer Academic Publishers, Dordrecht, Netherlands, pp. 187-198.

Kannaiyan, S and C. Sompom, 1987. Studies on sporulation, biomass production and nitrogen fixing potential of eleven cultures of the aquatic fern *Azolla. Indian J. Microbiol,* **27**: 22-25.

Kannaiyan, S., K. Kumar and G. Gopalaswamy, 2001. Stem nodulating *Sesbania rostrata* as a bio-resource for rice production, induction of nodulation and endophytic N_2 fixation in rice. *Mad. Agric. J.,* **88 (10-12)**: 547-560.

Kannaiyan, S., S.J. Aruna, P.K.S. Merina and D.O. Hall, 1997. Immobilized *cyanobacteria* as a bio-fertilizer for rice crops. *J. Appl. Phycol.,* **9**: 167-174.

Kannaiyan, S., K.K. Rao and D.O. Hall, 1994. Immobilization of *Anabaena azollae from Azolla filiculoides* in polyvinyl foam for ammonia production in photo bioreactor system. *World J. Microbiol. Biotechnol,* **10**: 55-58.

Kannaiyan, S., M. Thangaraju and G. Oblisami, 1981. Mineral nutrition of *Azolla.* Tamil Nadu Agric. Univ. Newslett., India **8**: 2.

Kannaiyan, S., M. Thangaraju, M. and G. Oblisami, 1982. Studies on the multiplication and utilization of *Azolla* bio-fertilizer for rice. In: Proc. National Symp. Biological Nitrogen Fixation. Indian Agri. Res. Inst, New Delhi, India, pp. 451- 460.

Kannaiyan, S., M. Thangaraju and Oblisami, G. 1984. Effect of *Azolla* application on grain yield of rice. *Agric Res. J. Kerala (India),* **22**: 183-185.

Kannaiyan, S. 1993b. Nitrogen contribution by *Azolla* to rice crop. *Proc. Indian Natl. Science. Acad.,* **59(5&6)**: 21-31.

Kattupitiya, S., P.B. New, C. Elmerich and I.R. Kennedy, 1995a. Improved nitrogen fixation in 2, 4-D created wheat roots associated with *Azospirillum lipoferum:* Colonization using reporter genes. *Soil Biol. Biochem.* **27**: 447-452.

Kattupitiya, S., J. Millet, M. Vesk, L. Viccars, A. Seman, L. Zhao, C. Elmerich, and I.R. Kennedy, 1995b. A mutant of *Azospirillum brasilense* Sp7 impaired in flocculation with modified colonization and superior nitrogen fixation in association with wheat. *Appl. Environ. Microbiol.* **61**: 1987 -1995.

Kennedy, I.R. 1994. Auxin induced nitrogen -fixing associations between *Azospirillum brasilense* and wheat. In: Nitrogen Fixation with Non-Legumes. (eds.) N.A. Hegazi, M. Fayez and M. Monib, Cairo Press, Cairo, Egypt, pp.513-523.

Kennedy, I.R. 1997. The establishment of new non-nodular endophytic systems. In: Biological Nitrogen Fixation: The Global Challenge and Future Needs. (eds.) I.R. Kennedy and E.C. Cocking. The Rockefeller Foundation, Bellagio Conference Centre, Lake Como, Italy, April 8-12, 1997, pp.65-67.

Kennedy, I.R. and Y.T. Tchan, 1992. Biological nitrogen fixation in non-leguminous field crops: Recent advances. *Plant Soil,* **141**: 93 -118.

Kennedy, I.R., L. Lily, C. Pereg-Gerk, R. Wood, K. Deaker and S. Katupitiya, 1997. Biological nitrogen fixation in non-leguminous field crops: Facilitating the evolution of an effective association between *Azospirillum* and wheat. *Plant Soil,* **194**: 65-79.

Kerby, N.W., S.C. Musgrave, S.Y. Shestakov and W.O.P. Stewart, 1986. Photo-production of ammonium by immobilized mutants strains of *Anabaena variabilis. Appl. Microbiol. Biotechnol,* **24**: 42-45.

Kijne, I.W. 1992. The *Rhizobium* infection process. In: Biological Nitrogen Fixation (eds.) G. Stacey, R.H. Bunis and H.J. Evans, Chapman and Hall, New York, pp.349-398.

Kitts, C.L. and R.A. Ludwing, 1994. *Azorhizobium caulinodans* respires with at least four terminal oxidases. *J. Bacteriol.,* **176**: 886-895.

Kloepper, J.W., B. Schippers and P.A.H.M. Bakker, 1992. Proposed elimination of the term endorhizosphere. *Phytopathology,* **82**: 726-727.

Kulasooriya, S.A., S.A. Hirimburegama and S.W. Abeysekava, 1987. Use of *Azolla* in Sri Lanka. In: *Azolla* utilization. Proc. Workshop on *Azolla* Use, Fuzhou, Fujian, China, Int. Rice Res. Inst., Los Banos, Philippines, p. 140

Kumar, K., M. Thangaraju, and S. Kannaiyan, 1995. Use of nitrification inhibitors and nitrogen use of efficiency in lowland rice soil In: Rice Management Biotechnology (ed.) S. Kannaiyan, Associated Publishing Co., New Delhi, India, pp. 101-106.

Kumar, K. 1994. Studies on the development of *Azolla* mutants and their nitrogen fixing potential, microbial decomposition and bio-fertilizer value for rice. Ph.D thesis, Tamil Nadu Agric. Univ., Coimbatore, Tamil Nadu, India, pp. 323.

Kumar, K. and S. Kannaiyan. 2000. Selection of mutants of *Azolla* for higher nitrogen potential, ammonia assimilation and nutrient content. *J. Microb. World,* **2 (1)**: 35-41

Kumar, K., S. Kannaiyan, and Cocking, E.C. 1996. Studies on the colonization of *Azorhizobium caulinodans* in rice roots and their role in symbiotic nitrogen fixation. Paper presented in the 37th Annual conference of the Association of Microbiologists of India, IiT, Chennai, 4-6, December, (Abstract), and p. 118.

Kwon, O.K and H. Beevers, 1993. Adverse effects of nitrate on stem nodules of *Sesbania rostrata* Brem. *New Phytol.,* **125(2):** 345-350.

Ladha, J.K., M. Garcia, S. Miyan, A.T. Padre, and Watanabe, 1998 Survival of *Azorhizobium Caulinodans* in the soil and rhizosphere of wetland rice under *Sesbania rostrata. Appl. Environ. Microbiol.,* **55**: 454- 460

Ladha, J.K., R. Pareek, and M. Becker, 1992. Stem nodulating legume -*Rhizobium* symbiosis and its agronomic use in low land rice. *Adv. Soil Sci.,* **20**:147-192

Lakshmikumari, M. S. K. Kavimandan, and N.S. Subba Rao, 1976. Occurrence of nitrogen fixing *Spirillum* in roots of rice, sorghum, maize and other plants. *Indian J. Exptl. Biol.,* **14**: 638-639

Latha, K.R., S. Kannaiyan, and S. Subramanian, 1988. Uptake of nitrogen in different growth phases of rice through *Azolla* biofertilizer. *Acta Agron. Hungarica,* **37(3-4)**: 239-243

Liu Chung Chu. 1987. Re-evaluation of *Azolla* utilization in agricultural production. In: *Azolla* Utilization. Proc. on the Workshop on *Azolla* use. Fuzhou, Fhjian, China, Int. Rice Res. Instt. Los banos, Manila. Philippines, pp. 67-76.

Long, S.R. 1996. *Rhizobium* symbiosis- Nod factors in perspective. *The Plant Cell,* **8:** 1885 -1898

Loudhapasitipom, L. and C. Kanareugsa 1987. *Azolla* use in Thailand. In: *Azolla* Utilization Proc. Workshop on *Azolla* Use, Fuzhou, Fujian, China Int. Rice Res. Instt., Los Banos, Manila, Philippines, pp.119-122.

Mabbayad, B.B. 1987. The *Azolla* programme of the Philippines. In: *Azolla* Utilization. Proc. Workshop *Azolla* Use, Fuzhou, Fujian, China, Int. Rice, Res. Instt., Los Banos, Manila, Philippines, pp.101-108.

Madhusoodanan, P.V and T.K. Dominic, 1995. On the acid tolerance of certain strains of cyanobacteria from Kerela. Natl Symp. Algal Biofertilizer, Tamil Nadu Agric. Univ. Coimbatore, India, (Abstract), p19.

Magalhaes, F. M., J. L. Baldani, J. Santo, J.R. Kuykendall and J. Dobereiner, (1983) A new acid tolerant *Azaspirillum* species. *Ann. Acad. Brar. Clienc.,* **55**: 417-430.

Mahesh, G. S. and Kannaiyan, 1994. Influence of certain growth regulators on the germination and fertilization of sporocarps *Azolla microphylla.* Paper presented in the National Seminar on *Azolla* and Algal Biofertilizer for Rice. Tamil Nadu Agric. Univ., Coimbatore, India, 1-3, March, 1994 (Abstract), p. 16.

Meelu, O.P. and R.A. Morris, 1988. Green manure management in rice-based cropping systems. In: Sustainable Agriculture: Green Manure Rice Farming. Int. Rice Res. Inst., Los Banos, Manila, Philippines, pp.209-222.

Militor, V., W.W.A. Erber, and G.A. Peschek, 1986. Increased levels of cytochrome C oxidase and sodium antiporter in the plasma membrane of *Anaystis nidulans* after growth in sodium enriched media. *FEBS Lett.,* **204**: 151-255.

Miller, D.M., J.V. Johnes, J.H. Yopp, D.R. Tindall, and W.O. Schmid 1976. Ion metabolism in a halophilic blue green alga *Aphanothece halophytica. Arch. Microbiol.*, **111**:145-149.

Mitra, A. K 1951. The algal flora of certain Indian soils. *Indian J. Agric. Sci.*, **21:** 375.

Moore, A. W. 1963. Occurrence of non-symbiotic nitrogen fixing microorganisms in nitrogen soils. *Plant Soil.*, **19**: 385-395.

Morris, RA, R. E. Furoc, R. Rajbandari, E. P. Marqueses and M.A. Dizon, 1987. Nitrogen accumulation in water logging tolerant green manure. *Agron. J.*, **79(5)**: 937-945.

Ndoye, I. and B.L. Dreyfus, 1988. N_2 fixation by *Sesbania rostrata* and *Sesbania sesban* estimated using ^{15}N and total N difference methods. *Soil Biol. Biochem.*, **20**: 209-213.

Ndoye, I., F. De Billy, J. Vasse, B. Dreyfus, and G. Trucjet, 1994. Root nodulation of *Sesbania rostrata. J. Bacteriol.*, **176**:1066 -1068.

Neyra, C .A. and J. Dobereiner, 1977. Nitrogen fixation in grasses. *Adv. Agron.*, **29:** 1-38.

Nie, Y. E. 1983. Studies on induced nodulation in plants lacking root nodules. *Nature.*, **6**: 326-360.

O'Callaghan, K. J., M.R. Davey and E.C. Cocking, 1997. Xylem colonization of the legume *Sesbania rostrata* by *Azorhizobium caulinodans. Proc. Royal. Soc. Lond.*, **B 264**: 1821- 826.

O'Hara, G.W., N. Boonkerd, and M.J. Dilworth, 1988. Mineral constraints to nitrogen fixation. *Plant Soil*, **108**: 93-110.

Okon, Y and C.A. Labandera-Gonzalez, 1994. Agronomic applications of *Azospirillum:* An evaluation of 20 years worldwide field inoculation. *Soil Biol. Biochem.*, **26:** 1591-1601.

Okon, Y. 1985 *Azospirillum* as a potential inoculant for agriculture. *TIBTECH.*, **3(9):** 223-228.

Palaneeswar, R., S.P. Sundaram, and A. Balasubramanian, 1992. Occurrence of acid producing *Azospirillum* in rice soil ecosystem. In: Biological Nitrogen Fixation and Biogas Technology. (eds.) S. Kannaiyan, K. Ramasamy, K. Ilamurugu and K. Kumar, Tamil Nadu Agricultural University, Coimbatore, Tamil Nadu, India, pp. 66-69.

Pandey, D. C. 1965. A study of the algae from paddy soils of Ballia and Ghazipur districts of Uttar Pradesh, India. 1. Cultural and ecological considerations. *Nova Hedwingia*, **9**: 299-334.

Pareek, R.P., J.K. Ladha, and Watanabe, I. 1990. Estimating N_2 fixation by *Sesbania rastrata and* S. *cannabina* (syn S. *aculeata*) in lowland rice soil by I dilution method. *Biol. Fertil. Soils.*, **10**: 77- 88.

Patel, L.B., B.S. Sidhu, and V. Beri, 1996. Symbiotic efficiency of *Sesbania rastrata* and S. *cannabina* as affected by agronomic practices. *Biol. Fertil Soils*, **21(3)**: 149-151.

Paul, P.L. and V.K. Mandhare 1991. Effect of direct incorporation of *Azolla* with nitrogen levels on yield of rice. 31st Annu. Conf. Assoc. Microbiol. India, Tamil Nadu Agric. Univ., Coimbatore, Tamil Nadu, India, Abstract, p. 135.

Paul, J. V. and S. Kannaiyan, 1994. Effect of the herbicide Butachlor on ammonia excretion by the colchicine mutants of *Anabaena azollae* (AS-OS). Paper presented in the National Seminar on *Azolla* and Algal Biofertilizer for Rice. Tamil Nadu Agric. Univ., Coimbatore, India, 1-3, March, 1994, Abstract, p. 3.

Prasad, B. N., R.K. Mehrotra and Y. Singh, 1978. On pH tolerance of some soil blue- green algae. *Acta Bot. Indica*, **6(2):** 130-138.

Purushothaman, D and K. Govindarajan 1984. Rhizosphere ecology and nitrogen fixation of *Azospirilhlm* in pearl millet *(Pennisetum americamum* (L.) Leeke. In: Working Group Meeting on Cereal Nitrogen Fixation. 9-12, October 1984, Int Crop. Res. Inst Semi-Arid Tropics, Patencheru, Andhra Pradesh, India, Abstract, p. 16.

Purushothaman, D., S. Gunasekaran and G. Oblisami, 1980. Nitrogen fixation by *Azospirillum* in some tropical plants. *Proc. Ind. Natl. Sci. Acad.*, **46**: 713-717.

Purushothaman, D., K. Jayanthamurali, and S. Gunasekaran 1987. Association of *Azospirillum* with the roots of upland rice. In: Biofertilizers: Potentials and Problems. (eds.) S. P Sen and P. Palit. Naya Prokash. Kolkata, India, pp. 171-182.

Quispel, A. 1974. The endophyte of root nodules in non-leguminous plants. In: The Biology of Nitrogen Fixation (ed.) A. Quipel, North -Holland Publishing Company, Amsterdam The Netherlands, pp. 520-599.

Rai, A.K and G. Abraham 1993. Salinity tolerance and growth analysis of the cyanobacterium, *Anabaena doliolum. Bull. Environ. Cont. and Toxicol.*, **51(5):** 724-73.

Rajagopal V., M. Jayanthi and M.V. Shantaram 1994. Growth studies on *Azolla* under Telangana conditions of Andhra Pradesh. India. Natl Seminar on *Azolla* and Algal Biofertilizer for Rice, Tamil Nadu Agric. Univ., Coimbatore, Tamil Nadu, India. Abstract. p. 15

Rao, V.R., P.B.B.N. Charyulu, D.N. Nayak and C. Ramakrishna 1978. Nitrogen fixation by free-living organisms in tropical rice soils. In: Limitations and Potentials for Biological Nitrogen Fixation in the Tropics. (ed). J. Dobereiner, R.H. Burris and A. Holluender, Plenum Press, New York, USA. pp. 354-368.

Reddy, P.M., J.K. Ladha, R.B. So, R. Hernandez, M.C. Ramos, O.R. Angeles, E.B. Dazzo, and E.J. de Bruijn. 1997. Rhizobial communication with rice roots: Induction of phenotypic changes, mode of invasion and extent of colonization. *Plant Soil.*, **194**: 81-98.

Reed, R and W.D.P. Stewart 1988. The responses of cyanobacteria to salt stress. In: Biochemistry of the Algae and Cyanobacteria. (eds.) L.J. Rogers and J.R. Gallon, Clarendon Press, Oxford. pp. 217-237.

Reed, R.H., L.J. Borowitzka, M.A. Mackay, J.A. Foster, S.R.C. Warr, D.J. Moore and W.D.P. Stewart, 1986. Organic solute accumulation in osmotically stressed cyanobacteria. *FEMS Microbiol Lett.*, **2**:83-86

Reinhold-Hurek, B., T. Hurek, I. Fendrik, B. Pot, M. Gillis, K. Kersters, S. Thielemans, and J. DeLey, 1987. *Azospirillum halopraeferens* sp. nov., a nitrogen-fixing organism associated with roots of kallar grass (*Leptochola fusca*) (L) *kunth*). *Int. J. Syst. Bacteriol.*, **37**: 43-51.

Reynders, K.L and Vlassak, 1976. Nitrogen fixing Spirillum species in Belgium soils. *Agriculture (Belgium)* **24**: 329-336.

Rinaudo, G., B.L. Dreyfus and Y. Dommergues 1983. Influence of *Sesbania rostrata* green manure on the nitrogen content of rice crop and soil. *Soil Biol. Biochem.*, **15**: 111-113.

Rinaudo, G., B. Dreyfus, Y.R. Land Dommergues 1982. *Sesbania rostrata* as a green manure for rice in West Africa. *BNF Technol. Trop. Agric.*, 441- 445.

Rinaudo, G., S. Orenga, M.P. Fernandez, H. Meugnier, and R.A.D. Bardin, 1991. DNA homologies among members of the genus *Azorhizobium* and other stem and root nodulating bacteria isolated from the tropical legume *Sesbania rostrata. Int. J. Syst. Bacteriol*, **41(1):** 114-120.

Robertson, B.K, B.L. Dreyfus and M. Alexander 1995. Ecology of stem nodulating *Rhizobium* and *Azorhizobium* in four vegetation zones of Senegal. *Microb. Ecol.*, **29**: 71- 81

Robertson, B.K. and M. Alexander, 1994. Mode of dispersal of stem nodulating bacterium. *Soil Biol. Biochem.*, **26(11):** 1535-1540.

Rodriguez, H. and R. Fraga 1999. Phosphate solubilizing bacteria and their role in plant growth promotion. *Biotechnol. Adv.*, **17**: 319-339.

Roger, R.A. and I. Watanabe, 1986. Technologies for using biological nitrogen fixation in wetland rice: Potentialities, current stage and limiting factors. *Fertil. Res.*, **9**: 39-77.

Rolfe, B.G. (and 13 others) 1998. Round table: Agriculture 2020: Eight billion people. In: Current Plant Science and Biotechnology in Agriculture: Biological Nitrogen Fixation for the 21st Century. (eds.) C. Elmerich, A. Kondorosi and W.E. Newton, Kluwer Academic Publishers, Dordrecht, The Netherlands, pp.685-692.

Sabry, S.R.S., S.A. Saleh, C.A. Batchelor, J. Jones, J. Jotham, G. Webster, S. Kothari, M.R. Devey, and E.C. Cocking, (1997) Endophytic establishment of *Azorhizobium caulinodans* in wheat. *Proc. Soc. Lond. B.*, **264**: 341-346.

Sadasivam, S., P.H. Lakshmi and S. Kannaiyan 1993. Phosphoenolpyruvate carboxylase activity, fixation of ^{14}C in amino acids and nitrogen transport in stem nodules of *Sesbania rostrata. Biologia Plant.*, **35(2):** 303-06.

Saraswati, R., Matoh, and J. Sekiva 1992. Nitrogen fixation of Sesbania rostrata: contribution of stem nodl1les to nitrogen acquisition. *Soil Sci. Plant Nutr.*, **38(4)780:** 775-780.

Sardeshpandey, J.S. and S.K. Goyal, 1981 a. Distributional pattern of blue green algae in rice field soils of Konkan region of Maharastra State. *Phykos,* **20 (1 &2):** 102-106.

Sardeshpandey, J.S. and S.K. Goyal 1981 b. Effect of pH on growth and nitrogen fixation by blue green algae. *Phykos,* **20(1-2):** 107-113.

Saxena, S and B.D. Kaushik 1991. Polysaccharides (biopolymers) from halotolerant cyanobacteria. *Indian J. Exptl Biol.,* **30**: 433-434.

Schaede, R. 1940. Die Knollchen der adventiven Wasser Wurzeln Yon *Neptunia oleracae* and ihre Bakteriensymbiose. *Planta.,* **31**: 1-21.

Schubert, H., S. Fulda and M. Hagemann, 1993. Effects of adaptation of different salt concentration on photosynthesis and pigmentation of the *Cyanobacterium, Synechocystis* sp. PCC 6803. *J. Plant Physiol.,* **142**: 291-295.

Selvakumar, G., G. Gopalaswamy and S. Kannaiyan 2002. Pigment analysis and ammonia excretion in herbicide tolerant cyanobacterial. *Indian J. Exptl. Biol.,* **40:** 934 -940.

Senthil Kumar, M. 2000. Studies on the induction of para nodules by the stem nodulating *Azorhizobium caulinodans* and its nitrogen fixing activity in rice and maize. M.Sc., (Ag) thesis, Tamil Nadu Agricultural University, Coimbatore, Tamil Nadu, India, pp. 73-92.

Shanmugasundaram, R. and S. Kannaiyan. 1992a. Studies on the sporocarp germination *of Azolla microphylla* and isolation of algal symbionts from megasporocarp. In: Biological Nitrogen Fixation and Biogas Technology (eds.) S. Kannaiyan, K Ramasamy, K Kumar, K. Ilamurugu, Tamil Nadu Agri. Univ. Coimbatore, India, pp. 44-48.

Shanmugasundaram, R. and S. Kannaiyan. 1994. Studies on the development of *Azolla* and algal bio-fertilizer for rice. Tamil Nadu Agric. Univ., Coimbatore, India, Abstract, p. 17.

Shanmugasundaram, R. and S. Kannaiyan. 1992 b. Effect of growth regulator and fungicide combination on germination, fertilization and viability of the dried sporocarps of *Azolla microphylla.* ***Phykos,*** **31**: 6 1-67.

Shi, D.J., M. Briers, D.O. Hall and R.J. Robins. 1987a. The effects of immobilization on the biochemical, physiological and morphological features of *Anabaena azollae.* Planta, **172**: 298-308.

Shi, D.J., D.O. Hall and P.S. Tang. 1987b. Photosynthesis, nitrogen fixation, ammonia photoproduction and structure of *Anabaena azollae* immobilized in natural and artificial systems. In: Progress in Photosynthesis Research. Vol II (ed.) J. Biggens, Martinus Nijhoff Publishers, Dordrecht, The Netherlands, pp. 641-644.

Singh, N.L., H.I. Dorycanta, G.A. Devi, N.S. Singh and M.S. Singh. 1997. Blue green algae from rice field soils of Nagai and. *Phykos,* **36(182):** 115-120.

Singh. P.K 1979. Use of *Azolla* in rice production in India. In: Nitrogen and Rice, Int. Rice Res. Inst., Los Banos, Manila, Philippines, pp. 407-418.

Singh, P.K. 1977a. Multiplication and utilization of fern *Azolla* containing nitrogen fixing algal symbiont, a green manure in rice cultivation. *IIRiso,* **26**: 125-137.

Singh, P.K 1977b. The use of *Azolla pinnata* as green manure for *Newsletter,* **2(7):** 7.

Singh, A.L and P.K. Singh. 1989. N_2 Fixation in Indian Rice Field *(Azolla* and Blue green algae). Agrobotanical Publisher, Bikaner, India, pp. 236.

Sinha, J.P. and D. Mukherjee. 1975. Blue-green algae from the paddy fields of Bankura districts of West Bengal. *Phykos,* **14(1-2):** 117-118.

Somasegaran, P and H.J. Hoben. 1994. Handbook for *Rhizobia*-Methods in Legume- *Rhizobium* Technology. Springer-Verlag, New York, USA, pp. 450

Srinivasan, G. 1980. *Azolla* manuring and grain yield of rice. *Int. Rice Res. Newslett.,* **6**: 25.

Stoltzfus, J.R, R. So, P.P. Malarvithi, J.K. Ladha and F.J. de Bruijn. 1997. Isolation of endophytic bacteria from rice and assessment of their potential for supplying rice with biologically fixed nitrogen. *Plant Soil,* **194**: 25-36.

Showers, M.P. and A.R.J. Eaglesham. 1983. A stem-nodulating physiological characteristics of both fast and slow growers. *J. Gen. Microbiol.*, **129**: 3651-3655.

Subba Rao, N.S. 1993. Bio-fertilizers in Agriculture and Forestry, III edn, Oxford & IBH Publishing Co., New Delhi, India.

Subba Rao, N.S. 1983. Nitrogen fixing bacteria associated with plantation and orchard plants. *Can. J. Microbiol.*, **29**: 863-866.

Subba Rao, N.S. 1988. Microbiological aspects of green manure in lowland rice soil. In: Sustainable Agriculture Green Manure in Rice Farming, Int. Rice Res Inst, Los Banos, Manila, Philippines. pp. 132-149.

Subhashini, R. K Kumar and S. Kannaiyan, 2000. Variation revealed by RAPD-PCR profiles of microsymbiont *Anabaena azollae* strains from four N_2- fixing *Azolla* cultures. *Acta Biotechnnol.*, **20(2)**: 169-174.

Subramani, S. and S. Kannaiyan, 1987. Effect of urea on decomposition of *Azolla. Int. Rice Res. Newslett.*, **12(4)**: 57.

Sussenguth, K., and R. Beyerle, 1936. Uber Bakterienknollchen am sprob von *Aeschynomene paniculata* wild. *Hedwigia,* **75:** 234-237.

Sundara Rao, W.V.B. and M.K. Sinha, 1963. .Phosphate dissolving organisms in soil and rhizosphere. *Indian J. Agric. Sci.,* **33:** 272-278.

Sundaram, S.P., S. Kannaiyan and S. Muthukumarasamy, 2001. Transformation studies in *Azospirillum halopraeferens.* www. microbiologyou. Com, p. 6.

Talley, S.N., B.J. Talley and D.W. Rains, 1977. Nitrogen fixation by *Azolla* in rice fields. In: Genetic Engineering for Nitrogen Fixation (ed.) AI Hollander, Plenum Publishing Company, New York, US, pp. 25- 2182.

Tchan, Y.T. and I.R. Kennedy, 1989. Possible N_2- fixing root nodules induced in non-legumes. Agricultural Science (Australian Institute of Alpines Sciences (AIAS) Melbourne New Series). **2**: 57-59.

Tchan, Y.T., A.M.M. Zeman and I.R Kennedy. 1991. Nitrogen fixation in paranodules of wheat roots by introduced free-living diazotrophs. *Plant Soil,* **137**: 43-47.

Thangaraju, M and Kannaiyan, S. (1990). Influence of *Azolla Sesbania* and urea super granules (USG) on rice yield and nitrogen uptake. *Int. Rice Res. Newslett.,* **15 (1)**: 25.

Thiruchenthilvasan, S. and Purushothaman, D. 1999. Rhizosphere colonization of genetically engineered *Azospirillum* strains and response of sorghum to inoculation. XI Southern Conference on Microbial Inoculants. 19th and 20th Feb. 1999. Guntur, India, Abstract, p. 1.

Tomekpe, K., Dreyfus, B.L. and Holsters, M. 1996. Root nodulation of *Sesbania rostrata* suppresses stem nodulation by *Sinorhizobium teranga* but not *Azorhizobium caulinodans. Can. J. Microbiol.,* **42(2)**: 187-190.

Tsien, H.C., B.L. Dreyfus and E.L. Schmidtt. 1983. Initial Stages in the morphogenesis of nitrogen-fixing stem nodules of *Sesbania rostrata. J. Bacteriol.,* **150**: 888-897.

Uma, D. and S. Kannaiyan. 1994. Effect of the Insecticide carbofuran on ammonia production by cyanobacteria under immobilized state in PU foam. Paper Presented in the National Seminar on *Azolla* and Algal Biofertilizers for Rice. 1-3, March 1994, Tamil Nadu Agric. Univ., Coimbatore, Tamil Nadu, India. p. 27

Urban, J.S., L.C. Davis and S.J. Brown. 1986. *Rhizobium trifolii* -0403 is capable of growth in the absence of combined nitrogen. *Appl. Environ. Microbiol.,* **52:** 1060-1067.

Vande Broek, A., J. Michiels, A. Van Gool and J. Vanderleyden. 1993. Spatial- temporal colonization patterns of *Azospirillum brasilense* on the wheat root surface and expression of the bacterial *nif H* gene during association. *Mol. Plant Microbe Int.,* **6:** 592-600.

Venkataraman, G.S. 1972. Algal Biofertilizers and Rice Cultivation. Today and Tomorrow's Printers and Publishers, New Delhi, India.

Visperas, R.M., R.E. Furoc, R.A. Morris, B.S. Vergara and G. Patna. 1987. Flowering response of *Sesbania rostrata* to photoperiod. Philipp. *J. Crop Sci.,* **12:** 147-149.

Vlachova, M., B.A. Metz, J. Schell and E.J. de Bruijn, 1987. The tropical legume *Sesbania rostrata* -Tissue culture, plant regeneration and infection with *Agrobacterium tumefaciens* and *rhizogenes* strains. *Plant Soil,* **50**: 213-223.

Vonshak, A., R. Guy and M. Guy. 1988. The response of the filamentous cyanobacterium, *Spirulina platensis* to salt stress. *Arch. Microbiol,* **150:** 417-420.

Wastyn, M., A. Achatz, M. Trnka and G.A. Peschek. 1987. Immunological and special characterization of partly purified cytochrome oxidase from the cyanobacterium, *Synechocystis* 6714. *Biochem. Biophys. Res. Commun.,* **149**:102-111.

Watanabe, I. 1987. Fertility and fertilizer evaluation for rice. In: *Azolla* Utilization. Proc. Workshop on *Azolla* use. Fuzhou Fujian China., IRRI. Los Banos, Philippines, pp. 208.

Webster, G., C. Cough, J. Vasse C.A. Batchelor, K.J. O'Callaghan, S.L. Kothari. M.R Davey, J. Denarie and E.C. Cocking. 1997. Interactions of rhizobia with rice and wheat. *Plant Soil,* **194**: 115-122.

Yahalom, E., Y. Okon, A. Dovrat, A. 1988. Early nodulation in legumes inoculated with *Azospirillum* and *Rhizobium. Symbiosis,* **6**: 69-80.

Yatazawa, M. and S. Yoshida. 1979. Stem nodules in *Aeschynomene indica* and their capacity of nitrogen fixation. *Plant Physiol.,* **45**: 293-295.

Yu, D. and I.R. Kennedy, 1995. Nitrogenase activity (C_2H_2 reduction) of *Azorhizobium* in 2, 4-D induced root structures of wheat. *Soil Biol. Biochem.,* **27:** 459-462.

Zahran, H.H. (2005) Rhizobial nitrogen fixation in agriculture: biotechnological perspectives. In: Microbial Biotechnology in Agriculture and Aquaculture (ed.) R.C. Ray, Science Publishers, New Hampshire, USA, in press.

Zeman A.M., Y.T. Tchan, C. Elmerich and L.R. Kennedy. 1992. Nitrogenase activity in wheat seedlings bearing para-nodules induced by 2, 4- dichlorophenoxyacetic acid (2, 4-D) and inoculated with *Azosprillum.* Res. *Microbiol.,* **143:** 847-855.

Zhang Zhuang Ta, Ke Yu-Si. Ling De-Quan, Duan Bing-Yuan and Liu Xi-lian. 1987. Utilization of *Azolla* in agricultural production in Guandong Province, China. In: *Azolla* Utilization. Proc. Workshop on *Azolla* Use. Fuzhou, Fujian, China, Int. Rice Res. Int., Los Banos, Manila, Philippines, pp. 141-145.

Soil Microbiology and Soil Biotechnology : Vol. 2
Editor : S. Kannaiyan
Associated Publishing Co., New Delhi, 2010, pp. 70-99

3

Azospirillum in Soil and its Colonization in Plant Roots

K. Kumar[1] *and S. Kannaiyan*[2]

Introduction

Azospirillum is a free-living, plant-growth-promoting bacterium (PGPB), capable of affecting growth and yield of numerous plant species, many of agronomic and ecological significance. Dobereiner and Day (1976) reported for the first time, the occurrence of nitrogen fixing bacterium *Spirillum lipoferum* (Later renamed as *Azospirillum lipoferum*) in the roots of digit grass in Brazil. The bacterium was found to fix nitrogen in association with roots of various plants. This startling new discovery opened new vistas for world's most staple food crops like rice, maize, sorghum, wheat and millets, which were previously believed to have no potential for biological nitrogen fixation. *Azospirillum lipoferum* and *A. brasilense* are primary inhabitants of soil, the rhizosphere and intercellular spaces of root cortex of graminaceous plants. Many other nitrogen-fixing azosprilla have now been described by the Brazilian workers such as *A. amazonense*, *A. halopraeferans* and *Herbaspirillum seropedicae* (Boddey and Dobereiner, 1988).

Isolates of these bacteria from Indian soils were made for the first time by Lakshmi Kumari *et al.* (1976). Subba Rao *et al.* (1979) tested its performance in pots and Subba Rao (1982-1986) investigated their agronomic potential in the field. Okon and Labandera-Gonzalez (1994) evaluated worldwide data accumulated over the past 20 years on field inoculation with *Azospirillum* and concluded that these bacteria are capable of promoting the yield of agriculturally important crops in different soils and climatic regions. The review of the extensive data collected revealed 60-70 per cent success with statistically significant increases in yield in 5-30 per cent of the trials. The leading theory concerning its growth promotion capacity lies in its ability to produce various phytohormones that improve root growth, adsorption of water and minerals that eventually yield larger, and in many cases, more productive plants (Dobbelaere *et al.*, 2001). The recent developments in the study of the distribution and survival of *Azospirillum* in soil and its mechanism of colonization in roots are discussed in

1. Professor, Dept. of Agrl.Microbiology, Tamil Nadu Agricultural University, Coimbatore-641003.
2. Former Chairman, National Biodiversity Authority, Chennai-600041.

detail in this chapter, for more effective utilization of this organism in the present context of global concern for environmentally conscious and economically viable crop production.

Distribution of *Azospirillum* in soil

Azospirillum spp. are of ubiquitous distribution in many parts of the world consisting of tropical, sub-tropical and temperate climatic conditions and also with various standing crops grown in a variety of soil types. Reynders and Vlassak (1976) from Belgium isolated this organism from more than 50 per cent of soil samples representing a wide range of soil types. Amer *et al.* (1977) reported that various geographical regions of the world were rich in *Azospirillum*. Neyra and Dobereiner (1977) reported that nearly 10 per cent of the soils and roots from temperate region have *Azospirillum*, while more than 50 per cent tropical soils were positive for *Azospirillum*.

In India, Rao *et al.* (1978) reported thc presence of *Azospirillum* in different rice soil types (alluvial, laterite, pokkali and kari). The occurrence of *Azospirillum* in the roots of cotton, maize, ragi, kudiraivali, guinea grass, hariyali grass (Purushothaman *et al.*, 1980), plantation crops, orchard plants (Subba Rao, 1983), pearl millet, arecanut, banana, cashew, coconut, cocoa, rubber and cardamom (Purushothaman and Govindarajan, 1984; Govindan and Purushothaman, 1985) and upland rice (Purushothaman *et al.*, 1987) was well documented.

A new acid tolerant species *viz.*, *Azospirillum amazonense* was reported by Magalhaes *et al* (1983) from Amazon region. Charyulu and Rao (1980) and Palaneeswar *et al.* (1992) observed the occurrence of *Azospirillum* with appreciable nitrogen fixing ability in highly acidic soils with extremely low pH. Reinhold-Hurek *et al.* (1987) reported *Azospirillum halopraeferens* characterized by its high salt tolerance. Salt tolerant *Azospirillum* isolates have also been reported by Sundaram *et al.* (2001). Thiruchenthilvasan and Purushothaman (1999) observed better colonization of sorghum rhizosphere by genetically engineered *Azospirillum* cultures. Two more species *viz.*, *A. doebereinerae* and *A. largomobile* have been included in the genus. *A. largomobile* was technically transferred, on the basis of some phylogenetic relationships, from the genus *Conglomeromonas* to *Azospirillum* (Dekhil *et al.*,1997). *A. doebereinerae* was found in association with roots of the gramineaous plant *Miscanthus* (Eckert *et al.* 2001). Kumar and Kumari Sugitha (2004) analysed the diversity endophytic diazotrophs associated with rice and found that *Azospirillum* was the dominant diazotroph in root followed by stem and leaf tissues of rice.

Characteristics of *Azospirillum* species

Earlier, taxonomists have recognized five species in the genus *Azospirillum viz.*, *A.brasilense, A. lipoferum, A. amazonense, A. halopraeferens* and *A. irakense* (Bashan and Levanony 1990; Bashan and Holguin 1997*a*). They are Gram negative, vibrioid in shape, motile with a long polar flagellum and occasionally with peritrichous flagella. Organic acids such as malate and succinate are the preferred carbon sources for *Azospirillum*. However the different species vary in their ability to use different carbon sources (Okon, 1985). Strains of the new species *viz.*, *A. doebereinerae* and *A. largomobile* are curved rods or S-shaped, 1.0-1.5 μm wide and 2.0-3.0 μm long, Gram-negative, and motile with a single polar flagellum. Nitrogen-fixation occurs under microaerobic, nitrogen-limited conditions. All these features are very similar to other *Azospirillum* spp. The trait that differentiates this species from others is its ability to use (or not) several sugars and some minute genetic details. Optimum growth occurs at 30°C, and at pH values between 6.0 and 7.0, but not at 37°C, as is the case for the other species (Eckert *et al.*, 2001).

Unique isolates

Azospirillum strains are routinely isolated from agricultural lands and crop plants, including traditional isolations from grasses and cereals (Nath *et al.*, 1997; Weber *et al.*, 1999). In recent years, using a combination of traditional isolation and *in situ* molecular detection, exemplified by phylogenetic oligonucleotide probes and PCR-fingerprinting techniques, strains of *Azospirillum* or sequences identical to sequences of *Azospirillum* have been found in many, previously unstudied, locations. Endophytic *A. lipoferum* was isolated from *Pennisetum purpureum*, *Miscanthus sinensis*, *Miscanthus sacchariflorus*, and *Spartina pectinata*, which are used for energy (Kirchhof *et al.*, 1997). PCRamplified *nifH* sequences from sediment in oligotrophic seagrass beds in the Bahamas, revealed several *nifH* sequences nearly identical to *A. brasilense* (Bagwell *et al.*, 2002). Using analyses of PCR amplicons containing the ribosomal intergenic spacer and partial 16S rRNA gene, thermotolerant *Azospirillum* strains were isolated from an aerated lagoon, where pulp and paper mill effluent are treated (Yu and Mohn, 2001). A high cell density strain of *Azospirillum* was found on the roots of black mangrove *Avicennia marina*. This saline tolerant bacterial strain has potential to improve beans and rice in coastal agricultural fields (Ravikumar *et al.*, 2002). Nitrogen-fixing microbial populations from leaf litter in an Oregon Douglas fir forest, assessed by restriction fragment length polymorphism (RFLP) of PCR products, showed that all the *nifH* sequences obtained from the forest litter were characterized by the *nifH* sequences of members of *Rhizobium*, *Sinorhizobium*, and *Azospirillum* genera (Widmer *et al.*, 1999).

Physiological mechanisms of root colonization

Attachment of *Azospirillum* to roots was determined to be a two-step process comprised of adsorption and anchoring. The primary adsorption phase is rapid (reaching a maximum within 2 h of exposure of the bacteria to roots), weak, reversible, and probably governed by bacterial proteinaceous compounds. The anchoring phase is stronger, takes several hours to form, is irreversible, and is based on bacterial extracellular surface polysaccharides involving a network of fibrillar material that permanently connects the bacteria to the surface and prevents removal from the site (Bashan and Holguin, 1997a). This attachment sequence in wheat has also been reconfirmed (Egorenkova *et al.*, 2000). Adsorption was evaluated by measuring the cell-surface charge and cell-surface hydrophobicity of 10 strains of *Azospirillum* spp. *Azospirillum* spp. had moderate cellsurface hydrophobicity and charge, lower than known values for human pathogens. Cell-surface hydrophobicity and charge of *Azospirillum* can be affected by external treatments of the bacterium cell, but a general pattern of modifying hydrophobicity and charge by external treatments of the cells of all strains was not found. Various chemicals, temperatures, and enzymatic treatments changed the cell-surface hydrophobicity and charge (Castellanos *et al*., 1997). *A. brasilense* Sp245 and its lipopolysaccharide-defective mutants were compared. Despite the similarity of adsorption dynamics of the parent strain and this mutant, attachment capacity of the mutant was lower than that of the parental strain. The mutant strain had a considerable decrease in hydrophobicity of the *Azospirillum* cell surface (Fedonenko *et al*., 2001). This shows that hydrophobicity and charge may play a role, perhaps small, in the primary adsorption of *Azospirillum* spp. to root surfaces.

When bacterial cells were subjected to starvation periods, many of the cell-wall characteristics changed. Nevertheless, no change was considered major, and apparently starvation is not a major limiting factor in the attachment of *Azospirillum* to surfaces (Castellanos *et al*., 2000). The effect of pH and cations was also studied. The presence of significant concentrations of Ca+2 or (PO4)-3 in the incubation medium reduced bacterial adsorption. The effect of the incubation medium's pH on adsorption of *Azospirillum* showed that strains from roots or rhizosphere of wheat had optimal adsorption at pH 6.0, while all other strains preferred pH 7.0 (Pinheiro *et al*., 2002). Additionally, lipopolysaccharides extracted from the outer membrane of *A. brasilense* Sp245 and its mutant cells induced deformation of the wheat seedling root hairs after exposure (Egorenkova *et al*., 2000; Fedonenko *et al*., 2001).

As some lectins were identified in the cell wall of *A. brasilense* and *A. lipoferum*, perhaps these compounds play a role in the adsorption phase (Castellanos *et al*., 1998). Adsorption of *A. brasilense* to roots of spring wheat was affected by the bacterial concentration and possible involvement of lectins. The number of attached cells increased with increasing inoculum size and time of contact. Saturation of root surface adsorption was observed after 3-24 h of incubation, depending on the strain. The longer the incubation period, the firmer was the attachment. Adsorption to roots was partially inhibited when the roots were treated earlier with *N*-acetyl-D-glucosamine, indicating the possible involvement of lectins. Root hair deformation induced by polysaccharides from *Azospirillum* capsular material was inhibited by *N*acetyl-D-glucosamine and chitotriose, specific haptens of the wheat germ lectin agglutinin (Yegorenkova *et al*., 2001).

A major outer membrane protein, acting as an adhesin and possibly involved in root adsorption and cell bacterial aggregation, was purified from *A. brasilense* Cd. This protein binds to root extracts of various plant species, but with stronger adhesion to extracts of cereals, in comparison with legumes and tomatoes. It also binds to roots of different cereal seedlings in an *in vitro* adhesion assay (Burdman *et al*., 2001). These studies indicate the participation of lectin molecules in the attachment process.

Involvement of polysaccharides and lectins in attachment

The role of cell-wall components, including lectin recognition between the plant and the bacterium, has been studied for decades without definite conclusions, contrary to the situation in rhizobia-legume symbiosis. Some advances in that topic have been made in Russia. *Azospirillum* is capable of secreting capsular polysaccharides, which is possibly involved in the adhesion of bacteria to surfaces and several other exopolysaccharides (EPS). Involvement of EPS in the aggregation phenomenon has been recently documented. Capsular polysaccharides, exopolysaccharides, and *O*-specific polysaccharides in *A. brasilense* strains Sp7 and Sp245, showed close antigenic identity. These extracellular polysaccharides (EPS) are likely to represent O-antigenic lipopolysaccharide fragments excreted by the bacteria into the culture medium. Identification as a capsule or EPS depends on the strength of the attachment of these polysaccharides to the cell surface. This could explain the absence of a specific capsular antigen in *A. brasilense* (Matora and Shchegolev, 2002).

Additionally, a new *O*-specific polysaccharide was isolated and characterized from the lipopolysaccharide of *A. brasilense* Sp245 (Fedonenko *et al.*, 2002). High-molecular-weight carbohydrate containing complexes (lipopolysaccharide–protein [LP] complex and polysaccharide–lipid [PL] complex) that were isolated from the capsular material of *A. brasilense* Sp245, are possibly responsible for protecting them from extreme conditions, such as high (46–48°C) and low (–20°C and –70°C) temperatures, extreme pH values (2 and 10), and desiccation, which are otherwise detrimental. Before exposure to extreme conditions, addition of these complexes to a suspension of decapsulated cells significantly increased survival rates (Konnova *et al.*, 2001). Maintaining the pH at 7.0 during incubation favors accumulation of LP and PL complexes in the capsule of *A. brasilense* (Kannova *et al.*, 2003). *In vitro*, some *Azospirillum* produced significant quantities of exocellular and capsular polysaccharide. After depletion of the carbon source from the culture medium, polysaccharides were consumed, especially by *A. lipoferum* strains, and this significantly increased nitrogenase activity, although oxygen uptake was low during this period. This allowed a mathematical model predicting biochemical activities of *Azospirillum* to be developed (Kefalogianni and Aggelis, 2002).

Interaction of EPS, LP complexes, PL complexes, lipopolysaccharides, and *O*-specific polysaccharides from *Azospirillum* with plant lectins is a promising avenue to explore (Skvortsov and Ignatov 1998). Also to be considered are bacterial lectins interacting with polysaccharides of plants. Cell-surface lectins (agglutination with fucose and glucose, and to a lesser extent, with mannose) were detected in 7 strains of *A. brasilense* and *A. lipoferum*. Cell-wall proteins extracted from 2 *Azospirillum* strains exhibited lectinlike activities (Castellanos *et al.*, 1998). A strong interaction between the *A. lipoferum* 59b lectin and 3 glucosidases was detected, but their exact role was not determined (Alenkina *et al.*, 2001). Under saline conditions, the normal composition of the EPS of *A. brasilense* Cd dramatically changed from 7 monosaccharides to mainly galactose. Root exudates induced changes in the EPS only under regular growth conditions, and led to higher amounts of arabinose and xylose (Fischer *et al.*, 2003). Attachment of *Azospirillum brasilense* to cereals roots also can be divided in two different steps (Rodriguez-Navarro *et al.*, 2007). Bacterial surface proteins, capsular polysaccharide and flagella appear to govern the first binding step while extracellular polysaccharide is involved in the second step.

The hypothesis that the lectin, wheat germ agglutinin (WGA), also found in wheat roots, is a signal molecule in the *Azospirillum*-plant interaction that diverts bacterial metabolism in the direction favorable for growth and development of the host plant and is essential for effective associative arrangement has been suggested. WGA induced modifications of the phospholipid fraction of *A. brasilense* membranes. Exposure to WGA decreased the total content of phosphatidylglycerol and phosphatidylinositol, increased phosphatidylethanolamine and cardiolipin levels, but did not induce a change in phosphatidic acid content. These specific changes in the phospholipids are thought to mediate transmembrane signaling and cell response to WGA (Antonyuk *et al.*, 1999). WGA, when added to *A. brasilense* Sp245 culture (final concentration 10–8 to 10–9 mol/L), enhanced IAA production, N_2-fixation, and ammonium excretion by the cells. Additionally, WGA promoted synthesis of proteins, both new and those already present in bacterial cells (Antonyuk and Ignatov, 2001). Changing the medium's C-source from malate to gluconate increased the content of glucosamine, which increased the affinity of polysaccharides for WGA (Konnova *et al.*, 2003).

Under *in vitro* conditions, N_2-fixation capability of *A. lipoferum* was efficiently increased in the presence of WGA, and a putative WGAbinding receptor was detected in the cell capsule. The stimulatory effect required the molecule *N*-acetyl-Dglucosamine dimer (GlcNAcdi), and was dependent on the number of GlcNAcdi links involved in the receptor-WGA interface. The effect on N2-fixation was attributed to the generated WGA-receptor complex stimulus leading to elevated transcription of the *nifH* and *nifA* genes and *glnBA* gene cluster (Karpati *et al.*, 1999). This indirect evidence is not, at the moment, strong enough to demonstrate significant involvement in primary attachment, and influence of this attachment on plant-bacteria interaction. However, these results provide a direction for future research.

Root colonization

If a positive effect of inoculation with *Azospirillum* sp. is expected, significant root colonization should happen first. Inappropriate root colonization usually resulted in marginal or no effects on plant growth (Hecht-Buchholz, 1998; Benizri *et al.*, 2001). Early studies with *Azospirillum* revealed a small cluster type of colonization when the bacteria were located on small roots, and in particular, in the root hair zone and bacteria were anchored to the root surface with numerous fibrilar materials and the partially digested mucigel layer (Bashan and Levanony, 1990; Bashan and Holguin, 1997*a*). Some strains of *A. lipoferum* and *A. brasilense*, but not others, are capable of colonizing the interior of wheat roots. This strain affinity is not present during the initial stages of root colonization (anchoring stage), since isolates from the root interior or an isolate with a proven ability to colonize the interior of wheat roots showed no greater ability to anchor to the roots than other *Azospirillum* strains isolated from the wheat rhizosphere or from the rhizosphere of other grasses. Thus, previous studies of this type of colonization were confirmed.

Cell aggregates and single bacteria were visible on the surface of young roots (root hairs and lateral roots). Large cellular clumps were observed at the emergence points of lateral roots or at intercellular spaces of root epidermal cells (Ramos *et al.*, 2002). Sometimes, the majority of the colonizing bacteria resided on the root surface and only a minority appeared in the root interior (Liu *et al.*, 2003). In the root hair zone, bacteria were more numerous, but principally located in depressions between epidermal cells. In all root zones, mucilage was present, and near the tip it appeared to have been partially digested, forming "halos" around the bacteria and revealing fibril-like strands attached to the bacteria (Pinheiro *et al.*, 2002). *A. brasilense* Cd grown under standard conditions was distributed over the entire wheat root system, except the elongation zone, as is well-known. However, bacteria under saline stress were mainly localized at the root tips and lateral roots (Fischer *et al.*, 2000).

Similarly, inoculation of black mangrove seedlings in seawater with either *A. halopraeferens* or *A. brasilense* produced high density colonization of the root surface. The colonization pattern was different for the 2 species. *A. halopraeferens* yielded mainly single cells embedded in a thick sheath, whereas *A. brasilense* produced primarily micro-aggregates. *A. brasilense* cells were anchored to the root surfaces and to themselves by a network of fibrillar material. *A. halopraeferens* was a better root surface colonizer, whereas the *A. brasilense* population was greater in the entire root (Puente *et al.*, 1999).

Differential wheat root colonization among different strains of *A. brasilense* (Sp7, Sp245, and Wa5) was evaluated. All strains colonized the root tip in large numbers. *A. brasilense* Sp245 showed the highest colonizing potential, and could be found in other parts of the root, including the inner root tissue, forming micro-colonies in intercellular spaces. The other 2 strains formed microcolonies exclusively on the root surface. Following colonization, from cotyledon to flowering stage, Schloter and Hartmann (1998) demonstrated that, while colonization of the total root by *A. brasilense* Sp7 and Wa5 dropped continuously, inside the roots, Sp245 remained constant during these growth stages. Competition among *Azospirillum* strains for root colonization was demonstrated. Coinoculation of all 3 *A. brasilense* strains resulted in similar colonization patterns, but the population of strain Wa5 decreased on the roots. Assmus *et al.* (1997) found that following co-inoculation of strains Sp7 and Wa3, Sp7 performed better than Wa3 for colonization niches. Competition with naturally occurring endophytic PGPB (*Corynebacterium flavescens* and *Bacillus pumillus*) on rice excluded root colonization with inoculated *A. brasilense* (Bacilio-Jiménez *et al.*, 2001).

The preferential colonization of the root isolates of *Azospirillum* in different rice varieties was studied by Govindarajan and Kavitha (2004).The population of the root isolate from rice var.ASD 18 varied from 39.82 X 10^5 to 1.92 X 10^5, in the seven different rice varieties tested (Table 1). Maximum colonization of this isolate was recorded in the rice var. ASD 18 followed by ADT 36. The results indicated that the homologous isolates of the different rice varieties colonized their respective host in significantly more numbers.

As reported earlier, wheat, maize, and rice can develop tumors (para-nodules) along primary and secondary roots when treated with low concentrations of various auxins, the most well-known is the herbicide 2,4-dichlor-phenoxyacetic-acid (2,4-D). As a consequence, they can develop an endophytic diazotrophic "symbiosis". Histologically, auxininduced tumors appear as cancerous root meristems. Auxinaffected root meristems do not recover and further develop into large nodule-like structures. Introduced diazotrophs (*Azospirillum* spp., *Azorhizobium caulinodans*, *Rhizobium* spp.) potentially inhabit para-nodules as a major colonization niche. Colonizing bacteria follow a "crack entry" at sites where developing tumors have emerged through the root cortex and epidermis, and therein establish high cell density inside intercellular spaces of cortical and meristematic tissues. Infection of tumor cells occurs with bacteria found inside cell-cytoplasm surrounded by membrane-like structures. Once they inhabit para-nodules, inoculated diazotrophs colonize endophytically with high cell numbers (Christiansen-Weniger, 1997 & 1998). Not all diazotrophs are capable of effectively colonizing paranodules. In wheat, only *Herbaspirillum seropedicae*, *Azorhizobium caulinodans*, and the mutant *A. brasilense* Sp7S strain displayed significant endophytic colonization of para-nodules, while *Acetobacter diazotrophicus*, *Azotobacter vinelandii*, *Derxia gummosa*, and other *Azospirillum* strains colonized the rhizoplane and not the interior of the paranodules (Kennedy *et al.*, 1997 & 1998).

Table 1: Selective colonization of *Azospirillum* isolates in the root region of rice varieties grown under axenic conditions.

Sl. No.	*Azospirillum* isolates	Population of *Azospirillum* ($X10^5$) g^{-1} of root tissue on d.w.b. Rice varieties						
		Pure line selection				CO RH 2 hybrid rice		
		ASD18	ADT36	CO46	CO47	A line	B line	R line
1.	Root isolate of ASD 18	39.82	34.22	32.93	19.87	3.81	2.46	1.92
2.	Root isolate of ADT 38	37.29	54.38	43.01	1.28	2.54	3.69	4.48
3.	Root isolate of CO 46	34.13	38.49	53.09	33.97	5.72	3.69	1.28
4.	Root isolate of CO 47	27.18	21.99	23.52	58.97	3.81	3.08	2.56
5.	Root isolate of CO RH 2 A line	10.11	7.33	9.41	5.77	50.71	37.52	37.76
6.	Root isolate of CO RH 2 hybrid B line	3.16	4.89	4.03	5.77	33.66	44.28	40.96
7.	Root isolate of CO RH 2 hybrid R line	1.26	6.72	8.06	10.89	24.13	31.50	57.60
	SEd	1.12	0.94	0.05	0.08	0.06	0.07	0.60
	CD (P=0.05)	2.40	0.20	0.11	0.17	0.13	0.15	0.12

The effect of saline stress on the colonization of wheat para-nodules by *A. brasilense* Cd established that para-nodules acted as bacterium-protected niches supporting higher bacterial populations than in plants without para-nodules. In para-nodules, most of the bacteria were present around the basal surface of the modified lateral root structures (Fischer *et al.*, 2000). No major breakthrough has occurred in the study of the attachment process and subsequent root colonization. Only refinements of previous studies have been reported. Perhaps a better understanding of the plant genes involved in the interaction will shed additional light on these crucial steps.

Cell aggregation as a factor of root colonization

Aggregation is one of the most basic phenomena of *Azospirillum* colonization of roots. *In vitro* aggregation/flocculation phenomena have been extensively studied (Bashan and Levanony, 1990; Bashan and Holguin, 1997), and some factors were recently summarized (Burdman *et al.*, 2001). The objective was to produce a superior inoculant for plants, based on this typical characteristic of *Azospirillum*. Growth of *A. brasilense* Cd in a high C: N medium, with fructose and ammonium chloride as C and N sources, induced visible flocculation after 24 h. Suspending aggregates in urea disrupted the aggregates. No cell aggregates were formed after 72 h when grown in low C: N medium. Aggregating cells, but not cells grown in a low C: N medium, accumulated large amounts of poly-β-hydroxybutyrate, and the cell envelope contained a well defined electron-dense layer outside the outer membrane.

The concentration of EPS produced by 4 different strains of *A. brasilense*, differing in their capacity to aggregate, was strongly correlated with the extent of aggregation (Burdman *et al.*, 1998). For cells grown in aggregate-induced medium, outer-membrane protein fractions exhibited high aggregation-specific activities that were strongly correlated with protein content. This suggests that this protein is also involved in the aggregation process of *A. brasilense* cells (Burdman *et al.*, 1999). The EPS and capsular polysaccharide of *A. brasilense* showed a positive correlation between aggregation and the relative quantity of arabinose. Arabinose was not detected in polysaccharides from mutant strains that were aggregation-impaired or in bacteria grown in nonaggregation-inducing medium. The only monosaccharides able to significantly inhibit aggregation at low sugar concentrations were arabinose and, to a lesser extent, galactose. It is possible that residues of arabinose present in the EPS are involved in aggregation of *A. brasilense* (Burdman *et al.*2000). Mutants that lack part of the surface polysaccharide, characteristic of wild-type cells, are impaired in flocculating activity (Blaha and Schrank, 2003).

Some evidence that lectins are involved in cell aggregation was presented. A mutant strain of *A. brasilense* Sp7 with impaired lectin activity exhibited poorer cell aggregation than its parent strain.Pretreatment of bacterial cells with haptens (L-fucose and Dgalactose) of a lectin located at the cell surface of the mutant inhibited aggregation of azospirilla. Interaction of lectins of several *Azospirillum* strains (*A. brasilense* 75, *A. brasilense*Sp7, and *A. lipoferum* 59b) with polysaccharide-containing complexes isolated from these strains was not specific, and no interstrain cross-interaction between the EPS and lectins of azospirilla was found (Nikitina *et al.*, 2001). These interactions show that the aggregation phenomenon is mediated by proteins and polysaccharides, but how this works is not clear. As aggregation can be manipulated easily, this may provide an opportunity to construct better *Azospirillum* inoculants.

Possible physiological mechanisms of plant growth promotion by *Azospirillum*

Despite intensive studies on physiology and molecular biology of these bacteria, the exact mode of action of the bacteria on plants is not much clearer than it was a decade ago. There are several facts that are beyond dispute; the bacteria fix nitrogen and produce several phytohormones in culture and in association with the plant, but the transfer of these products is limited and not always detected. Yet, the growth response is evident. The bacteria affect several plant metabolic pathways, including cell membrane activity. The most apparent outcome of inoculation is changes in the morphology of the root system (both positive and negative). Inoculated plants absorbed minerals and water better. Several possible mechanisms were suggested to explain these phenomena, some with more experimental data than others. Yet, there is no definite agreement on exactly how the bacteria affect plant growth, what the major mechanisms are, or if there is one major mechanism responsible for the observed effects on plant growth, and in particular, on plant yield. These questions are the driving force in the *Azospirillum* research field.

The Additive Hypothesis suggested 13 yr ago is apparently still valid. The hypothesis considers multiple mechanisms rather than 1 mechanism participating in the association. These mechanisms operate simultaneously or in succession, the contribution of an individual mechanism being less significant when evaluated separately. The sum of their activities, when induced under appropriate environmental conditions, results in the observed changes in plant growth (Bashan and Levanony, 1990). Today, the most common explanation for the effect of the bacteria on plants is the production of phytohormones that alter the metabolism and morphology of the plant, yielding better mineral and water absorption. The contribution of N2-fixation is more controversial. Finally, direct effects of the bacteria on plant cell membranes also have been proposed.

Production of phytohormones

Azospirillum spp. are known mainly for their ability to produce plant hormones as well as polyamines and amino acids in culture (Thuler *et al.*, 2003). Among these hormones, indoles, mainly indole acetic acid (IAA), and gibberellins may play a major role.

Indole Acetic Acid (IAA)

Confirmatory studies on IAA production by several strains of *Azospirillum* showed that production depended on the type of culture media and availability of tryptophan as a precursor. *A. brasilense* Cd produced the highest level of IAA among the strains tested (approx. 380 ìmol/L) (El-Khawas and Adachi 1999). The pH has a significant effect on the amount of IAA produced (Ona *et al.*, 2003). Assessment by chemical methods and with HPLC of possible precursors (indole, anthranilic acid, and tryptophan) for IAA formation in *A. brasilense* Sp245 revealed a high motive force for tryptophan synthesis from chorismic acid and for IAA synthesis from tryptophan, and this makes it unlikely that anthranilic acid and indole act as the precursors to IAA in a tryptophan-independent pathway (Zakharova *et al.*, 1999). Vitamins may also play a role in the regulation of IAA synthesis in *A. brasilense.* Very low levels of the B vitamins, especially pyridoxine and nicotinic acid, increased production of IAA in *A. brasilense* (Zakharova *et al.*, 2000).

Auxin production by *Azospirillum* sp. is believed to play a major role in plant growth promotion, although little new evidence with plants has been published in recent years. *A. brasilense* produced high quantities of extracellular IAA and tryptophan in culture medium supplemented with tryptophan, a precursor of IAA. Addition of filter-sterilized culture supernatants to rice roots grown in hydroponic tanks increased root elongation, root surface area, root dry matter, and development of lateral roots and root hairs, compared with untreated roots. Higher concentrations of the supernatant strongly inhibited root elongation, lateral root development, and caused nodule-like tumors on the roots (El-Khawas and Adachi 1999). Similarly, a cell-free supernatant of *A. brasilense* Cd applied to soybean plants induced the highest number of roots and increased root length (Molla *et al.*, 2001*b*). Inoculation of wheat with *A. brasilense* Sp245 and Sp7 wild strains led to a strong decrease in root length and increase in root hair formation, as is common for such inoculations. The effect on root morphology was further enhanced by adding tryptophan, and this could be mimicked by replacing *Azospirillum* cells with IAA (Dobbelaere *et al.*, 1999). A mutant of *A. brasilense* with low production of phytohormones, but with high nitrogenase activity did not enhance root growth over uninoculated controls. In contrast, a mutant with increased phytohormone production significantly affected root morphology. In general, increased plant biomass and

N2-fixation were recorded in strains having increased production of indole compounds (Kundu *et al.*, 1997).

Induction of para-nodules in rice roots by the auxins 2,4-D, naphthalene acetic acid (NAA), and IAA enhanced polygalacturonase activity in rice roots during formation of the para-nodules and endophytic colonization by *Azospirillum*. While inoculation with *Azospirillum* could augment polygalacturonase activity of rice roots to a small extent without any visible effect on root morphogenesis; auxin application, together with *Azospirillum* inoculation, enhanced polygalacturonase activity of rice roots to a high level, thus, yielding root change into para-nodules that later were colonized by *A. brasilense* (Sekar *et al.*, 1999).

One effect of *Azospirillum* inoculation on the metabolism of maize and common bean seedlings was increased respiration rates. *In vitro* inoculation of detached root segments reduced kinetic values (*K*m and *V*max) of ®-glucosidase, an enzyme that may be involved in phytohormone release from conjugates, and yielded lower total activity of ®-glucosidase, but higher affinity to the substrate of specific ®-glucosidases (Vedder-Weiss *et al.*, 1999).

Gibberellins

A beneficial effect of *Azospirillum* spp. on plants has been suggested to be partially caused by the production of gibberellins. Application of gibberellins had effects similar to *Azospirillum* inoculation in increasing root hair density. When *A. lipoferum* USA 5b, a gibberellin-producing strain, was cultured in the presence of glucosyl ester or glucoside of gibberellin A20, both conjugates were hydrolyzed. These in vitro results support the hypothesis that growth promotion in plants induced by *Azospirillum* inoculation results from a combination of both gibberellin production and gibberellin-glucoside/glucosyl ester deconjugation by the bacterium (Piccoli *et al.*, 1997). The effect of water potential or O_2 concentration on growth and gibberellin A3 (the main gibberellin identified from *Azospirillum*) production in *A. lipoferum* showed that gibberellin A3 produced by each culture was reduced severely at high water potentials or low O_2 concentrations. At the highest water potential concentration, gibberellin A3 was reduced by only 50%, despite a 90% reduction in cell numbers. This indicates an increase in the amount of gibberellin A3 produced per cell with increasing water potential (Piccoli *et al.*, 1999). The involvement of gibberellin A3 produced by *Azospirillum* spp. in promoting maize growth was also suggested (Lucangeli and Bottini, 1997).

Ethylene

During most phases of plant growth, ethylene production is minimal. Ethylene plays a major role in germination by breaking the dormancy of seeds. However, a high level of ethylene concentration inhibits subsequent root elongation. High levels of ethylene may be synthesized as a response to biological or environmental stresses, causing wilting and senescence (Glick *et al.*, 1999). Controlling ethylene levels, often by lowering them, prevents significant economic losses in agriculture. One of the precursors of ethylene synthesis is 1-aminocyclopropane-1-carboxylic acid (ACC). ACC deaminase is a key enzyme, commonly found in many soil microorganisms and PGPBs, capable of degrading ACC. Thus, lowering ethylene levels in plants can be considered as having potential for promoting growth (Glick *et al*., 1999).Wild *Azospirillum* spp. do not have ACC deaminase, but some strains can produce ethylene, nevertheless. Modulation of ethylene by PGPB can occur through degradation of the ethylene precursor ACC via ACC deaminase, yielding ammonia and á-ketobutyrate as by-products.

N_2 fixation

N_2-fixation was the first mechanism suggested to promote the growth of plants by *Azospirillum*. The majority of evidence collected during the last 3 decades concerning this mechanism has generated controversy. Some greenhouse and field experiments have shown repeatedly that the transfer of nitrogen (measured as transfer of $^{15}N_2$) fixed by *Azospirillum* spp. to the plant is minimal (Kennedy *et al*.,1997 & 1998; Bashan and Holguin 1997a, b). Yet, other studies showed that the bacteria cannot fulfill all of the nitrogen requirements of plants, but nevertheless contribute significant amounts of nitrogen. Furthermore, inoculation commonly and significantly reduced the required doses of nitrogen fertilization in numerous greenhouse and field experiments of many plant species (Bashan and Levanony, 1990; Bashan and Holguin, 1997a).

Several confirmatory reports about the contribution of fixed nitrogen by *Azospirillum* to plants, similar in nature to reports of previous years, illustrate the controversy. The ^{15}N isotope dilution technique indicated that there were significant biological N_2-fixation contributions to 2 genotypes of maize that showed similar increases in grain yield when they were inoculated with a mixture of *Azospirillum* spp. Strains or fertilized with the equivalent of 100 kg N/ha. These genotypes showed a large increase in total N in the plant. This suggests that the yield response resulted from increased nitrogen acquisition, but not from bacterial nitrate reductase; NR- mutants generally caused plant responses similar to those of the parent strains (Garcia de Salamone *et al*., 1997).

The ability of the bacteria to transfer fixed nitrogen from the atmosphere to wheat plants was tested using $^{15}N_2$-enriched atmosphere. Labeled fixed nitrogen was detected in plant growth media, in roots and shoots of wheat grown 26 d in $^{15}N_2$-enriched atmosphere, but the highest levels of ^{15}N were detected in wheat shoots. Ammonia or nitrate supply to plants did not repress $^{15}N_2$-fixation (Ruppel and Merbach 1997).

Para-nodules induced by low concentrations of the auxin 2, 4-D in rice seedlings were the preferential sites for colonization by a NH_4^+-excreting *A. brasilense* mutant. Nitrogenase activity in tumor structures inhabited by bacteria significantly increased, compared with untreated control plants (Christiansen-Weniger,1997). It is probable that within the para-nodule bacterial nitrogenase activity is less sensitive to increased oxygen tension in the roots. Host plants benefit from enhanced N2-fixation in their roots with para-nodules because fixed nitrogen is incorporated into host plant material. Host plants probably stimulate nitrogenase activity of endophytic azospirilla by providing a carbon source as energy (Christiansen-Weniger, 1998). The effects of inoculating wheat with a highly efficient *A. brasilense* strain under 3 nitrogen regimes revealed that inoculation stimulated plant growth, nitrogen accumulation, and nitrogen and NO3 -- accumulation in the tissues. At maturity, inoculated plants showed higher biomass, grain yield, and nitrogen content than the uninoculated ones, as well as a higher grain protein concentration. It was concluded that *A. brasilense* increased plant growth by stimulating nitrogen uptake by the roots (Saubidet *et al.*, 2002).

On the other hand, using an *in vitro* model (*A. brasilense* and wheat) within 70 h after inoculation, insignificant amounts of newly fixed N_2 were transferred from an ammonia-excreting strain of *A. brasilense* to the shoot tissue of wheat. By the addition of malate (a preferred carbon source for *Azospirillum*), transfer of nitrogen to the shoots increased 48-fold, indicating that 20% of shoot nitrogen had been derived from N2-fixation. Apparently, the inability of the host plant to release sufficient carbon into the rhizosphere is a significant constraint on the development of the *A. brasilense*-wheat association. Perhaps wheat plants with an increased release of photosynthate to the rhizosphere should be a priority for improving the association (Wood *et al.*, 2001).

N_2-fixation by aerobic bacteria is a very energy demanding process, requiring efficient oxidative phosphorylation, while O_2 is toxic for the nitrogenase complex. Azospirilla and other well-known N_2-fixing soil bacteria have evolved a variety of strategies to deal with and overcome this apparent "O_2 paradox". The question is whether the specific environmental adaptations of azospirilla are sufficient to allow optimal proliferation and N_2-fixation in their natural habitat. Could improving O_2 tolerance of the N_2-fixing process contribute to the development of more efficient strains for inoculation of plants? (Marchal and Vanderleyden, 2000). This is left for future research.

Proton efflux from roots

Improving plant growth by affecting proton and organic acid extrusion mechanisms of plants by inoculation with *Azospirillum* sp. was proposed over a decade ago (Bashan and Levanony, 1990 and Bashan and Holguin 1997a), and was shown to occur in cardon cactus plants (Carrillo *et al.*, 2002). Lowering the pH of the rhizosphere increases the availability of phosphorus and iron to plants, especially in arid lands with high calcium content and soil pH. A confirmatory study of the proton extrusion phenomenon in wheat showed that inoculation enhanced proton efflux and root elongation of the roots. Although the evidence is circumstantial, perhaps these 2 phenomena are related. This effect was directly dependent on the bacterial strain-plant combination, suggesting that compatible strains are necessary to induce this phenomenon (Amooaghaie *et al.*, 2002).

Effect of inoculation on plant growth and yield

Contrary to greenhouse studies on the effect of *Azospirillum* spp. on plant growth and yield that dominated older literature (Bashan and Holguin, 1997*a*), most contemporary reports involved field studies as indicators of maturity of the field and its transformation to commercialization (Bashan *et al.*, 2004). Three notable phenomena reported are (*i*) inoculation of plant species other than cereals. There are a significant number of those, demonstrating that over 100 species are capable of inoculation by this bacterium. Although *Azospirillum* was isolated initially from cereals and most of the initial inoculation has been done on the main cereal crops (Bashan and Levanony, 1990), there are more non-cereal species successfully inoculated with *Azospirillum* than cereals (Bashan and Holguin, 1997*a*). That review proposed that *Azospirillum* should be considered a general plantgrowth-promoting bacterium and not a cereal growth promoter. (*ii*) Apparently, in numerous cases, inoculation reduced the use of chemical fertilizers, especially nitrogen by 20%–50%, and provided superior results when organic fertilizers were incorporated. (*iii*) In many developing countries, inoculation increased the cost-benefit ratio (Kannaiyan, 2003).

Survival and persistence in soil

Azospirillum sp. is a common rhizosphere bacterium. Most species and strains prefer rhizosphere and root habitats and not the bulk soil, where they survive poorly. Yet, a fraction of the population persists in bulk soil (Bashan, 1999) and can be isolated directly from the soil (Egorenkova *et al.*, 2000). Effects of 3- and 19-y natural fallow treatment of tropical sandy topsoil on soil structure and *Azospirillum* diversity was assessed by aggregate size fractionation. Long-term fallow (19 y) under the grass *Pennisetum*, stimulated aggregation, whereas all clay particles were easily dispersed from the soil left fallow for 3 y. Coarse soil fractions supported potential sites for N_2-fixation, suggesting that these microhabitats were favorable for active N2-fixers. Yet, more than 70% of the N_2-fixing microorganisms and 90% of the recovered *Azospirillum* were isolated from the dispersed fine clay fraction.

Use of specific genetic probes on colonies within each soil fraction indicated an abundance of *A. irakense* only in the 3-y fallow soil fractions, and a selective effect of fallow on *A. brasilense*, *A. amazonense* genomic species in a 19-y fallow soil (Chotte *et al.* 2002). Edaphic factors affecting survival of 5 species of *Azospirillum* in bulk soil (*A. brasilense*, *A. lipoferum*, *A. amazonense*, *A. halopraeferens*, and *A. irakense*) were evaluated after the bacteria were inoculated into 2 natural, semi-arid soils (terra rosa and sandy loess), and 2 artificial soils constructed to simulate the native soils. Within 60 d, the populations of all 5 *Azospirillum* spp. declined significantly in a linear pattern in the native and analogous artificial soils.

Higher levels of $CaCO_3$ and fine and rough sand had significant detrimental effects on the survival of the 5 species, whereas increased organic matter content improved survival. In contrast, when the bacterial strains were incubated in the rhizosphere of tomato seedlings grown in the artificial soils, manipulation of these soil variables had only a marginal effect on bacterial survival; all *Azospirillum* spp. survived well in the tomato rhizosphere under conditions that were otherwise detrimental. This indicated that most cells of the 5 Azospirilla strains died in a linear pattern in 2 semi-arid soils, and that only major soil components affected *Azospirillum* survival in soil (Bashan and Vasquez, 2000). Spreading pig waste slurry as organic fertilizer seems to be a good management practice in corn-cultivated agrosystems, as it does not affect the presence of *A. brasilense* in soil.

A. brasilense was found only in unfertilized and organically amended plots, while fertilization with urea reduced its presence below the detection threshold of a molecular probe (Ceccherini *et al.*, 2001). All these studies enhance the notion that there is an effect on the community structure of azospirilla in bulk soil. Environmental factors affecting survival are organic matter content and clay fraction, as found in earlier studies. Yet, this does not provide sufficient evidence that azospirilla can pass from one season to the other as a saprophyte and inoculate crops in the next season or unintentionally promote weed growth in subsequent plantings, all fundamental and practical questions for the inoculant industry.

The effects of soil moisture on the survival of three diazotrophic bacteria species (*Azospirillum amazonense*, *Gluconacetobacter diazotrophicus* and *Azospirillum brasilense*) were tested (Oliveira *et al.*, 2004). Soil moisture had little influence on the survival of *A. brasilense*, which is considered a free-living species. On the other hand, increased soil moisture extended the survival of the endophytic *A. amazonense* (Figs. 1&2). These results indicate that nitrogen-fixing endophytic species are more affected by soil moisture than associative nitrogen fixing species.

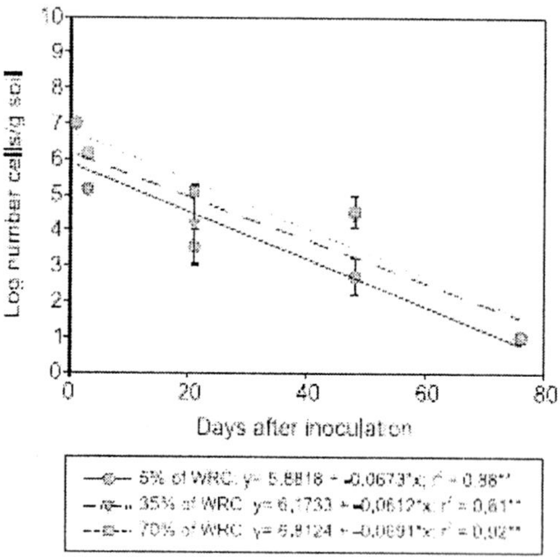

Fig. 1: Survival of *A. brasilense* strain BR 11001 in sterilized soil under three moisture levels. **p<0.001. Vertical bars denote standard error.

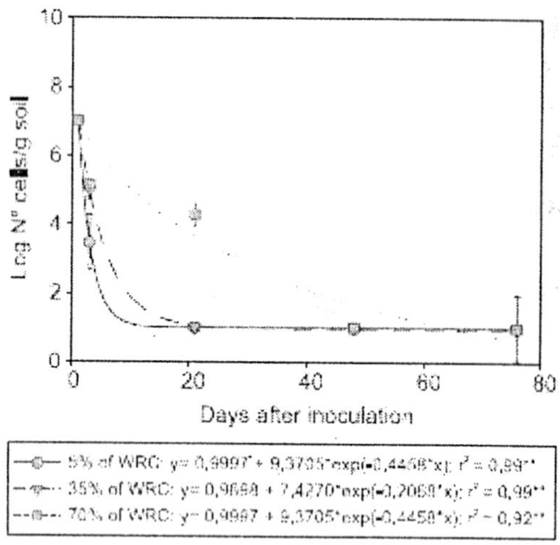

Fig. 2: Survival of *A. amazonense* strain BR 11115 in sterilized soil under three moisture levels. **p<0.001. Vertical bars denote standard error.

(Source: Oliveira *et al.*, 2004)

Influence of cell storage material on survival of the bacterium in soil

Like many prokaryotes, *Azospirillum* spp. produce high levels of intracellular storage substances, such as polyhydroxyalkanoates (previously described as PHB) under nutritional limitation. The C/N ratio in the bacterial growth medium and oxygen partial pressure, are the governing factors for concentrations of PHA in the cell. Under N_2-fixation and without mineral N, the bacteria can accumulate up to 75% of the cell dry wt as PHA. This energy and carbon storage compound is then used under stress conditions, such as limitations of carbon and energy, a capacity that might enhance the competence of *Azospirillum* in the rhizosphere, which is desirable in commercial agricultural inocula (Bashan and Holguin, 1997a; Bashan, 1998). Not all *Azospirillum* strains produced PHA. Out of 49 isolates of *Azospirillum* from West Bengal, only 13 were PHA producers. The majority of these belonged to *A. brasilense* and a few belonged to *A. amazonense* and *A. lipoferum*. When grown in a medium with nitrogen, the PHA content of these strains ranged from 1% to 14% of dry cell mass (Manna *et al.*, 1997). Three enzymes and their genes that are essential in the PHA biosynthetic and degradation pathways were identified in *A. brasilense* Sp7. A PHA– mutant of *A. brasilense* decreased its starvation resistance compared with the parent strain, but otherwise, motility, cell aggregation, root attachment, and EPS and capsular polysaccharide production were superior to the wild-type (Kadouri *et al.*, 2002; Edelshtein *et al.*, 2003).

In normally grown *A. brasilense* Cd cells, 2 unsaturated fatty acids, octadecanoate (18:1 *cis*-9) and hexadecanoate (16:1 *cis*-9), accounted for approximately 80% of the fatty acid content in *A. brasilense* Cd. The major lipids were phosphatidylethanolamine, phosphatidylglycerol, and phosphatidylcholine. Flocculation (creation of massive cell aggregates) resulted in the de novo synthesis of a glycolipid and cardiolipin. Flocculation also resulted in a decrease in cellular proteins and lipids and a proportional increase in cellular PHA and carbohydrates. PHA in flocculated cells reached 60%–65% of dry cell wt (Olubayi *et al.*, 1998).

Ecology of azospirilla and its interactions with soil microorganisms

It is well documented that azospirilla are mainly rhizosphere bacteria, and with the exception of some strains capable of living in bulk soil or as endophytes, most survive poorly in bulk soil (Bashan, 1999). Yet, environmental factors and other native soil microorganisms influence and interact with the inoculant. *Tn5-lacZ*-tagged *A. brasilense*, capable of growing at suboptimal temperatures, was compared with the parental strains. While the mutants fixed nitrogen, produced IAA and siderophores, and could colonize and promote wheat growth similar to the parental strains on their own, co-inoculation of the mutants with their respective parental strains showed that their colonization potential was very negatively affected (Rajeev *et al.*, 2000). However, 2 *A. brasilense* strains, capable of both growth promotion and growth at a suboptimal temperature of 22°C, were better inoculants for wheat than the regular, medium-high temperature strains (Kaushik *et al.*, 2002). This indicates that strains capable of growing at low temperatures and producing plant hormones are perhaps a better choice for wheat cultivation.

The positive relationship between nitrogenase activity and N_2-fixing populations in the top layers of 4 flooded tropical rice soils was demonstrated. Application of organic sources (cellulose and rice straw) to the topsoil (1–2 cm) produced higher nitrogenase activity associated with larger populations of *Azospirillum* sp., irrespective of soil type (Kanungo *et al.*, 1997). Microbiological analysis of field-grown rice indicated that cultivars with high nitrogen absorption efficiency harbored higher populations of N_2-fixing *Azospirillum* sp., *Azotobacter* sp., and anaerobic bacteria, compared with low nitrogen absorbing cultivars. Notwithstanding, the effect of nitrogen fertilizer management on N_2-fixation, nitrogen absorption efficiency of rice and the rhizosphere-associated nitrogenase activity and N_2-fixing bacteria were apparently interrelated (Kanungo *et al.*, 1998).

N_2-fixation capacity of 285 strains of *Azospirillum* isolated from soils from 7 geographic regions in Australia varied from zero or negligible nitrogenase activity compared with the normal activity of *Azospirillum*. Most of the isolates, when associated with wheat roots, reduced between 1–5 nmol C_2H_4 (mg dry root)$^{-1}$d and some strains showed considerably higher activity. The majority of these strains were *A. lipoferum*. N_2-fixation on wheat roots did not correlate well with N_2-fixation in pure culture. Some strains that fixed nitrogen vigorously in pure culture had low rates of N_2-fixation on roots, and *vice versa*, where isolates of *A. lipoferum* had a higher average nitrogenase activity than *A. brasilense* in nitrogen-free medium and in association with wheat roots (Han and New, 1998a & b).

Azospirillum spp. were found associated with natural arid vegetation of the North Sinai (Egypt). *Azospirillum* spp., and *Pseudomonas* spp., presumably dominating the diazotrophic community, were identified according to morphological, cultural, and physiological characteristics of the isolates. Acetylene-reducing activities of these isolates were highest with *A. brasilense* (Sedik, 1997). Similarly, *Azospirillum* was found to share with *Herbaspirillum* and *Acetobacterdiazotrophicus* the rhizosphere, and the interior of sugarcane *Saccharum officinarum* stems (Indira and Bagyaraj, 1997).. *Azospirillum* is usually inoculated into natural soils containing a multitude of microorganisms with a wide range of characteristics and microfauna. Interactions with some of them are inevitable.

Effects of a mixed inoculant of *Azospirillum* spp. and several AM fungi (*Glomus mosseae*, *G. deserticola*, and natural AM fungi from the test soil) upon mycorrhizal colonization in maize plants were evaluated for enzymatic activities (esterase, phosphatase, trehalase, and chitinase). The enzymes were indicators for detecting changes in the microbial functional groups in the soil. *Azospirillum* mixed with the AM fungi showed no negative effects on AM fungal establishment. As a result of mycorrhizal colonization and microbial inoculation, modifications of the microbial community structure and ecology were induced (Vázquez *et al*., 2000).

Anchoring of PGPB to AM fungal structures may have special ecological and biotechnological significance because it may facilitate colonization of new rhizosphere by the bacteria traveling on or moving with fungal hyphae, and may be an essential trait for development of mixed inocula. Mutants of *A. brasilense* deficient in EPS were shown to be strongly impaired in their capacity to attach to AM roots and AM fungal structures (Bianciotto *et al*., 2001). When barley was inoculated with the AM *Glomus intraradices*, addition of *A. lipoferum* 137 and the phosphate-solubilizing bacteria, *Agrobacterium radiobacter* and *Flavobacterium* sp., did not affect the intensity of development of mycorrhizal structures in the roots. However, while mycorrhization of the plants was conducive for better establishment of *A. lipoferum* 137 in the rhizoplane, it significantly decreased root colonization of *Flavobacterium* sp. (Belimov *et al*., 1999). Addition of *A. brasilense* to the AM fungi (*Hebeloma crustuliniforme* and *Pisolithus tinctorius*) used as a biocontrol inoculant for pine seedlings against the pine pathogens, *Rhizoctonia solani* and *Fusarium oxysporum,* increased fungal biomass and the shoot/root ratio of the seedlings (Dahm *et al*., 1998).

Synergistic and antagonistic effects on AM fungi (*G. fasciculatum*) co-inoculated with *A. brasilense* or *Rhizobium meliloti* on the photosynthesis in alfalfa plants might serve as a parameter to measure the effect of the microbial activity and their possible beneficial effects in the field. Beneficial effects of AM on photosynthesis were clearly revealed as enhancement of the electron transport activity per leaf area. When AM was co-inoculated with one of the PGPBs, the antagonistic effect of *A. brasilense* was higher than that of *R. meliloti,* though not strong enough to fully counterbalance the beneficial effect of AM. However, in the case of triple inoculation with both PGPB and AM, electron transport was found to be only slightly lower than in the case of single inoculation by AM, indicating that, in the presence of each other, PGPBs are no longer antagonistic to AM (Tsimilli-Michael *et al.*, 2000).

Two naturally occurring bacterial endophytes of rice seeds, *Corynebacterium flavescens* and *Bacillus pumilus*, competed with inoculated *A. brasilense* and excluded it from the rhizoplane (Bacilio-Jiménez *et al.*, 2001). Mixed cultures of *A. brasilense* Sp7 and *Penicillium corylophillum* allowed the efficient utilization of pectin by the fungus (93% pectin degradation within 9 d). The degraded components served as a carbon source for N2-fixation by *A. brasilense,* which is incapable of degrading pectin (El-Katatny *et al.*, 1997).

Azospirillum in the soil could indirectly fall prey to soil macrofauna. *A. brasilense* was introduced into oak leaf litter, or soil containing litter-dwelling diplopods and isopods or earthworms. In the presence of these animals, the number of bacteria in excrement was always 2–10 times lower than in food. Digestive fluid taken from the middle part of the gut of the diplopod *Pachyiulus flavipes* showed a strong antibacterial activity, indicating that *A. brasilense* appeared to be sensitive to digestion by the midgut fluid (Byzov *et al.*, 1997). All these studies add evidence for the rhizocompetence of *Azospirillum* as rhizosphere bacteria. Its capacity to accumulate large amounts of PHA, massively aggregate and flocculate, and produce cyst-like forms under stress conditions shows that *Azospirillum* has a potential as a formidable competitor in its ecological niche, the rhizosphere. Yet, its interaction with other soil microorganisms is unclear, and the available information allows only speculations, at best.

The changes in the rhizobacterial structure following *A. brasilense* inoculation of maize (*Zea mays*) plants was examined by PCR-denaturating gradient gel electrophoresis (DGGE) and automated ribosomal intergenic spacer analysis (ARISA), using two universal primers sets for the 16S rRNA gene, and an intergenic 16S–23S rDNA primer set, respectively (Lerner *et al.*, 2006). Similar results were obtained when using either ARISA or DGGE performed with these different primer sets, and analyzed by different statistical methods: no prominent effect of *A. brasilense* inoculation was observed on the bacterial communities of plant roots grown in two different soils and in different growth systems. In contrast, plant age caused significant shifts in the bacterial populations.

Effect of soil stress conditions on *Azospirillum*

Salt stress

Numerous cultivated soils worldwide are becoming more saline mainly from the use of marginal irrigation water, from excess fertilization, and various desertification processes. Inoculation with *Azospirillum* sp. under saline stress conditions is therefore commonplace. Prior findings showed that common agricultural *Azospirillum* strains tolerated high salinity (□ 2%). Salt resistance among species increased from *A. amazonense* (lowest) to *A. halopraeferens* (highest), the latter tolerating over 3% NaCl (seawater salinity). The common cellular mechanism of osmotic stress adaptation is intracellular accumulation of organic solutes (osmolytes). In sorghum, inoculation with *A. brasilense* diminished the adverse effects caused by osmotic stress (Bashan and Holguin, 1997a).

More detailed confirmatory *in vitro* studies demonstrated that *A. brasilense* Cd can tolerate up to 200 mmol/L NaCl in the medium without appreciable decline in growth. Higher concentrations of salt caused inhibition of bacterial growth. At 300 mmol/L NaCl, growth decreased 66% after 24 h. After 48 h at this concentration, bacteria reached maximum optical density, comparable to the optical density of control cultures after 12 h. *A. brasilense* responded to saline stress by elevating intracellular concentrations of glutamate at 24 h and K^+ at 48 h. Although several cellular functions are affected by saline stress, it seems that *A. brasilense* Cd has remarkable tolerance for saline conditions (Rivarola *et al.*, 1998). As already known, *Azospirillum* spp. can accumulate compatible solutes, such as glycine betaine, glutamate, proline, and trehalose, to allow adaptation to fluctuations in soil salinity/osmolarity. Addition of these osmo-protectants to bacterial cultures under saline stress usually increased cellular growth and nitrogenase activity (Choi and Gal 1998; Tripathi and Mishra 1998a). As inhibition of *A. lipoferum* growth by NaCl was relieved by exogenous glycine betaine, this suggested that *A. lipoferum* has a salinity-induced glycine betaine transport system (Tripathi and Mishra, 1998a).

Additionally, proline seems to play a major role in osmoadaptation. With increases in osmotic stress, the dominant osmolyte in *A. brasilense* shifted from glutamate to proline. Accumulation of proline occurs by uptake and synthesis. At higher osmolarity, *A. brasilense* Sp7 accumulated a high intracellular concentration of glycine betaine that was taken up via the glycine betaine transport system, probably similar to that of *A. lipoferum*. Except for *A. halopraeferens*, all other species of *Azospirillum* lacked the ability to convert choline to glycine betaine. Mobilization of the *betABT* genes of *E. coli* in *A. brasilense* enabled it to use choline for osmo-protection (Tripathi *et al.*, 1998b).

Saline stress alters *A. brasilense* Cd maize and wheat interactions, normal colonization, and N_2-fixation. Saline stress altered the early stages of plant development, which led to inadequate colonization and expression of *A. brasilense nif* gene promoters. While *nifA* expression increased in stressed bacteria, *nifH* transcription was diminished (Jofré *et al.*, 1998a & b). Yet, *A. halopraeferens* and *A. brasilense* were capable of colonizing roots of mangrove trees in seawater (Puente *et al.*, 1999). Attachment of *A. brasilense* Cd to maize and wheat roots was altered when the bacteria were grown under saline stress. Alteration in the adsorption phase of attachment appeared to be related to the disappearance of a 100 kD external membrane protein in the bacterium. Gene expression in *A. brasilense* was influenced by the presence of plant root exudates; wheat root exudates induced the reappearance of the 100 kD protein (Fischer *et al.*, 1999).

Unstressed bacteria grown under standard conditions were distributed over the entire root system of wheat, except the elongation zone. Bacteria subjected to saline stress were mainly found at the root tips and the lateral roots. Interestingly, salt treatment reduced surface colonization, but not colonization inside the root (Fischer *et al.*, 2000). *Azospirillum* inoculation at NaCl concentrations up to –1.2 MPa significantly increased chlorophyll, K, Ca, soluble saccharides, and protein contents as compared with control plants growing without NaCl (Hamdia and El-Komy, 1997), similar to alleviating water stress on wheat plants grown under drought conditions (El-Komy *et al.*, 2003).

Inoculating *A. brasilense* on wheat seedlings exposed to severe salt (NaCl) or osmotic (polyethylene glycol) stresses significantly reversed part of the negative effects; both stresses reduced relative elongation rate of shoots. Fresh wt, fresh wt/dry wt, water content, and relative water content were higher in shoots from inoculated plants than in stressed controls (Creus *et al.*, 1997). Turgor pressure at low water potential was higher in inoculated seedlings in 2 wheat cultivars under osmotic stress. This could result from better water uptake as a response to inoculation that, in turn, is reflected by faster shoot growth in inoculated seedlings exposed to these stresses. These results are similar to what was found in water-stressed wheat inoculated with *Azospirillum.* They showed better water status and effects on cell wall elasticity and (or) apoplastic water (Creus *et al.*, 1998).

Soil salinity plays a major role in diversity of indigenous azospirilla associated with rice along the coast of India. All the strains detected were *A. brasilense* or *A. lipoferum.* Decreasing diversity of *Azospirillum,* probably from higher salinity, is apparently more important than the clay fraction of the soil, a factor demonstrated as having a major role in several studies. If true, this association between soil salinity and range of *Azospirillum* should be considered when developing inoculants, especially for coastal saltaffected agricultural areas (Saleena *et al.*, 2002). Addition of NaCl resulted in a significant increase of diazotroph clones within the a-proteobacteria similar to *Azospirillum* sp. (Nelson and Mele, 2006).

The effect of salt on *Azospirillum*-plant interactions is a major factor to be considered in applied studies. In spite of its importance, meagre molecular studies have been conducted on the bacterium per se, and almost none as to interactionswith plants. The most fundamental omission in current knowledge is uncertainty on whether improved salt tolerance of the bacterium is needed to enhance the bacterium's effect on plants or if existing salt tolerance is adequate to ensure positive growth-promotion by inoculation.

Pesticides

In view of the high input of agrochemicals in contemporary crop production and the likelihood that *Azospirillum* inoculation is and will be used in regular crop production, not solely in organic agriculture, studies on the interactions of *Azospirillum* inoculants with common pesticides are essential. Apparently, as in the past (Bashan and Holguin, 1997*a*), this is an unattractive subject of *Azospirillum* field research. Few tests, mostly at a laboratory scale, have been conducted.

Insecticides

The effects of 4 insecticides, bromopropylate, diazinon, methidathion, and profenofos on the cells of *A. brasilense* were evaluated *in vitro* in chemically defined media. While bromopropylate and diazinon were completely harmless, methidathion at high doses significantly reduced N_2-fixation, intracellular levels of ATP, and growth in cells. However, these negative effects were not significant in cells grown in dialyzed soil media, indicating that *A. brasilense* can potentially tolerate high concentrations (300 mg/mL) of methidathion (Gomez *et al.*, 1998). Similarly, profenofos produced the same negative effect as methidathion, including inhibition of production of several vitamins. The effect was significantly greater in dialyzed soil media (Gomez *et al.*, 1999).

The insecticide carbofuran affects nitrogenase activity *in vitro*, where low levels (2 and 5 µg/mL) significantly stimulated nitrogenase activity, while higher levels (>5 µg/mL) caused significant inhibition. Interestingly, the usual inhibitory effect of combined high nitrogen levels on nitrogenase diminished gradually in the presence of carbofuran over a 10-day incubation period (Kanungo *et al.*, 1998). Of 4 insecticides (carbofuran, chlormephos, terbufos, and benfuracarb) currently used on maize at the time of sowing, only terbufos had a slight effect on growth of *A. lipoferum* in solid medium cultures. Physical contact of the insecticides with the commercial inoculant (Azogreen-m), bacteria inoculated directly on insecticide granules, or inoculant mixed with an insecticide all decreased survival of *A. lipoferum* (Revellin *et al.*, 2001).

Acaricides

A. lipoferum survived high concentrations of the organochlorine acaricide dicofol (1, 1-bis-(4-chlorophenyl)-2, 2, 2-trichloroethanol). The bacteria accumulated the acaricide in the cell envelope connected to neutral lipids and phospholipids of membranes (Mano and Langenbach, 1998).

Fungicides

Three fungicides (carbendazim, thiram, and Bordeaux mixture) and 2 insecticides (carbofuran and phorate) were evaluated for their effect on *Azospirillum* sp. in cowpea *in vitro* and in the field. Only thiram caused growth reduction of the bacterium *in vitro*. In the field, the rhizosphere bacterial population was reduced in thiram-treated plants, but not in plants treated with the other chemicals. A mix of thiram and insecticides reduced the *Azospirillum* population, but a gradual bacterial build-up was observed in the rhizosphere following the treatment (Raji and Pillai, 2000).

Herbicides

Cotton plants could be protected from harmful effects of 2, 4-D by inoculation with *A. brasilense*. The 2, 4-D degrading plasmid pJP4 was transferred into *A. brasilense* Sp7. Transconjugants degraded 2, 4-D in pure culture via cometabolism at concentrations up to 50 µg/mL. However, when the transconjugants were inoculated on cotton seeds, the plants were resistant only to low levels of the herbicide (0.5 µg/mL), which is not sufficient for protection of the cotton plant. Plants growing in soils with this concentration of herbicide, and inoculated with wild-type strains, died (Feng and Kennedy, 1997).

Toxic metals and antibiotics

Five strains of *Azospirillum* spp. were able to tolerate chromium to levels of 1000–1400 µg/mL *in vitro* (Govindarajan, 1999). *Azospirillum* spp. are highly sensitive to heavy metals; the order of toxicity is Hg > Cd > Pb. A mixture of these heavy metals (at 5 µg/mL each) was more toxic to *Azospirillum* cultures than the individual metals (Subramaniane and Govindarajan, 1998). The effect of heavy metals on 3 strains of *Azospirillum* was evaluated. At concentrations of 10 and 100 mg/L, Cd^{+2}, Zn^{+2}, and Cu^{+2} inhibited growth and biomass of *Azospirillum* in media containing these metals. At a concentration of 500 mg/L, growth was completely inhibited (Strzelczyk *et al.*, 1997). In the presence of 50 µmol/L

CdCl2, inoculation with *A. lipoferum* of barley seedlings partly decreased Cd toxicity, possibly through the improvement of mineral uptake. Cadmium causes severe inhibition of growth and nutrient uptake in barley plants.

Additionally, inoculation with the bacteria slightly enhanced root length and biomass of the seedling treated with Cd, and the amount of nutrients absorbed by the inoculated plants increased significantly. There was only some protection against Cd toxicity, but no uptake of Cd since Cd content in the inoculated plants was unchanged (Belimov and Dietz, 2000). *A. lipoferum* was found to be naturally resistant to antibiotics of the penicillin and cephalosporin families (Boggio and Roveri, 2003). In summary, the data available from the onset of research on *Azospirillum* preclude a firm conclusion about the interactions of this bacterium with pesticides.

Compost

Some compost may be toxic to plant growth because of elevated humic acids or inappropriate preparation. Inoculation of wheat seeds with *A. brasilense* or *A. lipoferum* prior to sowing in soil amended with 2 types of composts improved seed germination and development of plants. The bacteria possibly changed the humic acids since either bacterium could survive and grow in high humic acid solution as the sole source of carbon, modifying the composition during *in vitro* tests (Bacilio *et al*., 2003).

Conclusion

Of all bacterial genera associated with plant roots, *Azospirillum* is the most studied. The bacteria's plant growth promotion mechanisms are only partially understood. The main handicap is lack of a clear plant phenotype that appears after inoculation, like nodules in rhizobia. Consequently, screening of large numbers of isolates or mutants is necessary. There has been a proliferation of studies on the physiology and molecular biology of *Azospirillum* in an effort to circumvent the absence of a clear plant phenotype. Many genes involved in PGPB interactions such as N_2-fixation, hormone and polysaccharide production, chemotaxis, and motility genes, were identified, cloned and studied intensely, irrespective of direct effects on plants after *Azospirillum* inoculation, a fact that so far contributes only marginally to producing a better inoculant. However, a genetically manipulated "super bacterium" that provides plenty of nitrogen to the plant by fixing more N_2 or excreting ammonium without compromising its fitness, is still far from development. Once this happens, the transfer of *Azospirillum* genes into plants might be considered.

References

Alenkina, S.A., V.E.Nikitina and M.V. Borisova-Golovko, 2001. Investigation of the relationship between the lectin activity of azospirilla and their alpha-glucosidase, ®-glucosidase and ®-galactosidase. *Microbiology (Moscow, Russ. Ed.)*, **70**: 557-559.

Amer H., N. A. Hegazi, and M.Nonib, (1977). A survey on the occurrence of nitrogen fixing *Spirillum* in Egypt. *Non symbiotic Nitrogen fixation Newsletter*, **5**: 101-102.

Amooaghaie, R., A. Mostajeran and G. Emtiazi, 2002. The effect of compatible and incompatible *Azospirillum brasilense* strains on proton efflux of intact wheat roots–*Azospirillum* and proton efflux of wheat root. *Plant Soil*, **243**: 155-160.

Antonyuk, L.P. and V.V.Ignatov. 2001. The role of wheat germ agglutinin in plant-bacteria interactions: A hypothesis and the evidence in its support. *Russ. J. Plant Physiol.*, **48**: 364-369.

Antonyuk, L.P., M.V. Sumaroka and V.V. Ignatov, 1999. Effect of wheat lectin on the membrane phospholipid content of *Azospirillum brasilense*. *Appl. Biochem. Microbiol.*, **35**: 53-55.

Assmus, B., M.Schloter, G. Kirchhof, P. Hutzler and A. Hartmann, 1997. Improved *in situ* tracking of rhizosphere bacteria using dual staining with fluorescence-labeled antibodies and rRNAtargeted oligonucleotides. *Microb.Ecol.*, **33**: 32-40.

Bacilio-Jiménez, M., S.Aguilar-Flores, M.V.del Valle, A.Pérez, A. Zepeda and E.Zenteno. 2001. Endophytic bacteria in rice seeds inhibit early colonization of roots by *Azospirillum brasilense*. *Soil Biol. Biochem.*, **33**: 167-172.

Bacilio, M., P.Vazquez and Y. Bashan, 2003. Alleviation of noxious effects of cattle ranch compost on wheat seed germination by inoculation with *Azospirillum* spp. *Biol. Fertil. Soils*, **38**: 261-266.

Bagwell, C.E., J.R.La Rocque, G.W. Smith, S.W.Polson, M.J.Friez, J.W.Longshore and C.R. Lovell. 2002. Molecular diversity of diazotrophs in oligotrophic tropical seagrass bed communities. *FEMS Microbiol. Ecol.*, **39**: 113-119.

Bashan, Y. 1998. Inoculants of plant growth-promoting bacteria for use in agriculture. *Biotechnol. Adv.*, **16**: 729-770.

Bashan, Y. 1999. Interactions of *Azospirillum* spp. in soils: a review. *Biol. Fertil. Soils*, **29**: 246-256.

Bashan, Y., and G.Holguin. 1997*a*. *Azospirillum*-plant relationships: environmental and physiological advances (1990-1996). *Can. J. Microbiol.*, **43**: 103-121.

Bashan, Y., and G.Holguin, 1997*b*. Short - and medium term avenues for *Azospirillum* inoculation. In: *Plant growth-promoting rhizobacteria–present status and future prospects.* (eds.) A. Ogoshi, K. Kobayashi, Y. Homma, F. Kodama, N. Kondo and S. Akino. Faculty of Agriculture, Hokkaido University, Sapporo, Japan. pp. 130-149.

Bashan, Y., and H.Levanony, 1990. Current status of *Azospirillum* inoculation technology: *Azospirillum* as a challenge for agriculture. *Can. J. Microbiol.*, **36**: 591-608.

Bashan, Y., and P. Vazquez. 2000. Effect of calcium carbonate, sand, and organic matter levels on mortality of five species of *Azospirillum* in natural and artificial bulk soils. *Biol. Fertil.Soils*, **30**: 450-459.

Bashan,Y., Holguin,G. and L.E. de-Bashan.2004. Azospirillum-plant relationships: physiological . molecular, agricultural and environmental advances (1997-2003). *Can. J. Microbiol.*, **50**: 521-577.

Belimov, A., and K.J. Dietz, 2000. Effect of associative bacteria on element composition of barley seedlings grown in solution culture at toxic cadmium concentrations. *Microbiol. Res.*, **155**: 113-121.

Belimov, A.A., N.V. Serebrennikova and V.V.Stepanok. 1999. Interaction of associative bacteria and an endomycorrhizal fungus with barley upon dual inoculation. *Microbiology (Moscow, Russ. Ed.)*, **68**: 104-108.

Benizri, E., E.Baudoin and A.Guckert. 2001. Root colonization by inoculated plant growth-promoting rhizobacteria. *Biocontrol Sci. Technol.*, **11**: 557-574.

Bianciotto, V., S. Andreotti, R. Balestrini, P.Bonfante and S.Perotto, 2001. Extracellular polysaccharides are involved in the attachment of *Azospirillum brasilense* and *Rhizobium leguminosarum* to arbuscular mycorrhizal structures. *Eur. J. Histochem.*, **45**: 39-49.

Blaha, C.A.G. and I.S. Schrank, 2003. An *Azospirillum brasilense* tn5 mutant with modified stress response and impaired in flocculation. *Antonie Leeuwenhoek*, **83**: 35-43.

Boggio, S.B. and O.A. Roveri, 2003. Catalytic properties of an endogenous beta-lactamase responsible for the resistance of *Azospirillum lipoferum* to beta-lactam antibiotics. *Microbiol. SGM*, **149**: 445-450.

Boddey, R.M. and J. Dobereiner. 1988. Nitrogen fixation associated with grasses and cereals. Recent results and perspectives for future research. *Plant and Soil*, **108**: 53-65.

Burdman, S., E. Jurkevitch, B.Schwartsburd, M. Hampel and Y. Okon, 1998. Aggregation in *Azospirillum brasilense*: effects of chemical and physical factors and involvement of extracellular components. *Microbiology (Reading, U.K.)*, **144**: 1989-1999.

Burdman, S., E.Jurkevitch, B. Schwartsburd and Y.Okon. 1999.Involvement of outer-membrane proteins in the aggregation of *Azospirillum brasilense*. *Microbiology (Reading, U.K.)*. **145**: 1145-1152.

Burdman, S., E.Jurkevitch, M.E. Soria-Diaz, A.M.G. Serrano and Y.Okon, 2000. Extracellular polysaccharide composition of *Azospirillum brasilense* and its relation with cell aggregation. *FEMS Microbiol. Lett.*, **189**: 259-264.

Burdman, S., G. Dulguerova, Y. Okon and E.Jurkevitch, 2001. Purification of the major outer membrane protein of *Azospirillum brasilense*, its affinity to plant roots, and its involvement in cell aggregation. *Mol. Plant Microbe Interact.*, **14**: 555-561.

Byzov, B.A., H. Claus, E.B. Tretyakova, D.G. Zvyagintsev and Z.Filip. 1997. Effects of soil invertebrates on the survival of some genetically engineered bacteria in leaf litter and soil. *Biol. Fertil. Soils*, **23**: 221-228.

Carrillo, A.E., C.Y. Li and Y.Bashan, 2002. Increased acidification in the rhizosphere of cactus seedlings induced by *Azospirillum brasilense*. *Naturwissenschaften*, **89**: 428-432.

Castellanos, T., F. Ascencio, and Y.Bashan, 1997. Cell-surface hydrophobicity and cell-surface charge of *Azospirillum* spp. *FEMS Microbiol. Ecol.* **24**: 159-172.

Castellanos, T., F. Ascencio and Y.Bashan, 1998. Cell-surface lectins of *Azospirillum* spp. *Curr. Microbiol.*, **36**: 241-244.

Castellanos, T., F. Ascencio and Y.Bashan, 2000. Starvationinduced changes in the cell surface of *Azospirillum lipoferum*. *FEMS Microbiol. Ecol.*, **33**: 1-9.

Ceccherini, M.T., M. Castaldini, A. Azzini, S. Fancelli, M.Bazzicalupo and N.Miclaus, 2001. Occurrence of *Azospirillum brasilense* in soils amended with swine manure. *Ann. Microbiol.*, **51**: 29-37.

Charyulu, P.B.B.N. and V.R. Rao, (1980). Influence of various soil factors on nitrogen fixation by *Azospirillum* spp. *Soil Biol. Biochem.*, **12**: 343-346.

Choi, Y.J. and S.W. Gal, 1998. Effects of osmoprotectants on the growth and nitrogenase activity of *Rhizobium* and *Azospirillum* under osmotic stress. *Agric. Chem. Biotechnol.*, **41**: 53-59.

Chotte, J.L., A. Schwartzmann, R. Bally and L.J. Monrozier, 2002. Changes in bacterial communities and *Azospirillum* diversity in soil fractions of a tropical soil under 3 or 19 years of natural fallow. *Soil Biol. Biochem.*, **34**: 1083-1092.

Christiansen-Weniger, C., 1997. Ammonium-excreting *Azospirillum brasilense* C3: *gusA* inhabiting induced tumors along stem and roots of rice. *Soil Biol. Biochem.*, **29**: 943-950.

Christiansen-Weniger, C., 1998. Endophytic establishment of diazotrophic bacteria in auxin-induced tumors of cereal crops. *Crit. Rev. Plant Sci.*, **17**: 55-76.

Creus, C.M., R.J. Sueldo and C.A. Barassi, 1997. Shoot growth and water status in *Azospirillum* -inoculated wheat seedlings grown under osmotic and salt stresses. *Plant Physiol. Biochem.* (Paris), **35**: 939-944.

Creus, C.M., R.J. Sueldo and C.A. Barassi, 1998. Water relations in *Azospirillum* – inoculated wheat seedlings under osmotic stress. *Can. J. Bot.*, **76**: 238-244.

Dahm, H., E. Strzelczyk, A.Ciesielska and K.Redlak, 1998. The effect of ectomycorrhizal fungi and bacteria on pine seedlings. *Acta Mycol.*, **33**: 25-36.

Dekhil, S.B., M.Cahill, E.Stackebrandt and LI.Sly. 1997. Transfer of *Conglomeromonas largomobiliz* subsp *largomobiliz* to the genus *Azospirillum* as *Azopirillum largomobile* comb nov., and elevation of *Conglomeromonas largomobiliz* subsp *parooensis* to the new type species of *Conglomeromonas*. *Conglomeromonas parooensis* sp nov. *Syst. Appl. Microbiol.*, **20**: 72-77.

Dobbelaere, S., Croonenborghs, A., Thys, A., Vande Broek, A., and Vanderleyden, J. 1999. Phytostimulatory effect of *Azospirillum brasilense* wild type and mutant strains altered in IAA production on wheat. *Plant Soil*, **212**: 155-164.

Dobbelaere, S., Croonenborghs, A., Thys, A., Ptacek, D., Vanderleyden, J., Dutto, P., Labandera-Gonzalez, C., Caballero-Mellado, J., Aguirre, J.F., Kapulnik, Y., Brener, S., Burdman, S., Kadouri, D., Sarig, S., and Okon, Y. 2001. Responses of agronomically important crops to inoculation with *Azospirillum*. *Aust. J. Plant Physiol.*, **28**: 871–879.

Dobereiner, J. and Day, J.M. 1976. Associative symbiosis in tropical grasses: characterization of microorganisms and nitrogen fixing sites. In: *Proceedings of the First International Symposium on Nitrogen Fixation.* (eds.). W E Newton and C J N Washington State University Press, Pullman. pp. 518-536.

Eckert, B., O.B. Weber, G. Kirchhof, A. Halbritter, M. Stoffels and A.Hartmann. 2001. *Azospirillum doebereinerae* sp. nov., a nitrogen-fixing bacterium associated with the C4-grass *Miscanthus. Int. J. Syst. Evol. Microbiol.*, **51**: 17-26.

Edelshtein, Z., D. Kadouri, E. Jurkevitch, A.Vande Broek, J. Vanderleyden and Y.Okon, 2003. Characterization of genes involved in poly-beta-hydroxybutyrate metabolism in *Azospirillum brasilense. Symbiosis*, **34**: 157-170.

Egorenkova, I.V., S.A Konnova, I.M.Skvortsov and V.V. Ignatov, 2000. Investigation of the initial stages of interaction of the bacterium *Azospirillum brasilense* with wheat seedling roots: Adsorption and root hair deformation. *Microbiology (Moscow, Russ. Ed.)*, **69**: 103-108.

El-Katatny, M.S., H.M. El-Komy and A.M. Attia, 1997. Pectin decomposition by mixed cultures of *Azospirillum* spp., and *Penicillium corylophillum* and its role in *Azospirillum*–host plant association. *Microbiol. Res.*, **152**: 143-149.

El-Khawas, H., and K. Adachi, 1999. Identification and quantification of auxins in culture media of *Azospirillum* and *Klebsiella* and their effect on rice roots. *Biol. Fertil. Soils*, **28**: 377-381.

El-Komy, H.M., M.A.Hamdia and G.K.A.El-Baki, 2003. Nitrate reductase in wheat plants grown under water stress and inoculated with *Azospirillum* spp. *Biol. Plant.*, **46**: 281-287.

Fedonenko, Y.P., I.V. Egorenkova, S.A. Konnova, and V.V. Ignatov, 2001. Involvement of the lipopolysaccharides of azospirilla in the interaction with wheat seedling roots. *Microbiology (Moscow, Russ. Ed.)*, **70**: 329-334.

Fedonenko, Y.P., G.V.Zatonsky, S.A. Konnova, E.L. Zdorovenko and V.V.Ignatov, 2002. Structure of the O-specific polysaccharide of the lipopolysaccharide of *Azospirillum brasilense* Sp245. *Carbohydr. Res.*, **337**: 869-872.

Feng, L., and I.R. Kennedy, 1997. Biodegradation and plant protection from the herbicide 2, 4-D by plant-microbial associations in cotton production systems. *Biotechnol. Bioeng.*, **54**: 513-519.

Fischer, S.E., M.J. Miguel and G.B.Mori, 2003. Effect of root exudates on the exopolysaccharide composition and the lipopolysaccharide profile of *Azospirillum brasilense* cd under saline stress. *FEMS Microbiol. Lett.*, **219**: 53-62.

Fischer, S., V. Rivarola, and G.Mori, 1999. Saline stress alters the gene expression of *Azospirillum brasilense* Cd. *Microbios*, **100**: 83-88.

Fischer, S., V. Rivarola, and G.Mori, 2000. Colonization of wheat by *Azospirillum brasilense* Cd is impaired by saline stress. *Plant Soil*, **225**: 187-191.

Garcia de Salamone, I.E., J.Döbereiner, S.Urquiaga, and R.M. Boddey, 1997. Biological nitrogen fixation in *Azospirillum* strain-maize genotype associations as evaluated by the 15N isotope dilution technique. *Biol. Fertil. Soils*, **23**: 249-256.

Glick, B.R., C.L. Patten, G.Holguin, and D.M. Penrose, 1999. Biochemical and genetic mechanisms used by plant growth promoting bacteria. Imperial College Press, London, UK.

Gomez, F., V.Salmeron, B. Rodelas, M.V.Martinez-Toledo and J.Gonzalez-Lopez, 1998. Response of *Azospirillum brasilense* to the pesticides bromopropylate and methidathion on chemically defined media and dialysed-soil media. *Ecotoxicology*, **7**: 43-47.

Gomez, F., M.V. Martinez-Toledo, V. Salmerón, B.Rodelas, and J.González-Lopez, 1999. Influence of the insecticides profenofos and diazinon on the microbial activities of *Azospirullum brasilense. Chemosphere*, **39**: 945-957.

Govindan, M and D.Purushothaman, 1985. Association of nitrogen fixing bacteria with certain plantation crops. *Natl.Acad. Sci. Lett.*, **8**: 1631-1665.

Govindarajan. K. 1999. Studies on chromium tolerance in *Azospirillum*. *Madras Agric. J.*, **86**: 22-24.

Govindarajan. K. and K. Kavitha. 2004. Studies on Azospirillum associated with rice Varieties. In : *Biofertilizers Technology for rice based cropping System.* (eds.) S.Kannaiyan. K.Kumar and K.Govindarajan. Scientific Publishing Co.. Jodhpur. Rajasthan. India. pp. 221-224.

Hamdia, M.A.. and H.M. El-Komy. 1997. Effect of salinity, gibberellic acid and *Azospirillum* inoculation on growth and nitrogen uptake of *Zea mays*. *Biol. Plant.*, **40**: 109-120.

Han. S.O.. and P.B. New. 1998*a*. Variation in nitrogen fixing ability among natural isolates of *Azospirillum*. *Microb.Ecol.*, **36**: 193-201.

Han. S.O.. and P.B. New. 1998*b*. Isolation of *Azospirillum* spp. from natural soils by immunomagnetic separation. *Soil Biol. Biochem.*, **30**: 975-981.

Hecht-Buchholz. C. 1998. The apoplast-habitat of endophytic dinitrogen-fixing bacteria and their significance for the nitrogen nutrition of nonleguminous plants. *Z. Pflanzenernaehr. Bodenkd.*, **161**: 509-520.

Indira, B.N.. and D.J. Bagyaraj. 1997. Associative diazotrophs of sugarcane (*Saccharum officinarum* L.) cultivars. *Indian J.Microbiol.*, **37**: 95-98.

Jofré, E., S. Fischer. V. Rivarola. H.Balegno. and G.Mori. 1998*a*. Saline stress affects the attachment of *Azospirillum brasilense* Cd to maize and wheat roots. *Can. J. Microbiol.*, **44**: 416-422.

Jofré. E., Rivarola. V.. Balegno. H.. and Mori. G. 1998*b*. Differen-tial gene expression in *Azospirillum brasilense* Cd under saline stress. *Can. J. Microbiol.* **44**: 929-936.

Kadouri. D.. S. Burdman. E. Jurkevitch. and Y.Okon. 2002. Identification and isolation of genes involved in poly â-hydroxybutyrate biosynthesis in *Azospirillum brasilense* and characterization of a *phbC* mutant. *Appl. Environ. Microbiol.*, **68**: 2943-2949.

Kannaiyan. S. 2003. Inoculant production in developing countries-problems. potentials and success. In: *Maximising the use of biological nitrogen fixation in agriculture* (eds.) G.Hardarson and W.J.Broughton. Kluwer Academic Publishers. The Netherlands. pp. 187-198.

Kanungo. P.K.. B.Ramakrishnan and V.R. Rao. 1997. Placement effects of organic sources on nitrogenase activity and nitrogen-fixing bacteria in flooded rice soils. *Biol. Fertil. Soils,* **25**: 103-108.

Kanungo. P.. B. Ramakrishnan and V.R. Rao. 1998. Nitrogenase activity of *Azospirillum* sp. isolated from rice as influenced by a combination of NH_4^+-N and an insecticide. carbofuran. *Chemosphere*. **36**: 339-344.

Karpati. E.. P.Kiss. T. Ponyi. I. Fendrik. M.de Zamaroczy and L. Orosz. 1999. Interaction of *Azospirillum lipoferum* with wheat germ agglutinin stimulates nitrogen fixation. *J. Bacteriol.*, **181**: 3949-3955.

Kaushik. R.. A.K. Saxena. and K.V.B.R. Tilak. 2002. Can *Azospirillum* strains capable of growing at a sub-optimal temperature perform better in field-grown-wheat rhizosphere. *Biol. Fertil. Soils*. **35**: 92-95.

Kefalogianni. I.. and G.Aggelis, 2002. Modeling growth and biochemical activities of *Azospirillum* spp. *Appl. Microbiol. Biotechnol.*, **58**: 352-357.

Kennedy, I.R.. L.L. Pereg-Gerk, C. Wood, R.Deaker. K. Gilchrist and S.Katupitiya. 1997. Biological nitrogen fixation in nonleguminous field crops: facilitating the evolution between *Azospirillum* and wheat. *Plant Soil*, **194**: 65-79.

Kennedy. I.R.. S.Katupitiya. D. Yu. K.Gilchrist. R.Deaker. L.Pereg-Gerk and C.Wood. 1998. Prospects for Facilitated evolution of effective N_2-fixing associations with cereals: comparative performance of *Azospirillum brasilense* Sp7-S with various free-living diazotrophs in para-nodulated wheat. In: *Developments in plant and soil sciences. Vol. 79. Proceedings of the 7th International Symposium on Nitrogen Fixation with Non-Legumes.* 16–21 October. 1996. Faisalabad. Pakistan. (eds.) K.A. Malik. M. Mirza. and J.K. Ladha. Kluwer Academic Publishers. Dordrecht. pp. 109-124.

Kirchhof. G., M. Schloter. B. Assmuss and A.Hartmann, 1997. Molecular microbial ecology approaches applied to diazotrophs associated with non-legumes. *Soil Biol. Biochem.* **29**: 853-862.

Konnova. S.A.. O.S. Brykova, O.A. Sachkova. I.V. Egorenkova. and V.V. Ignatov. 2001. Protective role of the Polysaccharidecontaining capsular components of *Azospirillum brasilense*. *Microbiology (Moscow, Russ. Ed.)*. **70**: 436-440.

Konnova, S.A., Y.P. Fedonenko, O.E. Makarov and V.V. Ignatov, 2003. Effect of cultivation conditions on the composition of extracellular polysaccharide-containing substances in the bacterium *Azospirillum brasilense*. *Biol. Bull. Russ. Acad. Sci.*, **30**: 354-360.

Kumar, K., and T.C.Kumari Sugitha, 2004. Diazotrophic biodiversity in rice ecosystem *Paper presented in the International Conference on Microbial Ecology*, Held at Cancun, Mexico, 21-26 August, 2004.

Kundu, B.S., P. Sangwan, P.K. Sharma, and A.S. Nandwal, 1997. Response of pearlmillet to phytohormones produced by *Azospirillum brasilense*. *Indian J. Plant Physiol.*, **2**: 101-104.

Lakshmi kumari, M., S.K Kavimandan and N. S. Subba Rao. 1976. Occurrence of nitrogen fixing *Spirillum* in roots of rice, sorghum, maize and other plants. *Indian J. Exptl. Biol.*, **14**: 638-639.

Lerner, A., Y. Herschkovitz, E. Baudoinb, S.Nazaret, Y.Moenne-Loccoz, Y. Okon and E. Jurkevitch, 2006. Effect of Azospirillum brasilense inoculation on rhizobacterial communities analyzed by denaturing gradient gel electrophoresis and automated ribosomal intergenic spacer analysis. *Soil Biol. & Biochem.*, **38**: 1212-1218.

Liu, Y., S.-F. Chen and J.-I. Li, 2003. Colonization pattern of *Azospirillum brasilense* yu62 on maize roots. *Acta Bot. Sin.*, **45**: 748-752.

Lucangeli, C., and R.Bottini, 1997. Effects of *Azospirillum* spp. on endogenous gibberellin content and growth of maize (*Zea mays* L.) treated with uniconazole. Symbiosis, **23**: 63-72.

Magalhaes, F. M., J. I, Baldani, Santo, J. Kuykendall, J .R and Dobereiner, J. (1983) A new acid tolerant *Azospirillum* species. *Ann. Acad. Brar. Clienc.*, **55**: 417-430.

Manna, A., S. Pal, and A.K. Paul. 1997. Occurrence of poly-3- hydroxybutyrate in *Azospirillum* sp. *Folia Microbiol.*, **42**: 629-634.

Mano, D.M.S. and T.Langenbach, 1998. 14C-dicofol association to cellular components of *Azospirillum lipoferum*. *Pestic. Sci.*, **53**: 91-95.

Marchal, K. and J.Vanderleyden, 2000. The "oxygen paradox" of dinitrogen-fixing bacteria. *Biol. Fertil. Soils*, **30**: 363-373.

Matora, L.Y. and S.Y. Shchegolev, 2002. Antigenic identity of the capsule lipopolysaccharides, exopolysaccharides, and O-specific polysaccharides in *Azospirillum brasilense*. *Microbiology (Moscow, Russ. Ed.)*, **71**: 178-181.

Molla, A.H., Z.H. Shamsuddin and H.M. Saud, 2001*b*. Mechanism of root growth and promotion of nodulation in vegetable soybean by *Azospirillum brasilense*. *Commun. Soil Sci. Plant Anal.*, **32**: 2177-2187.

Nath, D.J., Bhattacharjee, R.N., Devi, M.R., and Patgiri, S.R. 1997. Widespread occurrence of *Azospirillum* spp. in the North-Eastern Region of India. *Adv. Plant Sci.*, **10**: 189-194.

Nelson, D.R. and P.M. Mele, 2006. Subtle changes in rhizosphere microbial community structure in response to increased boron and sodium chloride concentrations. *Soil Biol.,& Biochem.*, **39**: 340-351.

Neyra, C. A. and J. Dobereiner, 1977. Nitrogen fixation in grasses. *Advan. Agron.*, **29**: 1-38.

Nikitina, V.E., E.G. Ponomareva, S.A. Alenkina, and S.A. Konnova, 2001. The role of cell-surface lectins in the aggregation of azospirilla. *Microbiology (Moscow, Russ. Ed.)*, **70**: 408-412.

Okon, Y. 1985. *Azospirillum* as a potential inoculant for agriculture. *TIBTECH*, **3**(9): 223-228.

Okon, Y and C.A. Labandera-Gonzalez, 1994. Agronomic applications of *Azospirillum*: An evaluation of 20 years worldwide field inoculation. *Soil Biol. Biochem.*, **26**: 1591-1601.

Oliveira, A.L.M., L.E. Canuto, E.E. Silva, M.R. Veronica and J.I. Baldani, 2004. Survival of endophytic diazotrophic bacteria in soil under different moisture levels. *Brazilian J. Microbiol.*, **35**: 295-299.

Olubayi, O., R. Caudales, A. Atkinson, and C.A. Neyra, 1998. Differences in chemical composition between nonflocculated and flocculated *Azospirillum brasilense* Cd. *Can. J. Microbiol.*, **44**: 386-390.

Ona, O., L., Smets, P., Gysegom, K., Bernaerts, J.V., Impe, E., Prinsen, and J. Vanderleyden, 2003. The effect of pH on indole-3-acetic acid (IAA) biosynthesis of *Azospirillum brasilense* sp7. *Symbiosis*, **35**: 199-208.

Palaneeswar, R., S.P. Sundaram, and A. Balasubramanian. 1992. Occurrence of acid producing *Azospirillum* in rice soil ecosystem. In: *Biological nitrogen fixation and Biogas technology*. (eds). S. Kannaiyan. K. Ramasamy. K. Ilamurugu and K. Kumar. Tamil Nadu Agricultural University. Coimbatore. Tamil Nadu. India.

Piccoli. P., C.D. Lucangeli. G. Schneider. and R.Bottini. 1997. Hydrolysis of (17. 17-2H2) gibberellin A20-glucoside and (17. 17-2H2) gibberellin A20-glucosyl ester by *Azospirillum lipoferum* cultured in a nitrogen-free biotin-based chemically-defined medium. *Plant Growth Regul.* **23**: 179-182.

Piccoli, P., O. Masciarelli. and R.Bottini. 1999. Gibberellin production by *Azospirillum lipoferum* cultured in chemicallydefined medium as affected by oxygen availability and water status. *Symbiosis*, **27**: 135-145.

Pinheiro, R.D., L.H. Boddey, E.K. James. J.I. Sprent and R.M. Boddey. 2002. Adsorption and anchoring of *Azospirillum* strains to roots of wheat seedlings. *Plant Soil.* **246**: 151-166.

Puente, M.E., G. Holguin, B.R. Glick and Y.Bashan. 1999. Root-surface colonization of black mangrove seedlings by *Azospirillum halofraeferens* and *Azospirillum brasilense* in seawater. *FEMS Microbiol. Ecol.*, **29**: 283-292.

Purushothaman, D., and K.Govindarajan. (1984). Rhizosphere ecology and nitrogen fixation of *Azospirillum* in pearl millet (*Pennisetum americanum* (L.) Leeke).In: *Working group meeting on cereal nitrogen fixation*.9-12. October.1984.ICRISAT. Andhra Pradesh. India p. 16 (*Abstr.*).

Purushothaman. D., K.Jayanthamurali and S.Gunasekaran. (1987). Association of *Azospirillum* with the roots of upland rice. In: *Biofertilizers: Potentials and Problems*. (eds) S.P. Sen and P. Palit. Naya Prokash. Calcutta. India. pp. 171-182.

Rajeev, K., A.K. Saxena. and K.V.B.R.Tilak. 2000. Selection of Tn5::*lacZ* mutants isogenic to wild type *Azospirillum brasilense* strains capable of growing at sub-optimal temperature. *World J. Microbiol. Biotechnol.*, **16**: 567-570.

Raji. P. and M.V.R. Pillai. 2000. Effect of plant protection chemicals on *Azospirillum* in cowpea (*Vigna unguiculata* (L) Walp). *Legume Res.*, **23**: 177-179.

Ramos. H.J.O., L.D.B. Roncato-Maccari. E.M. Souza. J.R.L. Soares-Ramos. M. Hungria. and F.O. Pedrosa. 2002. Monitoring *Azospirillum*-wheat interactions using the *gfp* and *gusA* genes constitutively expressed from a new broad-host range vector. *J. Biotechnol.*, **97**: 243-252.

Rao. V. R., P. B.B.N. Charyulu. D.N. Nayak and C.Ramakrishna. (1978). Nitrogen fixation by free-living organisms in tropical rice soils. In :*Limitations and potentials for biological nitrogen fixation in the tropics*. (ed.). J. Dobereiner. R H Burris and A Holluender. Plenum Press. New York. USA. pp. 354.

Ravikumar. S., G. Ramanathan. N. Suba. L.Jeyaseeli. and M.Sukumaran. 2002. Quantification of halophilic *Azospirillum* from Mangroves. *Indian J. Mar. Sci.*, **31**: 157-160.

Reinhold-Hurek, B., T. Hurek, I. Fendrik. B. Pot. M. Gillis. K.Kersters. S. Thielemans. and J.DeLey. (1987). *Azospirillum halopraeferens* sp. nov., a nitrogen fixing organism associated with roots of kallar grass (*Leptochola fusca*) (L.) kunth.). *Int. J. Syst. Bacteriol.*, **37**: 43-51.

Reynders. L and K. Vlassak. (1976). Nitrogen fixing *Spirillum* species in Belgium soils. *Agriculture (Belgium)*. **24**: 329-336.

Revellin. C., J.J. Giraud. N. Silva. P. Wadoux and G. Catroux. 2001. Effect of some granular insecticides currently used for the treatment of maize crops (*Zea mays*) on the survival of inoculated *Azospirillum lipoferum*. *Pest Manag. Sci.*, **57**: 1075-1080.

Rivarola. V., S. Castro. G. Mori. E. Jofre. A. Fabra. R. Garnica. and H. Balegno. 1998. Response of *Azospirillum brasilense* Cd to sodium chloride stress. *Antonie Leeuwenhoek.* **73**: 255-261.

Rodelas. B., J. Gonzalez-Lopez. M.V. Martinez-Toledo. C. Pozo. and V. Salmeron. 1999. Influence of *Rhizobium Azotobacter* and *Rhizobium Azospirillum* combined inoculation on mineral composition of faba bean (*Vicia faba* L.). *Biol. Fertil. Soils.* **29**:

Rodriguez-Navarro. D N., M. S.Dardanelli and J.E. Ruiz-Saiýnz. 2007. Attachmentof bacteria tothe roots of higher plants *FEMS Microbiol Lett*., pp. 1-10.

Ruppel, S., and W.Merbach, 1997. Effect of ammonium and nitrate on $^{15}N_2$-fixation of *Azospirillum* spp. and *Pantoea agglomerans* in association with wheat plants. *Microbiol. Res.*. **152**: 377-383.

Saleena, L.M., S. Rangarajan and S.Nair, 2002. Diversity of *Azospiril'um* strains isolated from rice plants grown in saline and nonsaline sites of coastal agricultural ecosystem. *Microbiol. Ecol.*, **44**: 271-277.

Saubidet, M.I., N. Fatta and A.J. Barneix, 2002. The effect of inoculation with *Azospirillum brasilense* on growth and nitrogen utilization by wheat plants. *Plant Soil*, **245**: 215-222.

Schloter, M. and A.Hartmann, 1998. Endophytic and surface colonization of wheat roots (*Triticum aestivum*) by different *Azospirillum brasilense* strains studied with strain-specific monoclonal antibodies. *Symbiosis*, **25**: 159-179.

Sedik, M.Z. 1997. Biodiversity of diazotrophs associated with natural vegetation of North Sinai. *Bull. Fac. Agric. Univ. Cairo*, **48**: 143-158.

Sekar, C., N.N. Prasad and M.D. Sundaram, 1999. Enhancement of polygalacturonase activity during auxin induced para nodulation and endorhizosphere colonization of *Azospirillum* in rice roots. *Indian J. Exp. Biol.*, **38**: 80-83.

Skvortsov, I.M., and V.V. Ignatov, 1998. Extracellular polysaccharides and polysaccharide-containing biopolymers from *Azospirillum* species: properties and the possible role in interaction with plant roots. *FEMS Microbiol. Lett.*, **165**: 223-229.

Strzelczyk, E., A. Pokojska-Burdziej and C.Y. Li, 1997. Effect of Cd^{+2}, Zn^{+2} and Cu^{+2} on auxins production by *Azospirillum* strains. *Acta Microbiol. Pol.*, **46**: 263-270.

Subba Rao, N. S. (1983). Nitrogen fixing bacteria associated with plantation and orchard plants. *Can. J. Microbiol.* **29**: 863-866.

Subramaniane, S., and Govindarajan, K. 1998. Studies on heavy metal tolerance in *Azospirillum*. *Natl. Acad. Sci. Lett.(India)*, **21**: 274-277.

Sundaram, SP., S. Kannaiyan, and S. Muthukumarasamy, (2001). Transformation studies in *Azospirillum halopraeferens*. www. microbiologyou. com p. 6.

Thiruchenthilvasan, S., and D. Purushothaman, (1999). Rhizosphere colonization of genetically engineered *Azospirillum* strains and response of sorghum to inoculation. *XI Southern Conference on Microbial Inoculants*. 19th and 20th Feb.1999. Guntur- 522 005, India. p. 1 *(Abstr.)*.

Thuler, D.S., E.I.S.Floh, W. Handro, and H.R. Barbosa, 2003. Plant growth regulators and amino acids released by *Azospirillum* sp in chemically defined media. *Lett. Appl. Microbiol.*, **37**: 174-178.

Tripathi, A.K., and B.M. Mishra, 1998*b*. Isolation and cloning of an *Azospirillum lipoferum* locus that complements *Escherichia coli pro*U mutant. FEMS *Microbiol. Lett.*, **162**: 241-247.

Tripathi, A.K., B.M. Mishra, and P. Tripathi, 1998a. Salinity stress responses in the plant growth promoting rhizobacteria, *Azospirillum* spp. *J. Biosci., (Bangalore)*, **23**: 463-471.

Tsimilli-Michael, M., P. Eggenberg, B. Biro, K. Koves-Pechy, I. Voros and R.J. Strasser, 2000. Synergistic and Antagonistic effects of arbuscular mycorrhizal fungi and *Azospirillum* and *Rhizobium* nitrogen-fixers on the photosynthetic activity of alfalfa, probed by the polyphasic chlorophyll a fluorescence transient O-J-I-P. *Appl. Soil Ecol.*, **15**: 169-182.

Vázquez, M.M., S. César, R. Azcón and J.M. Barea, 2000. Interactions between arbuscular mycorrhizal fungi and other microbial inoculants (*Azospirillum, Pseudomonas, Trichoderma*) and their effects on microbial population and enzyme activities in the rhizosphere of maize plants. *Appl. Soil Ecol.*, **15**: 261-272.

Vedder-Weiss, D., Jurkevitch, E., Burdman, S., Weiss, D., and Okon, Y. 1999. Root growth, respiration and beta-glucosidase activity in maize (*Zea mays*) and common bean (*Phaseolus vulgaris*) inoculated with *Azospirillum brasilense*. *Symbiosis*, **26**: 363-377.

Weber, O.B., V.L.D. Baldani, K.R.S. Teixeira, G. Kirchhof, J.I. Baldani, and J. Döbereiner, 1999. Isolation and characterization of diazotrophic bacteria from banana and pineapple plants. *Plant Soil*, **210**: 103-113.

Widmer, F., B.T. Shaffer, L.A. Porteous and R.J.Seidler, 1999. Analysis of *nifH* gene pool complexity in soil and litter at a Douglas fir forest site in the Oregon Cascade Mountain Range. *Appl. Environ. Microbiol.,* **65**: 374-380.

Wood, C.C., N. Islam. R.J. Ritchie and I.R. Kennedy, 2001. A simplified model for assessing critical parameters during associative 15N2 fixation between *Azospirillum* and wheat. *Aust. J. Plant Physiol.*, **28**: 969-974.

Yegorenkova, I.V., S.A. Konnova, V.N. Sachuk and V.V. Ignatov, 2001. *Azospirillum brasilense* colonization of wheat roots and the role of lectin-carbohydrate interactions in bacterial adsorption and root-hair deformation. *Plant Soil,* **231**: 275-282.

Yu, Z., and W.W. Mohn, 2001. Bacterial diversity and community structure in an aerated lagoon revealed by ribosomal intergenic spacer analyses and 16S ribosomal DNA sequencing. *Appl. Environ. Microbiol.,* **67**: 1565-1574.

Zakharova, E.A., A.A. Shcherbakov, V.V. Brudnik, N.G. Skripko, N. S. Bulkhin and V.V. Ignatov, 1999. Biosynthesis of indole- 3-acetic acid in *Azospirillum brasilense*: Insights from quantum chemistry. *Eur. J. Biochem.*, **259**: 572-576.

Zakharova, E.A., A.D. Iosipenko and V.V. Ignatov, 2000. Effect of water-soluble vitamins on the production of indole-3-acetic acid by *Azospirillum brasilense. Microbiol. Res.*, **155**: 209-214.

Soil Microbiology and Soil Biotechnology : Vol. 2
Editor : S. Kannaiyan
Associated Publishing Co., New Delhi, 2010, pp. 100-104

4

Liquid Bioinoculants

Geeta Singh[1] *and T.S. Marwaha*[2]

Introduction

To harness the beneficial effect of microbial bio-inoculants, it is a common practice to provide a sufficient number of viable, effective microorganism to induce rapid colonization of the rhizosphere. Inoculation of seeds is an efficient and convenient way of introducing effective PGPR to soil and subsequently the rhizosphere of the crop plant. However, it's full potential is yet to be realized. The efficacy and quality of bio-inoculants is dependent on many factors all of which affect the number of viable PGPR available in it. The major factors affecting the viable counts of PGPR per seed following inoculation are desiccation, temperature, seed coat toxicity. The shelf life of a bio-inoculants is also influenced by nature of carrier material.

Conventionally PGPR bio-inoculants are solid formulations, where peat is the most suitable favoured material used. Experimental evidences indicate that origin of peat degree of sterility, its pH and salt content all contributes to the survival of the added microbial bioinoculant in it (Steinborn and Roughley, 1975). Due to scarcity of peat and to overcome the constraints encountered in solid peat based bioinoculants, alternative carrier materials/ formulations for optimization of fundamental delivery techniques is required. The common forms of bioinoculant include agar cultures, peat, liquid, freeze dried and granules. In this direction liquid formulations are found to be a low cost technology. Two main functions of carrier materials have been identified namely they must support the growth of the organisms and support high viable numbers over an acceptable period of time.

Inoculants formulation and survival

Physiological preconditioning is the prerequisite for extending the survival of microbial bioinoculants. Physiological and morphological changes during the maturation of the peat inoculants have been shown to affect survival (Fend *et al.*, 2002). Changes in the rhizobial cells such as thickening of cell wall, the disappearance of polyhydroxy butyrate (PHB) granules and appearance of heat shock proteins were found to be correlated with improved survival but the mechanism is not clear. Poor survival on seeds may be overcome by applying

1. Geeta Singh and : 2. T.S. Marwaha, Senior Scientists, Division of Microbiology, IARI, New Delhi-110 012

the inoculant directly to the soil. For effective liquid formulation of PGPR, morphological forms like spore/cyst possess advantage. These cells remain dormant for extended periods under conditions of low nutrient availability and germinate under suitable conditions. The encystment process in PGPR namely *Azotobacter* and *Azospirillum* confers a survival advantage due to low endogenous respiration rate and ability to withstand desiccation (Sociolofsky and Wyss, 1962). *Azotobacter* and *Azospirillum* have been isolated from 15 year old dry soil samples.

Nature of carbon source regulates cyst formation. In case of *Azotobacter* spp. Butanol, crotonate and hydroxy butyrate induces abundant cyst formation when vegetative cells accumulate cell wall or capsular material and intra cellular polymers such as poly β hydroxy butyrate (PHB) they are transformed into cyst form. Similarly diazotroph *Azospirillum* spp. undergo abundant encystment following manipulations in carbon and nitrogen sources. Combination of fructose and nitrate was effective in production of exocellular polymers particularly linked polysaccharides synthesis of these exocellular polymers precedes cyst formation in *Azospirillum* spp. (Sadasivam and Neyra, 1985). The transition from vegetative cells to refractile cyst form is accompanied with synthesis of abundant poly β hydroxybutyrate granules within a central body surrounded by a thick layer of exopolysaccharides (Maltseva and Laski, 1982). Dried flocks of *Azospirillum* retain their viability for more than six months as confirmed by pellicle formation and acetylene reduction in semisolid malate medium (Sadasivam and Nyra, 1985). Other workers (Lamm and Neyra, 1981) reported that cysts remain viable for more than 15 months. Cyst enriched cultures of *Azospirillum* are highly resistant to desiccation compared to their negative counter parts. Cysts of *Azotobacter* spp. and *Azospirillum* spp. are morphologically distinct from their vegetative cells. In case of *Azospirillum* spp. the cysts are thick walled, ovoid refractile cells of 2.5 μm diameter and 3-15 μm lengths. (Lamm and Neyra, 1981). Gram positive spore forming PGPR like *Bacillus* spp. endure stresses like high temperatures, low nutrient availability and desiccation due to the presence of dipicolinic acid (Halvorson and Howitt, 1961) and anhydrous condition of the spore (Rode and Foster, 1960) the ability to form spores is the reason that viable Bacilli are isolated from soils stored for more than 300 years.

Exopolysacharides (EPS)

EPS biosynthesis is focal point of the ability of some bacteria to express stress tolerance. EPS synthesis is co-ordinated expression of a set of genes that responds to changes in water potential of cells and its environment (Hypoosmotic and Hyperosmotic). It is well documented that production of exopolysaccharides by bacterial cells is a means to overcome desiccation stress (Bushby and Marshall, 1977; Hartel and Alexander 1986; Osa-Afiana and Alexander 1982; Roberson and Fisestone, 1992). An EPS matrix may slow the rate at which bacterial colony equilibrates with surrounding environment. Thereby slowing the rate at which a bacterial colony equilibrates with the surrounding soil. This permits the bacterial survival by increasing the time available for metabolic adjustments. Due to their hygroscopic nature EPS maintain higher water content in the colony microenvironment which permits easy diffusion of nutrients to the microorganisms (McAneney *et al.*, 1982).

Physiological tolerance to desiccation

Survival in case of non spore formers in nutrient poor environment is through physiological adaption by slowed metabolic rate. Conditioning of PGPR cells to induce the production of osmoprotectants such as trehalose or protective EPS is a useful way to overcome desiccation stress experienced on seed surfaces. For maintenance of biological integrity during desiccation the cells accumulate osmoprotectants. These include potassium ions, glutamate, glutamine protein, glycine-betaine and sugars like trehalose, sucrose and glucosylglycerol (Pott, 1994). Compatible solutes help maintain the stability of cellular proteins during osmotic stress. Some desiccation tolerant cells accumulate large amounts of the disaccharides trehalose and sucrose (Crowe *et al.*, 2001). Trehalose a non reducing disaccharide is an effective protectant against various kinds of stress (Crowe *et al.*, 1992). The possible mechanism in such systems is that hydrogen bonding occurs between the carbohydrate and protein preserving the cell structure (Tsevtkov *et al.*, 1989). Similarly hydrogen bonding occurs between trehalose and phospholipids in membranes replacing the hydrating water molecules (Gaber, *et al.*, 1986). Yeasts (*Saccharomyces cerevisiae*) and thermophilic fungus *Chaetomium thermophilum* are also known to accumulate trehalose as stress protectant (Hounsa, 1998; Thevelein, 1996; Jepsen and Jensen, 2004). Trehalose accumulation through denovo synthesis occurs in *Bradyrhizobium japonicum*. In slow growing strains trehalose accounted for greater than 80% of total mono and disaccharides. An apparent relationship exists between high accumultion of trehalose and survival in water culture (Streeter, 1985).

Additives to improve PGIPR survival on seeds

A large mortality of the applied bioinoculant occurs between inoculation and sowing (Roughley *et al.*, 1993). Shelf life of liquid bioinoculant is a crucial factor determining its performacne. Selection of a liquid medium and the additives is governed by the morpho-physiological status of the suspended cells and osmotic considerations. Extensive research has been carried out on the use of bacterial nutrients as suspending agents for freeze drying and storage of cells (Dye, 1982). It was found that survival was related to assimilability and solubility of the compound. Sucrose was superior to glucose, xylose, trytophan, salicin and saline. Vincent (1958) reported 24-44% of cells suspended in a 10% sucrose solution survived primary drying whereas only 0.1% survived when suspended in water. Some of the additives found to improve the performance of liquid formulations are enlisted below.

PVP: Binds toxic compounds present in seed exudates high water binding capacity, slows down the rate of drying of bioinoculant after application. PVP coalesce into ridges of seed coat, providing a thick protective layer. This is sticky therefore enhances adherence of bioinoculant to seed surface.

Glycerol: Glycerol is a carbon source for rhizobia. Glycerol also has a high water-binding capacity and may protect cells from the effects of desiccation by slowing the drying rate. Its flow characteristics appear to promote rapid and even coating of seeds.

Trehalose: This compound is widely reported to enhance cell tolerance to desiccation and to osmotic and temperature stress. Trehalose acts by stabilizing both enzymes and cell membranes, and is a compatible osmoticum as well. Trehalose is readily manufactured by *Bradyrhizobium* given the correct conditions.

Glucose: Glucose promotes the production of the storage product glycogen. In the absence of glycogen, osmotically stressed cells are reported to release their supplies of trehalose, an undesirable outcome. Glucose also enhances exopolysaccharide production, which could protect cells during the rapid drying they experience at inoculation. Glucose had no apparent effect on cell survival.

Mannitol: Mannitol has many potential roles in a liquid inoculant. It is a carbon source for all strains of *Bradyrhizobium,* and in conjunction with arabinose it is reported to support the greatest production of trehalose in *B. japonicum* TAI. 102. Mannitol is reported to protect cells during the rapid drying they experience at inoculation both as a solute and in enhancing polysaccharide production. When the mannitol was removed from the media survival of *B. japonicum* was adversely affected. Its effect on survival appears to be maximum at approximately 1.0 g/L.

Conclusion

Liquid bioinoculant preparation requires physiological preconditioning and optimization of cell/spore/cyst in a osmotically balanced liquid formulation. The efficacy of the formulation is determined by the plant growth promoting attributes and extended shelf life.

References

Brockwell, J., Gault, R.R., Herridge, D.F., Morhorpe, L.J. and Roughley, R.J. 1988. Studies on alternative means of legume inoculation: micro-biological and agronomic, appraisals of commercial procedures for inoculating soybeans with *Bradyrhizobium japonicum. Australian Journal Agricultural Research,* **39**: 965-972.

Crowe, J.H., Crowe, L.M., Oliver, A.E., Tsvetkova, N., Wolkers, W. and Tablin, F. 2001. The trehalose myth revisited: introduction to a symposium on stabilization of cells in the dry state. *Cryobiology,* **43**: 89-105.

Crowe, J.H., Carpenter, J.F. and Crowe, L.M. 1998. The role of nitrification in anhydrobiosis. *Annual Reviews of Physiology,* **6**: 73-103.

Crowe, J.H, Hoekstra, F.A. and Crowe, L.M. 1992. Anhydrobrosis. *Annual Review of Physiology,* **54**: 579-599.

Dye, M. 1982. A note on some factors affecting the survival of *Rhizobium* cultures during freeze drying and subsequent storage. *Journal of Applied Bacteriology,* **52**: 461-464.

Fred, E.B., Baldwin, I.L. and McCoy, E. 1932. Root Nodule Bacteria and Leguminous Plants. University of Wisconsin Press, Madison.

Gaber, B.P., Chandrasekhar, I. and Patabiraman, N. 1986. The interaction of trehalose with the phospholipid bilayer: a molecular modeling study in Leopold, A.C., (Ed.), Membranes, Metabolism and Dry Organisms. Cornell University Press, Bhaca. pp. 231-241.

Halvorson, H. and Howitt, C. 1961. The role of DPA in bacterial spores. *In* H.O. Halvorson (ed.), Spores II. Burgess Publishing Co., Minneapolis.

Hounsa, C.G., Brandt, E.V., Thevelein, J., Hobmann, S. and Prior, B.A. 1998. Role of trehalose in survival of *Saccharomyces cerevisiae* under osmotic stress. *Microbiology,* **144**: 671-680.

Lamm, R.B. and Neyra, C.A. 1981. Characterization and cyst production of azospirilla isolated from selected grasses growing in New Jersey and New York. *Can. J. Microbiol.*, **27**: 1320-1325.

Maltseva, N.N. and Laski. 1982. Synthesis and monosaccharide composition of exopolysaccharides in certain soil bacteria. *Mikrobiol. Zh.*, **44**: 33-36.

Potts, M. 1994. Desiccation tolerance of prokaryotes. *Microbiology Reviews*, **58**: 755-805.

Rode, L.J. and Foster, J.W. 1960. Mechanical germination of bacterial spores. *Proc. Nat. Acad. Sc.*, USA. **46**: 118-128.

Roughley, R.J., Gemell, L.G., Thompson, J.A. and Brockwell, J. 1993. The number of *Bradyrhizobium* sp. (lupinus) applied to seed and its effect on *rhizosphere* colonization, nodulation and yield of lupin. *Soil Biology & Biochemistry*, **25**: 1453-1458.

Sadasivan, L. and Neyra, C. 1985. Flocculation in *Azospirillum brasilense* and *A. lipoferum:* Exopolysaccharides and cyst formation. *J. of Bacterial.*, No. 2. **163**: 716-723.

Socolofsky, M.D. and Wyss, O. 1962. Resistance of the Azotobacter cyst. *J. Bacteriology*, **84**: 119-124.

Steinborn, J. and Roughley, R.J. 1975. The effect of the ratio of magnesium: Calcium on numbers of Rhizobium in liquid medium. *Journal of Applied Bacterial.* **39**: 213-216.

Streeter, J.G. 1985. Accumulation of alpha. Alpha. Trehalose by *Rhizobium* bacteria and bacteroids. *Journal of Bacteriology*, **164**: 78-84.

Thevelein, J.M. 1996. Regulation of trehalose metabolism and its relevance to cell growth and function, in: Brambl, R., Marzluf, G.A. (Eds.). The Mycota III. Biochemistry and Molecular Biology., Springer, Berlin, pp. 395-420.

Tsevtkov, T.D., Tsonev, L.I., Tsevtkova, N.M., Koynova, R.D. and Tenchov, B.G. 1989. Effect of trehalose on the phase properties of hydrated and lyophilized dipalmitorylphosphatidylcholine multilayers. *Cryobiology*, **26**: 162-169.

Vincent, J.M. 1958. Survival of root-nodule bacteria. In: Hallsworth, E.G., (Ed.), Nutirition of the Legumes, pp. 108-123.

Soil Microbiology and Soil Biotechnology : Vol. 2
Editor : S. Kannaiyan
Associated Publishing Co., New Delhi. 2010. pp. 105-112

5

Mineral Phosphate Solubilization by Blue Green Algae

Mahesh Yandigeri[1], *Shrikrishna D. Jadhav and Sunil Pabbi**

Introduction

Phosphorus ('P') has long been considered as an essential constituent of all living organism. The amount of phosphorus in the earth's crust is of the order of 1015 million tonnes. All the phosphorus currently cycling through the biosphere was originally derived from minerals that composed the parent material of soils. There are 200 known forms of phosphorus minerals occurring in nature. It is considered as 'master key' element in crop production (Pierre, 1938) and the supply of 'P' depends directly on the concentration of inorganic phosphate (*Pi*) in the soil solution and indirectly on soil, which help in its maintenance. In terms of availability, this nutrient is found in insufficient amounts in most of the Indian soils. According to Hasan (1994), in India, about 49.3 per cent areas have low, 48.8 per cent have medium and only 1.9 per cent areas have high phosphorus status. Phosphorus plays an important role in plant growth and reproduction, comprising ~0.2 per cent dry weight in plants. Among the commonly used chemical phosphatic fertilizers, single super phosphate (SSP) represents the popular one among farmers. Only 20-25 per cent of applied phosphatic fertilizers is utilized by plants and rest is rendered unavailable due to chemical fixation by the formation of complexes with metal cations and clay minerals. Under acidic conditions phosphate ions are fixed up with iron and aluminium, whereas in alkaline conditions, it reacts with calcium to form tricalcium phosphate (TCP). The two reactions *viz.*, fixation and immobilization results into forms unavailable for plant uptake. The immobilized organic form can account for 2-80 per cent of the 'P' content of surface soils and nearly 90 per cent of the applied phosphatic fertilizer is fixed in the soil, rendering them unavailable for plant consumption. These characteristics cause widespread deficiency of 'P' for agricultural production. Phosphorus utilization is not only intimately linked with crop requirement and soil properties, but also with the crop-soil-fertilizer complex taken as a whole. Judicious management of phosphorus should therefore, take into account soil factors, fertilizers, their

1. National Bureau of Agriculturally Important Microorganisms, Kusmaur, Mau Nath Bhanjan, UP 275101.

* Centre for Conservation and Utilisation of Blue Green Algae, Division of Microbiology, Indian Agricultural Research Institute, New Delhi-110012.

time and method of application and cropping system. Phosphorus plays a key role in the nutrition of plants as it promotes the development of deeper roots. Being a constituent of ATP, it is involved in a number of processes like cell division, energy transduction through photosynthesis and biological oxidation and nutrient uptake. Both organic and inorganic forms of phosphorus are found in the soils. The inorganic forms occur in minerals as insoluble iron or alumnium phosphate. In soil, phosphorus may be found as fluorapatite, hydroxyapatite or chlorapatite as iron or aluminium phosphate or in combination with the clay fraction. The calcium phosphates are dominant in neutral to alkaline soils whereas iron and aluminium phosphates occur in acidic soils. The chief sources of organic phosphorus compounds are vegetation and decaying plant residues entering the soil. This comprises phytins, phospholipids, nucleic acids, phosphrylated sugars and coenzymes. Most of the phosphorus assimilated by microorganisms is mineralized in due course into organic forms and becomes part of available soil phosphorus pool. The addition of phosphate solubilising microorganisms help not only in augmenting the phosphorus availability but also in mobilizing chemically fixed phosphorus of the soil, thereby permitting better use of P available in the soil and reducing the dependence on super-phosphates.

Phosphorus solubilization by microorganism

It has been known since early 1900s that some microorganisms can solubilize phosphate from inorganic and organic sources and are actively engaged in 'P' transformation in soil and transport to plants. Phosphate solubilising microorganisms (PSMs) are found in all soils with their concentration varying with soil, climate and its history. Microbial involvement in solubilization of inorganic phosphates was first shown by Stalstrom (1903), who observed solubilization of TCP when it was incubated with milk cultures and soil infusions. Sackett *et al.*, (1908) gave conclusive evidence to show that soil bacteria dissolved rock phosphate, bone meal, di and tricalcium phosphates and calcium carbonate. A large microbial population can dissolve insoluble inorganic phosphates known to occur in soil, which include bacteria, fungi, actinomycetes and cyanobacteria (Blue Green Algae). Solubilization of insoluble phosphates by PSMs varies with different phosphorus sources, types of microorganisms, source of isolation and culture condition and physico-chemical properties of the soil. The P solubilising properties of cyanobacteria on bound phosphates has been exploited for utilization of low cost, low grade rock phosphate fertilizer, where $(PO_4)^{-3}$ remains bound in different forms of apatite like hydroxyl apatite, carbonoapatite, chloroapatite etc. Phosphorus is considered necessary for growth and nitrogen fixation by these organisms. It plays a major role in most cellular processes, especially those involved in generation and transformation of metabolic energy. It occurs in cells/living organisms in the form of either free phosphate ions or as organic phosphate constituents. The extra-cellular organic compounds liberated by algae and oxygen released due to photosynthesis are likely to bring about some important changes in chemical properties of soil such as availability of phosphorus, iron and manganese etc.

A number of phosphate solubilising cyanobacteria have been isolated using different substrates (Table 1) from different locations. In different field experiments, the available phosphorus content of the soil was observed to increase with the incorporation of blue green algae, *Azolla* or urea-N throughout the incubation. The increase of available phosphorus when blue green algae and *Azolla* were incorporated in soil might be related to solubilising effect of organic acids produced as a result of their decomposition. Moreover, during their decomposition organic phosphorus is released to the soil solution.

Table 1: Occurrence and isolation of various phosphate solubilising blue green algae using different substrates.

Blue Green Algae	Substrate	Author	Year
Anabaena and *Nostoc*	Hydroxyapatite	Cameron and Julian	1988
Hapalosiphon, Scytonema, Nostoc, Westiellopsis, Calothrix	Mussoorie rock phosphate	Roychoudhury and Kaushik	1989
Anabaena and *Phormidium*	Rock phosphate	Natesan and Shanmugasundaram	1992
Anabaena, Nostoc, Cylindrospermum, Tolypothrix	Soils	Mandal *et al.*	1992
Anabaena	Phosphomono-and di-esters	Kumar *et al.*	1992
Anabaena, Nostoc, Aulosira, Fischerella, Westiellopsis, Oscillatoria, Tolypothrix	Mussoorie rock phosphate	Latha *et al.*	1992
Anabaena PCC 7119	Phosphate	Fernandez-Valiente and Avendano	1993
Phormidium laminosum	Phosphates	Prieto *et al.*	1997
Synechococcus	Phosphates	Li Hong *et al.*	1998
Anabaena variabilis and *Westiellopsis prolific*	Mussoorie rock phosphate and Tri- calcium phosphate	Yandigeri and Pabbi	2005

It has also been shown that growth of a mixture of blue green algae comprising *Anabaena, Nostoc, Cylindrospermum* and *Tolypothrix* in soil caused an increase in organic phosphorus with concomitant decrease in Olsen-phosphorus, Al-P, Fe-P and Ca-P, but with little change in reductant soluble Fe-P and occluded Al-P fractions. Phosphorus assimilated by blue green algae and incorporated into their cells during growth is released upon bacterial decomposition in the form of soluble organic-P compound or condensed polyphosphates. These compounds are later mineralized or hydrolyzed to orthophosphates resulting in an increase in available P in soils.

Mechanism of P solubilization

Among the various mechanisms reported by which phosphate may be liberated from insoluble mineral by blue green algae include production of organic acids, synthesis of chelators, dissimilatory reduction of ferric ion and enzymatic solubilization or simultaneous operation of more than one of them. There is fall in pH during the growth of phosphate solubilising microorganisms as a result of production of organic acids. The organic acid released in culture filtrates react with insoluble phosphate and amount of soluble phosphate released depends on strength and type of acid. Aliphatic acids are found to be more effective in

P solubilization than phenolic and citric acids. Fumaric acid has highest P solubilising ability. Tribasic and dibasic acids are found to be more effective than monobasic and aromatic acids.

$$Ca_3(PO_4)_2 + 2H_2CO_3 \longrightarrow 2CaHPO_4 + Ca(HCO_3)_2$$

These acids may also form unionized association compounds with calcium, thus increasing soluble phosphate concentration. Calcium gets chelated to a small extent with hydroxy aliphatic acids (Lactic acid) while it is held more strongly with dibasic acids like malic and tartaric acids and most strongly with tribasic acids like citric acid. The principal acids produced are lactic, glycolic, citric and succinic acids. Paper chromatographic analysis of culture filtrates have indicated the production of these acids with their proportion varying with the organism. Hence, acidification of medium and production of organic acids appear to be the prominent mechanism involved in solubilization of phosphorus. The microbial mineral phosphate solubilising (MPS) phenotype has historically been associated with the production of low molecular weight organic acids (Goldstein, 1986). The metabolic and genetic basis for high efficiency solubilization of poorly soluble calcium phosphates have also been identified. It has been shown that solubilization is the result of acidification of the periplasmic space by direct oxidation of glucose (nonphosphorylating oxidation) or other aldose sugars. The direct (or nonphosphorylating) oxidation pathway is one of the four major metabolic routes for microbial dissemination of aldose sugar (Figure 1). Over expression of genes involved in biosynthesis/regulation of organic acid production from a different organism may lead to metabolic deregulation resulting in enhanced acid production. This has been practically demonstrated by cloning genes responsible for conferring mineral phosphate solubilising (MPS) ability from *Synechocystis* sp. PCC 6803 and two mps genes were selected based on ability of transformants to show a zone of clearance on solid medium containing dicalcium phosphate (DCP) as sole source of phosphorus. These transformants also solubilized rock phosphate.

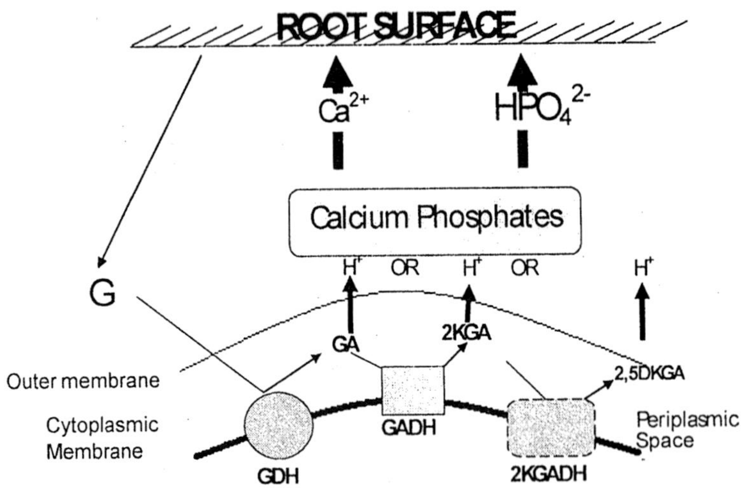

Fig.1: Ecophysiological role for the direct oxidation pathway in the release of Pi from calcium phosphates at or near the rhizosphere

2KGADH: 2ketogluconate dehydrogenase, 2,5DKGA: 2,5-diketogluconic acid, 2KGA: 2-ketogluconic acid, GADH: Gluconate dehydrogenase, GDH: Glucose dehydrogenase, GA: Gluconic acid, G: Glucose

The synthesis of chelators by phosphate solubilising organism, so as to drive the following dissolution reaction forward without changing pH of the growth medium has also been suggested as a mechanism for P solubilization. A chelator smaller than 6000-8000 molecular weight appeared to be synthesized by algae in response to phosphate deprivation, calcium deprivation or both, which allowed the blue green algae to utilize hydroxy-apatite.

$$Ca_{10}\ (OH)_2(PO_4)_6 = 10Ca^{+2} + 2OH^- + 6PO_4^{-3}$$

The involvement of phosphatases in the extra-cellular solubilization of rock phosphates has also been demonstrated. PSM's are known to produce phosphatase enzymes along with acids, which cause the solubilization of P in aquatic environments. The inorganic phosphates are brought back into medium by action of mineral and organic acids, whereas the organic phosphorus compounds are decomposed and mineralized by many enzymes complexes produced specially by heterotrophic bacteria and algae. The term phosphatase is commonly used for an enzyme which catalyzes a variety of hydrolysis of phosphomonoesters. The schematic reaction mechanism is as follows:

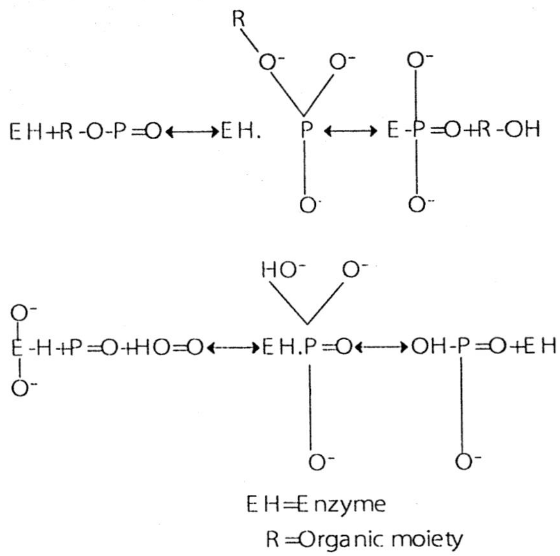

Fig. 2: Schematic reaction mechanism of phosphatases

Phosphatases have been typically classified as alkaline and acid phosphatase according to their optimum pH values for maximum hydrolyzing capacity. The alkaline phosphatase has a pH optimum well above 7 and acid phosphatase below 7. Alkaline phosphatase, acting maximally at pH 9-10.5 promote degradation of complex phosphorus compounds into orthophosphates and an organic moiety and thus, believed to have an important function in the nutrient dynamic of aquatic environment. A single phosphatase may act on different substrates, R-group being the variable. Thus, one enzyme may act on cleavage of ethyl phosphate, glycerophosphate, phenyl phosphate etc. The cycling of organic P has a large effect on P-availability and long term ecosystem productivity. The phosphorus concentration in soil solution may also be increased by hydrolytic cleavage of soil organic form notably the phosphomonoesters through the action of extracellular phosphohydrolases. It is also believed that above mechanisms operate simultaneously. Once solubilized, the phosphate is taken up by the growing algal cells for their nutrition. After completing their growth cycle, when cells undergo lysis the cell bound phosphate is released in the growth medium and becomes available to plants on mineralization. Most cyanobacteria are unable to take up large organic P-containing molecules but can cleave *Pi* moiety of a variety of substrates using extra cellular phosphatase with broad substrate specificity. So far, three distinct phosphatases have been identified in different cyanobacterial species. Of these, the best characterized is a typical alkaline phosphatase (A) of *Synechococcus* PCC 7942. Therefore, depending on substrate specificity of these phosphatases, there may be efficient scavenging of P from a variety of available sources. A study by Whitton (1991) on surface phosphomonoesterase activity of macroscopic colonies in paddy fields showed that *Aphanothece stagnina* tended to have low activity, whereas *Gloeotrichia* always showed high activity. Among different species of *Nostoc*; *Nostoc commune* had high activity whereas other species tended to have low activity as phosphomonoesterase activity is enhanced in all these species. Under conditions of P-limitation, *Aphanothece* typically grew under P-rich conditions whereas *Gloeotrichia* grew under conditions where growth was P-limited. Both cell bound phosphomonoesterase and phosphodiesterase activity has also been described of *Anabaena* cultures in P limited medium and to generate inorganic phosphates for transport into the cells by hydrolyzing the phosphate ester bonds (organic phosphates). A number of bound forms of phosphates in MRP can be made available by alkaline phosphatase activity of cyanobacterial cells. Phosphatase activity in the culture filtrates of *Anabaena* and *Phormidium* supported the argument that phosphatases are involved and secreted upon phosphate starvation/ depletion and these might be involved in solubilization of rock phosphate. The shifting of pH of the culture filtrate towards alkalinity also strengthened the possible involvement of alkaline phosphatase in extra-cellular solubilization of rock phosphate. Alkaline phosphatase activity among the planktonic communities in the northern Red Sea was reported to be highly stable and considered as potential indicator for the inorganic P-availability. A good correlation was found between depth, distribution of alkaline phosphatase activity and *Synechococcus* suggesting that this blue green alga is a significant contributor to the enzyme activity. *Synechococcus* sp. strain PCC 7942 has been shown to possess at least two alkaline phosphatases; one alkaline phosphatase is a phosphate repressible (pho A, 145 kDa) while the other is most likely not repressed by phosphates (pho V, 61.3 kDa). The presence of second alkaline phosphatase in *Synechococcus* sp. strain PCC 7942 could be important in providing the cell with sufficient phosphates under stress conditions.

Genes involved in P-solubilization

Many workers have worked and reported on cloning of mineral phosphate solubilising genes. Majority of mineral phosphate solubilizers belong to gram negative bacteria, which solubilize the phosphate. The ability of some gram negative bacteria to dissolve poorly soluble calcium phosphate (Mps phenotype) is periplasmic oxidation of glucose to gluconic acid via quinoproteins glucose dehydrogenase (GDH), a component of the direct oxidation pathway. The first step in this pathway is the oxidation of glucose to gluconic acid via the membrane bound glucose dehydrogenase. This enzyme contains 2,7,9- tricarboxyl- 1H- pyrrolo[2,3-f]-quinoline-4,5-dione (Pyrroloquinoline-quinone, PQQ) as prosthetic group, so that the functional holoenzyme is a quinoprotein. Addition of PQQ (Pyrroloquinoline-quinone) to whole cells, the prosthetic group of this enzyme results in production of gluconic acid from glucose. Among enteric bacteria, *Klebisella pneumoniae* has been shown to contain the genes required for PQQ biosynthesis, whereas *E. coli* cannot produce gluconic acid as it cannot synthesize PQQ, the cofactor for apo-glucose dehydrogenase enzyme that *E. coli* normally possess. PQQ dependent enzymes methanol dehydrogenase (MDH) and soluble dehydrogenase (s-GDH) share similarity in several lines of biochemical and kinetic patterns and it has been proposed that MDH and s-GDH react with their substrate through an identical mechanism comprising general base catalyzed hydride transfer from substrate to PQQ and subsequent tautomerization of the PQQ intermediate to reduced PQQ.

Glucose dehydrogenase is a member of the largest group of quinoproteins, those that use the cofactor 2, 7, 9-tri carboxyl-1H-pyrrolo (2, 3-F) quinoline-4, 5-dione (PQQ) (Liu *et al.*, 1992; Duine, 1991). It was found that the MPS trait was induced/repressed by low or high levels of exogenous *Pi* (Goldstein, 1986). This underlined the fact that genes play an important role in phosphate solubilization by prokaryotes. Efforts have been made to characterize the regulation of MPS trait in both recombinant *E. coli* and *Erwinia herbicola*. At levels less that 1mM soluble orthophosphate (*Pi*), *E. herbicola* solubilizes dicalcium phosphate or hydroxyapatite in the external culture media but repression of MPS trait by addition of 30 mM of soluble phosphate was observed. The phosphate stress induced switching on of the MPS activity by externally added *Pi* is a classic example of physiologically regulated gene expression in micro organisms capable of MPS activity. Such observation made it possible to propose the existence of mps genes in *E. herbicola*. These data thus suggest that P availability could regulate mineral phosphate solubilization in some species but have no effect on others.

The presence or synthesis of PQQ is not vital to a microorganism under growth conditions where the metabolic pathways do not require the involvement of quinoproteins. The administration of PQQ or its biosynthesis is important only for gram negative bacteria on those substances requiring a metabolic pathway in which quinoproteins operate. Glucose dehydrogenase (GDH) is a member of the largest group of dehydrogenase enzymes, utilizing quinoproteins as cofactors. Quinoproteins PQQ (2, 7, 9-tricarboxy-1H-pyrrolo {2, 3-f} quinoline-4, 5-dione) is utilized by GDH for synthesis of holoenzyme. These mps genes had been cloned from *Pseudomonas cepacia* (Babu-Khan *et al.*, 1995) and *E. herbicola*. The genes for solubilising mineral phosphates are present in many cyanobacteria, but are unable to express as PQQ synthase the cofactor for glucose dehydrogenase is absent. They can synthesize apo-GDH enzyme.

References

Babu-Khan, S., Yeo, T.C., Martin, W.L., Duron, M.R., Rogers, R.D. and Goldstein, A.H. 1995. Cloning of mineral phosphate solubilising gene from *Pseudomonas cepacia*. *Appl. Enviorn. Microbiol.*, **61**: 972-978.

Cameron, H.J. and Julian, G.R. (1988). Utlization of hydroxyapatite by cyanobacteria as their sole source of phosphate and calcium. *Plant and Soil*, **181**: 123-124.

Duine, J.A. 1991. Quinoproteins: enzymes containing the quinoid cofactor pyrroloquinoline quinine, topaquinone or tryptophan-tryptophan quinine. *Eur. J. Biochem.*, **200**:271-284.

Fernandez-Valiente, E. and Avendano, M.C. (1993). Sodium stimulation of phosphate uptake in the cyanobacterium *Anabaena* PCC 7119. *Plant and Cell Physiology*, **34**:201-207.

Goldstein, A.H. 1986. Bacterial solubilization of mineral phosphates: historical perspective and future prospects. *Am. J. Alt. Agric.*, **1(2)**:57-65.

Hasan, R. 1994. In: *Phosphorus research in India*. Dev. G. (ed.), Proc. Group Discussion at IARI-PPICIP Gurgaon, India. Pub. 7.

Kumar, T.A., Mahasneh, I.A. and Tiwari, D.N. (1992). Alkaline phosphatase activities of *Anabaena* from deep-water rice. *World J. Microbiol. Biotechnol.*, **8**: 585-588.

Latha, A., Shanthi, K. and Kannan, N. (1992). Rock phosphate solubilization by free-living nitrogen fixing blue-green algae. In: Proc. Nat. Symp. on Cyanobacterial Nitrogen Fixation (Ed. B.D. Kaushik), Associated Publishing Co., New Delhi, pp. 415-421.

Li Hong, Marcel, J.W. Veldhuis, Anton F. Post (1998). Alkaline phosphatase activities among planktonic communities in the northern Red Sea. *Marine Ecology Progress Series*, **173**: 107-115.

Liu, S.T., Lee, L.Y., Tai, C.Y., Hung, C.H., Chang, Y.S., Wolfarm, J.H., Rogers, R. and Goldstein, A.L. 1992. Cloning of an *Erwinia herbicola* gene necessary for gluconic acid production and enhanced mineral phosphate solubilization in *Escherichia coli* HB101: Nucleotide sequence and probable involvement in biosynthesis of coenzyme pyrrolo quinoline quinine. *J. Bacteriol.*, **174 (18)**: 5814-5819.

Mandal, B., Das, S.C., and Mandal, L.N. (1992). Effect of growth and subsequent decomposition of blue-green algae in the transformation of phosphorus in submerged soils. *Plant and Soil*, **143**: 289-297.

Natesan, R. and Shanmugasundaram, S. (1982). Extracellular phosphate solubilization by the cyanobacterium *Anabaena* ARM 310. *J. Biosci.*, **14:** 203-208.

Pierre, W.H. 1938. Phosphorus deficiency and soil fertility. *Soils and Men*. Year book of Agric. USDA, Washington, D.C., pp. 377-396.

Prieto, B., Pardo, M.A., Garbisu, M.J. Llama and Serra, J.L. (1997). Phosphate uptake by phosphorus- starved cells of cyanobacterium *Phormidium laminosum*. *World J. Microbiol. Biotechnol.*, **13**: 699- 709.

Roychoudhury, P., and Kaushik, B.D. (1989). Solubilization of mussorie rock phosphate by cyanobacteria. *Curr.Sci.*, **58(10)**: 569-570.

Sackett, W.G., Patten, A.J. and Brown, C.W., 1908. The solvent action of soil bacteria upon insoluble phosphates of raw bone meal and natural rock phosphate. *Zbl. Bakteriol.*, **28**: 688.

Stalstrom, V.A. 1903. Beitrag Zur Kennturs der Ein-Wisking Steriler und in Garung bofindlicher Organischer Stroffe auf dil loslichkeit der phosphorsen des tricalcium phosphate. *Zbl. Bakt.*, **11**: 724-732.

Whitton, B.A. 1991. Diveristy, ecology and taxonomy of the cyanobacteria. In: *Photosynthetic Prokaryotes*. Mann, N.H. and Carr, N.G., (ed.). New York: Plenum Press, pp. 1-51.

Yendigiri, M.S. and Pabbi S. 2005. Response of diazotrophic cyanobacteria to alternative sources of phosphorus. *Indian J. of Microbiology*. **45(2)**: 131-134.

Soil Microbiology and Soil Biotechnology : Vol. 2
Editor : S. Kannaiyan
Associated Publishing Co., New Delhi, 2010, pp. 113-135

6

Arbuscular Mycorrhizal Fungi

D.J. Bagyaraj[1] *and J. Arpana*[2]

Introduction

Many microorganisms form mutualistic symbiosis with plants. Among these the most widespread mutualistic symbiosis is the arbuscular mycorrhiza, formed between arbuscular mycorrhizal (AM) fungi and vascular flowering plants. These associations occur in terrestrial ecosystems throughout the world and have a global impact on plant phosphorus nutrition. The arbuscular mycorrhiza is an endosymbiosis in which the fungus inhabits the root cortical cells and obtains carbon provided by the plant while it transfers mineral nutrients from the soil to the cortical cells. Development of symbiosis involves the differentiation of both symbionts to create novel symbiotic interfaces within the root cells. AM fungi play an essential role in plant growth, plant protection and soil quality. They are ubiquitous in the soil and can form symbiosis with most terrestrial plants including major crops, cereals, vegetables and horticultural plants. In agriculture several factors, such as host crop dependency to mycorrhizal colonization, tillage system, fertilizer application and inoculum potential of AM fungi etc. can affect plant response to the association. Due to their obligate symbiotic status AM fungi need to associate with plant for growth and proliferation. Consequently, the cultivation of AM fungi strains and the maintenance of reference collections require methodologies and infrastructure quite different from those used with other microbial collection and inoculum production. Interest in AM fungi propagation for agriculture is increasing due to their role in the promotion of plant and soil health.

These beneficial effects of AM symbiosis occur as a result of a complex molecular dialogue between the two symbiotic partners. Identifying the molecules involved in the dialogue is a prerequisite for a greater understanding of the symbiosis.

Recently, molecular approaches are being used to understand the underlying dialogue and concomitant molecular changes occurring in the plant and the fungus during the establishment of a functioning AM symbiosis. Co-inoculation of AM fungi with other beneficial soil micro-organisms is more useful in improving plant growth thus suggesting the development of suitable "microbial consortia" for inoculating different crop plants.

1. D. J. Bagyaraj, Chairman, email: djbagyaraj@vsnl.com
2. J. Arpana, Senior Research Fellow, email: arpanamohan@yahoo.co.in
CNBRCD, #41, RBI Colony, Anand Nagar, Bangalore- 560 024

The AM symbiosis, formed between soil fungi and vascular plants have a long history with fossils providing evidence of AM fungi in the roots of the earliest land plants more than 400 mya. Sequence data and fossils of spores and hyphae point to the existence of AM fungi even earlier, more than 460 mya, and it is suggested that the AM fungi assisted plants for their colonization in land. Hence it is clear that the ability to form an AM symbiosis occurred early in the evolution of plants and today the capacity to form these associations in distributed widely throughout the plant kingdom and includes angiosperms, gymnosperms, pteridophytes and some bryophytes. Within the angiosperms, at least 80% of species are able to form AM symbiosis.

Plant roots provide an ecological niche for many of the microorganisms that abound soil. German botanist Albert Bernard Frank in 1885 introduced the Greek word mycorrhiza, which literally means "fungus root". Most plant roots form mycorrhizal associations of one kind or the other with certain fungi in soil. These mycorrhizal fungi perform the function of root hairs. Mycorrhizal association generally enhances the growth and vigor of the host plants. In addition the mycorrhizal plants have greater tolerance to toxic heavy metals, root pathogens, drought, high soil temperatures, soil salinity, adverse soil pH and to transplantation shock. Because of their widespread occurrence in nature and their numerous benefits to plants, these fungi are currently attracting much attention in agricultural, horticultural and forestry research. Though there are different mycorrhizal associations, the most common type occurring in all ecological situations, especially in tropics, is the arbuscular mycorrhiza (AM) (Barea and Jeffries 1995). Increased plant growth because of AM colonization is well documented (Bagyaraj and Varma 1995).

Types of mycorrhizae

Mycorrhizal associations vary widely in structure and function. Despite the many exceptions, it is possible to state broad generalizations about latitude (or altitude), soil properties and structure and function of the different mycorrhizal types that colonize the dominant vegetation in a gradient of climatic zones. Mycorrhizas are classified into five different categories delineated by the interacting organisms and the mycorrhizal morphology, they are

1. Arbuscular mycorrhizae
2. Ectomycorrhizae
3. Ectendomycorrhizae
4. Ericoid mycorrhizae
5. Arbutoid mycorrhizae
6. Monotropoid mycorrhizae
7. Orchid mycorrhizae

Arbuscular mycorrhizae

This is the most common type of mycorrhiza and can be found in almost all the plant communities, natural and cultivated. These mycorrhizal associations are evolutionary very old and it is believed that types of mycorrhizae and non-mycotrophy evolved more recently. Infact, it has been speculated that arbuscular mycorrhiza may have been involved in the successful invasion of land by vascular plants and played a controlling influence on the evolution of roots. It is a hugely diverse group of autotrophs that engage in this type of symbiosis. The fungal partners are members of the Glomeromycota (a newly designated phylum whose members were formerly placed in the Zygomycota). These AM fungi colonize the roots of most families of Angiosperms as well as Gymnosperms, Pteridophytes and Bryophytes (liverworts). Non-mycotrophy in the Angiosperms appears to be restricted primarily to the families Amaranthaceae, Brassicaceae, Chenopodiaceae and Zygophyllaceae, and many hemiparasitic plants. These mycorrhizas are characterized by the penetration of the cortical cell by the fungus and formation of intracellular arbuscules. These are small tree like, hyphal filled, invaginations of the cortical cells—-which provide intimate contact between the plasmalemmae of the two symbiotic partners and are, presumably, the point of material exchange between host and fungus. They persist in active form for a very short period, 1 to 15 days. With the exception of species of *Gigaspora* and *Scutellospora*, the fungi form vesicles within the roots. These are lipid filled, terminal swellings of hyphae with a storage/ perennating function.

Ectomycorrhizae

The fungi involved in the formation of ectomycorrhizas are mostly Basidiomycetes and are often species that have quite large and conspicuous fruiting bodies. Toadstools and earthballs are conspicuous examples. Fungi from other major fungal groups have also been shown to form sheathing ectomycorrhizae including two species belonging to *Endogone*, Zygomycetes closely related to the fungi that form arbuscular mycorrhizae. Most ectomycorrhizal (EM) fungi can synthesize plant growth substances, which cause the morphology of the root to be changed. The changes induced differ between species. At one extreme, branching is totally suppressed and, at the other, prolific branching takes place, with a single root having a dense aggregate of as many as 2000 mycorrhizal tips. Most ectomycorrhizae are enveloped in a dense weft of hyphae while others appear smooth and seem to lack radiating hyphae. All gradations between these two forms can be found. The first sign of mycorrhiza formation is the development of fungal mycelium around a short root. This gradually thickens over the root surface developing into a sheath (sometimes called the mantle). The sheath may be only one or two cells thick but can exceed 100μm. The entire root, including the apex, is often enclosed by the sheath so that there is no direct contact between younger roots and the soil. Root hair development is normally totally suppressed. Thus, all nutrients and water absorbed by the roots have to pass through the fungal sheath. Depending on the fungal species, various secondary structures in addition to radiating adsorptive hyphae may be found associated with the sheath.

Behind the zone of cell division, hyphae from the inner side of the sheath penetrate between the root cortical cells producing a network of hyphae called the Hartig net. The size and shape of the cells of the Hartig net differ greatly depending on the EM fungal species as does the depth of penetration, which is never beyond the endodermis. The Hartig net is a three-dimensional branching structure that provides a large area of contact with the cortical cells allowing efficient transfer of mineral nutrients to the plant and metabolites to the endophyte. Hyphal penetration of cortical cells does not normally occur and is limited to the cell wall. EM fungi can be found on about 90% of the trees in temperate forests and each tree can be associated with many fungal species concurrently.

Ectendomycorrhizae

Ectendomycorrhizae possess qualities characteristic of both ectomycorrhizae and arbuscular mycorrhizae. As with ectomycorrhizae, both a hartig net and mantle structures are produced, although the mantle may be reduced (and in some cases lacking entirely). But unlike EM, intracellular penetration of healthy plant cells by these fungi also occurs, a character consistent with AM. Ectendomycorrhizae can be formed with the roots of many angiosperm and gymnosperm species, and the fungal symbionts can be members of the Basidiomycota, Ascomycota, or Zygomycota phyla. In fact, the same fungal species can form either ectomycorrhizae or ectendomycorrhizae depending upon the identity of the plant that it is associated with.

Arbutoid mycorrhizae

Similar to ectendomycorrhizae, arbutoid mycorrhizae possess characteristics of both ectomycorrhizae and arbuscular mycorrhizae. There is a well developed mantle, a hartig net, and prolific extramatrical mycelium. In addition, there is also intracellular penetration and hyphal coils produced in autotrophic cells. These mycorrhizae are formed with members of the Ericales; namely, *Arbutus* and *Arctostaphylos* species. The fungal symbionts are exclusively Basidiomycete species, which may form EM with other autotrophic hosts.

Ericoid mycorrhizae

These are mycorrhizae formed between autotrophs in the Ericaceae and fungi in the Ascomycota. Intracellular penetration of root cells occurs and there is no mantle or hartig net development.

Monotropoid mycorrhizae

As the name suggests, monotropoid mycorrhizae are formed between plants in the Monotropaceae family and a specific subset of fungi in the Russulaceae or the Boletaceae. Monotropaceae is an achlorophyllous family within the Ericales. Plants of this family are, therefore, even more dependent upon their mycorrhizal partners than their autotrophic relatives. The monotropoid mycorrhizas have a thick compact mycelial sheath, a Hartig net one cell deep in the root cortex, and, thus, limited intracellular penetration. The penetration of epidermal and outer cortical cells is achieved by the development of fungal pegs that become extensively invaginated. The final structure resembles the specialized transfer cells, which are formed in a variety of plant species when there is a need for rapid, short-distance transport of solutes. Several different species from different major fungal groups have been found forming this type of mycorrhiza in North America and Britain. Preliminary studies of freshly isolated endophytes from *Monotropa hypopitys* L. have shown them to have considerable proteolytic capability. Access to protein could, in the organic rich substrates in which these plants normally grow, provide the carbon and nitrogen required for growth.

Orchid mycorrhizae

These are endomycorrhizae formed between plants of the Orchidaceae and basidiomycetous or rarely ascomycetous, fungi often of the form-genus *Rhizoctonia*. Where the perfect state is known, these have been shown to belong to genera of several fungal families.

The fungi form intracellular coils or less regular hyphal aggregates within host tissue. These structures are known as peletons and usually persist for a limited period before collapsing and degenerating. The peletons in superficial tissues persist for a longer time than those in deeper tissues. Orchid mycorrhizae have no sheath or comparable structure but hyphae penetrate the soil around the infected organ. Achlorophyllous orchids, presumed to be epiparasitic on their fungal partner, have the facultative pathogen *Armillaria* as their endophyte and the fungus may simultaneously be mycorrhizal with a chlorophyllous host.

For all or part of their life cycle, orchids are obligately dependent on their mycorrhizal partner. Under natural conditions, orchid seeds will not successfully germinate and the protocorm will not develop without infection by the mycorrhizal fungus. During the non-photosynthetic portion of its life cycle an orchid obtains carbon compounds from the fungus.

Outline of some of the features of different types of mycorrhiza associations.

Mycorrhizal type	Host plant group	Characteristics	Fungal associate
Arbuscular mycorrhizae	Herbaceous plants, grasses; some trees	Formation of arbuscules within cortical cells of host root	Glomeromycetes
Ectomycorrhizae	Coniferous and deciduous trees	Formation of a sheath or mantle of fungal tissue around the root surface and a Hartig net of fungal penetration between the cortical cells to the endodermis	Basidiomycetes Ascomycetes
Ectendomycorrhizae Ericoid mycorrhizae	Ericaceae	Hyphal coils within the host root cortical cells	Basidiomycetes Ascomycetes Zygomycetes
Arbutoid mycorrhizae	Arbutus	Hyphal coils within the host root cortical cells	Basidiomycetes
Orchidaceous mycorrhizae	Orchids	Fungal propagule carried in the seed of the plant	Basidiomycetes

It is very well documented that arbuscular mycorrhizal fungi improve fitness and growth of plants that are important in agriculture, horticulture and forestry. Hence more importance is given to this particular symbiosis compared to other types of mycorrhizal associations and to harness maximum benefit from them. The increased plant growth is attributed to enhanced uptake of diffusion limited nutrients, hormone production, biological nitrogen fixation, drought resistance and suppression of root pathogens. Several workers have reported that AM fungi can act as biocontrol agents for alleviating the severity of disease caused by root pathogenic fungi, bacteria and nematodes. The benefit accrues primarily because mycorrhizal fungi form a critical linkage between plant roots and the soil.

Taxonomy and classification of AMF

A taxonomic framework must be established prior to any biodiversity study, especially when comparisons among land use systems are required. Taxonomy and classification of AMF have changed radically over the last few years. Gerdemann and Trappe (1974) included all AMF species within the Order Endogonales in the family Endogonaceae (Division Zygomycota). Morton and Benny (1990) transferred all AMF species to the Order Glomerales (Division Zygomycota) with three families and 6 genera. Molecular and morphological data set were used together to transfer members of the genus *Sclerocystis* to *Glomus* (Almeida and Schenck, 1990; Redecker *et al.*, 2000) and to erect two new genera pertaining to two distinct families (Morton and Redecker, 2001). SchÜbler *et al.* (2001) moved all AMF species to a newly monophyletic Division, the Glomeromycota, placing this group of organisms at the same level of Basidiomycota and Ascomycota. They also included in the new classification scheme three new orders and several families. Recently, a new genus named *Pacispora* was proposed by Oehl and Sieverding (2004); spore formation in this genus is similar to those of *Glomus* but they differentiate inner germinal walls. This genus is included within the Order Diversisporales in the molecular phylogeny proposed by SchÜbler *et al.* (2001). Although molecular techniques have been applied to clarify phylogenetic classification among AMF species, identification of species has been based mainly on morphological characteristics of spores. Therefore, we propose a classification scheme to be followed on diversity surveys of AMF, merging results from Morton and Benny (1990), Morton and Redecker (2001) and SchÜbler *et al.* (2001), Oehl and Sieverding (2004) (Table 1).

Table 1: Taxonomic framework proposed for diversity studies of arbuscular mycorrhizal fungi and morphological characters defining genera in Glomerales (compiled from Morton and Benny (1990), Oehl and Sieverding (2004) and http://invam.caf.wvu.edu).

Division Glomeromycota Schüßler, Scharzott & Walker ·

Order Glomerales Morton & Benny

Suborder Glomineae Morton & Benny

Family Glomeraceae Pirozysnki & Dalpé

Glomus Tulasne & Tulasne (85 species) -

Spores formed blastically on a subtending hyphae, singly, in loose aggregates or in a sporocarp. Vesicles are thin walled and ellipsoid. Intraradical hyphae rarely coiled, with cross-connecting branch hyphae. Mycorrhizae stains darkly. Arbuscules with flared or cylindrical trunks with incremental narrowing of branch hyphae. Spores with spore wall formed by a variable number of layers all originating from the subtending hyphae, no germinal walls differentiated. Germination through the lumen of the subtending hyphae or through the spore wall.

Family Acaulosporaceae Morton & Benny

Acaulospora Gerd. & Trappe emend. Berch (31 species)

Spores formed laterally from the neck of a sporiferous saccule which leaves one scar on the spore surface. Vesicles vary in shape with knobs and concavities. Intraradical hyphae straight or coiled near the entry points. Mycorrhizae stains weakly. Arbuscules with flared or cylindrical trunks with incremental narrowing of branch hyphae. Spores with spore wall formed by three layers and two inner germinal walls, each with two thin layers that can be adherent. The innermost germinal wall has a beaded surface. Germination through a flexible, plate-like germination orb.

Entrophospora Ames & Schneider (4 species)

Spores formed within the neck of a sporiferous saccule which leaves two scars on the spore surface. Vesicles, arbuscules, intraradical hyphae and mycorrhizae staining as in *Acaulospora*. Spores with spore wall formed by two layers. Other spore subcellular structures and germination identical to that in *Acaulospora*.

Family Archaeosporaceae Morton & Redecker

Archaeospora Morton & Redecker (3 species)

Spores formed terminally from a subtending hypha or as a branch from a structure resembling a sporiferous saccule. Arbuscules and intraradical hyphae stains lightly. Vesicles and auxiliary cells are not differentiated. Spores with spore wall formed by 3-4 layers and no true bi-layered germinal wall formed. Dimorphic species found forming acaulosporoid and glomoid spores.

Family Paraglomeraceae Morton & Redecker

Paraglomus Morton & Redecker (2 species)

Spores formed terminally from a subtending hyphae (like in *Glomus*). Arbuscules and intraradical hyphae stains lightly. Vesicles and auxiliary cells are not differentiated. Spore subcellular structures and germination like in *Glomus*.

Family Pacisporaceae Walker, Blaszkowski, Schüßler & Schwarzott

Pacispora Sieverding & Oehl (7 species)

Spores formed terminally from a subtending hypha (like in *Glomus*). Spore wall usually formed by three distinct layers and germinal wall composed also by three layers. The second layer of the germinal wall usually reacts to Melzer's reagent. Walls of subtending hyphae continuous with the first and second layer of the spore wall. Spore germination from the germinal wall directly through the spore wall.

Suborder Gigasporineae Morton & Benny

Family Gigasporaceae Morton & Benny

Gigaspora Gerd. & Trappe (5 species) -

Spores formed terminally on a bulbous sporogenous cell; auxiliary cells finely papillate or echinulate. No vesicles produced. Intraradical hyphae frequently coiled, especially near entry points, often knobby or with projections. Arbuscules with swollen trunks with abrupt narrowing of branch hyphae. Spores with spore wall formed by two permanent layers, no inner germinal walls differentiated. At germination, a thin layer interspersed with warts differentiate and germ tube grows throughout the spore wall.

Scutellospora Walker & Sanders (30 species)

Spores formed terminally on a bulbous sporogenous cell; auxiliary cells almost smooth to knobby. No vesicles produced. Arbuscules and intraradical hyphae similar in morphology to organisms of *Gigaspora*. Spores with spore wall formed by two permanent layers and 1-3 inner germinal walls, each with two layers. Germ tube grows from flexible, plate-like germination shield that differentiates on the surface of the last germinal wall.

Development of AM symbiosis

Despite the wide array of plant and fungal species involved, the development of the AM symbiosis progresses, at least at a morphological level, along broadly similar lines. The interaction is accompanied by significant alterations in the cellular layers of both the symbionts to create the novel symbiotic interfaces of the arbuscular mycorrhiza over which carbon and phosphates are exchanged. AM-fungi are obligate symbionts, which in the absence of a plant are present in the soil as multinucleate spores surrounded by thick cell walls. After spore germination, hyphae grow through the soil and start to branch. Already at this time point, before the two organisms get into direct contact with each other, an interaction between the AM-fungus and the host plant takes place. Although spore germination, first hyphal growth and branching also occur in the absence of a host plant, these preinfection processes can be stimulated by signal molecules synthesized and secreted by the roots as root exudates and volatiles. If hyphae get into direct contact to a host plant root, they form an appressorium. Appressoria are attachment structures formed on root surfaces. From here the fungal hyphae grow intercellularly through the root until they reach inner cortical cell layers. Here, the fungus starts to penetrate the cortical cell wall and forms tree-like intracellular hyphae structures, the arbuscules. At no time point, the fungus enters the plant cytoplasm. In the arbuscule-containing cells a periarbuscular membrane is formed that surrounds the arbuscule. The fungal cell wall becomes progressively thinner as the arbuscule develops. This leads to the establishment of an extensive intracellular interface between the two symbionts, which is the place where the phosphate and possibly also carbon is transferred between the plant and the fungus. Despite its central role in mycorrhiza functioning the life span of an arbuscule is short, already after a few weeks it collapses and becomes degraded. The former arbuscule-containing plant cell is not damaged by the arbuscule degradation and can host further arbuscules. Apart from arbuscules, *Glomus intraradices* fungi form a second intaradicle structure, the vesicles. These are thin-walled, lipid-filled structures that usually form in intercellular spaces. Their primary function is thought to be connected to energy storage; however, vesicles do not occur in all AM-fungi.

The basic functions of mycorrhizae

Mycorrhizal association can alter morphology, where in root hair development is suppressed as the function of the root hair development is replaced by fungal hyphae. These hyphae have 2 major benefits. They are of smaller diameter than root hairs and can penetrate more easily and to a greater distance from the root into the soil, thus exploring a greater volume of soil and presenting a greater surface area for nutrient absorption than could the root-root hair system alone. The second benefit is that it is energetically more efficient to produce a long, thin hyphum than a root hair. Further in AM type number of lateral roots produced is greater than non-mycorrhizal plant, suggesting dual benefits of mycorrhizal plant. Using radio tracer phosphate, it was shown that the phosphate depletion zone was greater (up to 110 mm) in mycorrhizal plant compared to non-mycorrhizal plant (10-20mm) indicating that AM hyphae exploit a larger volume of soil than root hairs.

Thus the role of AM fungi in the uptake of diffusion limited nutrients like P,Zn, Cu etc. is well documented. In addition to their role in conferring resistance to soil-borne diseases alleviating the effects of salt and water stress, synergistic interaction with beneficial soil organisms, enhancing the abundance of soil aggregation and in improving plant growth, leading to the "big plant – little plant" syndrome is well documented (Jumpponen *et al.*, 1998). Further, the derived benefit of a mycorrhiza association is predominantly nutrient acquisition if the host plant root system is poorly branched.

i. Mycorrhizal role in P uptake

Due to the immobility of phosphorus in the soil matrix, it has been generally suspected that one of the key benefits of the mycorrhizal symbiosis to the host plant is increased phosphorus exploitation of the soil. Unlike other more mobile nutrients, which will flow from nearby sources into the sinks created by root exploitation, P will remain relatively fixed in place even in the presence of steep gradients. Consequently, a root must be in very close proximity to a P source in order to exploit it directly. In this regard, connections to an extensive hyphal networks of mycorrhizal fungi in the soil provide a host plant much greater access to P deposits. Mycorrhizae may also help to mobilize P by decomposing organic sources which immobilize P, such as tree litter.

ii. Mycorrhizal role in N uptake

In contrast to P, nitrogen is a highly soluble and mobile nutrient. As such, the benefit to the plant of a mycorrhizal interaction is not as obvious with regards to N. With arbuscular mycorrhizae, so far no research has demonstrated a measurable increase in N uptake in mycorrhizal vs. non-mycorrhizal plants, and more detailed analyses have turned up mostly mixed results. (Smith and Read 1997).

iii. Effect of AM on root pathogens

Most of the studies on VAM-root pathogens suggest that VAM decreased or mitigated the disease severity. Consistent reduction of disease symptoms has been described for fungal pathogens such as *Phytophthora parasitica, P.cactorum, P.vignae, Gaeumannomyces graminis var.tritici, Fusarium oxysporum, Chalara (Thielaviopsis) basicola, Rhizoctonia solani, R.bataticola, Sclerotium rolfsii, Pythium ultimum, P. splendens, Dothiorella gregantia, Botrytis fabae, Ganoderma pseudoferreum* and *Aphanomyces* spp., bacteria such as *Pseudomonas syringae* and *P. solanacearum* and nematodes such as *Meloidogyne arenaria, M. incognita, M.hapla, M. javanica, Tylenchulus semipenetrans, T.vulgaris, Pratylenchus brachyurus, P.zeae, Helicotylenchus dihystera* and *Radopholus similis* (Bagyaraj 1996).

iv. Mechanisms of suppression of root pathogens by AM fungi

Studies conducted so far suggest that the mechanisms of suppression may be due to morphological, physiological and biological alterations in the host.

It has been demonstrated that AM colonization induces remarkable changes in root system morphology, as well as in the meristematic and nuclear activities of root cells (Atkinson *et al.* 1994). This might affect rhizosphere interactions and particularly pathogens-infection development. The most frequent consequence of AM colonization is an increase in branching, resulting in a relatively larger proportion of higher order roots in the root system. Thickening of the cell walls through lignification and production of other polysaccharides in mycorrhizal plants preventing penetration and growth of pathogens like *Fusarium oxysporum* and *Phona terrestris* have been demonstrated. A stronger vascular system observed in mycorrhizal plants will increase the flow of nutrients, impart greater mechanical strength and diminish the effect of vascular pathogens (Bagyaraj, 1996).

The physiological and biochemical alterations brought about by the AM fungi are through improvement of the nutrient status of the host plant. This results in more vigorous plants thus resistant to pathogen attack. It also alters the host physiology and in turn the root exudation pattern. Decreased root exudation in mycorrhizal plants, because of membrane phospholipid content, possibly helps in reducing the infection by root pathogens. The mycorrhizal fungi activate the production of certain compounds involved in specific plant defense mechanisms like phytoalexins, chitinases, pathogenesis related (PR) proteins etc. in the host plants.

Mycorrhizal plants also harbour higher populations of microorganisms in the rhizosphere thus making it difficult for the pathogen to compete and gain access to the root.

v. Use of AM fungi in rootings of cuttings, air layering and in overcoming transplant shock

Some horticultural plants are propagated through cuttings. In such cases, rooting of the cuttings is very important. Enhanced rooting of cuttings through inoculation with AM fungi has been reported in apple. More rooting in tamarinds and cashew plants propagated through air layering was reported. Such plants not only withstood transplant shock but also established better when planted in polybags. AM inoculated avocado and cashew plants withstood transplant shock better than uninoculated stock (Bagyaraj *et al.*, 2006).

vi. AM fungi impact on C sequestration and soil-aggregate stability

Soil-aggregates stability is an important soil physical property that can be affected by AM fungi. Recently, a glycoprotein produced by AM fungi that promotes soil aggregation, "glomalin," has been discovered. Furthermore, higher than normal carbon-dioxide concentrations help to promote soil aggregation by increasing the production of glomalin (Rilling *et al.*, 1999). These findings could have important future implications in the use of mycorrhizal fungi to promote the production of soil stable aggregates, improve water infiltration, and soil C sequestration in agricultural systems.

vii. Mycorrhizal uptake of water and production of plant growth promoting substances

Apart from the above beneficial activities rendered by AM fungi, it also helps in uptake of water away from the rhizosphere soil and provides drought resistance to host plants. AM fungi are also known to secrete certain plant growth promoting substances like Indole acetic acid (IAA) and cytokinins.

Relative Mycorrhizal Dependency

Gerdemann (1975) defined Relative Mycorrhizal Dependency (RMD) as the degree to which a plant is dependent on the mycorrhizal condition to produce maximum growth or yield at a given level of soil fertility. Based on this, plants can be grouped into three categories as highly dependent, moderately dependent, less dependent and not dependent on mycorrhiza. To begin with, it will be worthwhile to work on plants, which are highly dependent on mycorrhiza.

Host specificity

The AM fungi show a preferential colonization to hosts, and thereby, the extent to which a host is benefited depends on the fungal species involved in the symbiosis (Abbott and Robson, 1982). The existence of inter and intraspecific variations among the plant species involved in relation to their phosphorus requirement and the ability of the host to translocate the native soil phosphorus further determines the efficacy of these fungi. Thus it is essential to screen for an efficient AM fungus for a particular host in order to harness the maximum benefit from the fungus.

Selection of efficient AM fungi

Where inoculation with selected AM fungi is required, it is essential to characterize the native AM fungal population first apart from the soil characteristics, particularly soil P status which also influences the efficacy of the fungus to a great extent. The selection process may then be carried out with host plant and inoculum of various strains/species of AM fungi. Such experiments should be carried out under conditions, which are nearest possible to the field conditions. Such pot experiments/micro plot experiments will be realistic and effective only when the exact host plant is used.

Appropriate technology for nursery raised crops

AM inoculum of suitably selected strains can be used for inoculation in the nursery bed. Growers only need to incorporate inoculum in the nursery beds or seedling trays at the appropriate rate by hand. Seedlings thus raised will be colonized by the introduced fungus and then can be planted out in the field.

Such inoculation increased yield of chillies, tomato, capsicum and other vegetables. Increased growth of the seedlings of plantation crops like cashew was observed when inoculated with AM fungi. It was also observed that in AM inoculated plants there was increased number of flowers and vase life of cut flowers of chrysanthemum and china asters (Bagyaraj *et al.*, 2006).

Application methods and time of application

It is important to standardize the dose and time of application apart from the method of application in order to harness the maximum benefit from this association. Our experience reveals that inoculation at the time of sowing/planting will result in better colonization of the roots as they emerge and hence help in better establishment of the host. The experiments conducted with micro-propagated banana, syngonium and spathiphyllum have revealed that the best time for inoculation would be just before planting out hardened plantlets.

Methods of application of AM fungi generally include hand placement, placing below the seed material or in case of pot experiments uniformly mixing the inoculum with the substrate. The importance of method of application of inoculum arises with the need to initiate colonization in the early stages of plant growth. Based on the plant parameters it was concluded that placing AM inoculum 8 cm below the surface of the soil (7 cm below the seed) at a point or as a layer is the best method of inoculum placement for seedlings raised in polybags. For seedlings raised on nursery beds placing the inoculum 2 cm below the seed was found to be the best approach.

Development of multiagent inoculum

Despite the potential of AM fungi in enhancing nutrient uptake, stress tolerance, the monoculture of these organisms is not often enough to achieve maximum benefit for the plant. It is, therefore, necessary to build up a multiagent inoculum consisting of growth promoting microbial consortia. The various beneficial interactions among plant mycorrhizae other growth promoting microorganisms *i.e.* phosphate solubilizing fungi and bacteria, N_2–fixing bacteria such as *Azotobacter*, *Rhizobium* have been well documented (Bagyaraj, 1984).

Propagation cycle of arbuscular mycorrhizal fungi

The major biological characteristic of AMF is their obligate biotrophic nature. This means that each of their life cycle steps requires the association with a living plant. As with most of the filamentous fungi, AMF propagation can occur either by spore differentiation and germination or by mycelium extension through soil and roots. Spores are differentiated by budding intercalary or apically on hyphae. Sexual reproduction has not yet been observed for these symbiotic fungi; therefore they are considered asexual.

Fungal filaments grow through soil particles and come in contact with young plant roots, the fungus threads its way through root surface, and then grow between and inside cortical cells. The wide dispersal of the fungal network through its filaments gives the plant access to a much larger volume of soil than the root system itself. The establishment of mycorrhizal network in roots and soil constitute a soil-root fungal continuum, which is required for beneficial symbiotic exchanges between fungi and plant.

Inoculum propagation

The main obstacle in the production of efficient and reliable AM fungal inoculum lies in their symbiotic behavior, the fungi obligatory requiring a host plant for growth. Traditionally, mycorrhizal fungi are propagated through pot-culture. Starting fungal inoculum, usually made of spores and colonized root segments, are incorporated to a growing substrate for seedling production (Brundett *et al.*, 1996). The fungi spread in the substrate and colonize root seedlings. Both colonized substrates and roots can then serve as mycorrhizal inoculum. Soilless similar culture systems such as aeroponic cultures enable the production of cleaner spores and facilitate uniform nutrition of colonized plants (Jarstfer and Sylvia, 1999). The successful propagation of some AM fungal strains on root-organ culture allowed the cultivation of monoxenic strains that can be used either directly as inoculum or as starting inoculum for large-scale production (Fortin *et al.*, 2002).

(i) Pot-culture propagation

The inoculum propagation process entails the following stages.

Isolation of AM fungi pure culture strain

Pure culture strains can be obtained originally from a single spore that germinate and colonize roots of a host plant. AM fungal strains can also be generated from colonized root segments isolated directly from field plants. Monospecific cultures will then be obtained through subsequent pot-culture generation, using isolated spores or fine root segments as starting inoculum. A technical problem usually encountered with AMF is that spores can easily fall into dormancy and germination rates decrease dramatically (Gemma and Koske, 1988). A cold-temperature treatment can be used to break dormancy (Juge *et al.,* 2002). Research culture collection can provide users with reliable fungal cultures appropriate to start AM fungus propagation, accompanied with detailed information on species origin, spore morphology, and sometimes strain molecular biology and biochemistry.

Choice of a host plant

The most important criteria required for the host plant is its high mycorrhizal potential (*i.e.*, its capacity to be colonized by the AMF strain and to promote its growth and sporulation), a tolerance to growth under growth chamber and greenhouse conditions, and an extensive root system made of solid but non-lignified roots. Rhodes grass (*Chloris gayana*), Leek (*Allium porrum* L.), Sudan grass (*Sorghum bicolor* (L.) Moench), corn (*Zea mays* L.), and bahia grass (*Paspalum notatum* Flugge) are the most frequently used plant host for inoculum propagation.

Optimum growing conditions

Pasteurized, steamed, or irradiated growing substrates are required in order to avoid culture contamination which could affect the quality of the inoculum. A well-aerated substrate is recommended, such as coarse textured sandy soil mixed with vermiculite or perlite or turface. Inadequate mineral nutrient composition may affect fungal development. Optimum P levels vary with the host plant and cultivated fungal strains and an excess of available phosphorus can inhibit AM fungi propagation. Potassium, nitrogen, magnesium, and a selection of microelement ratios may also affect inoculum development, especially when inert growing substrates are used and plant fertilization is performed artificially (Dixon *et al.*, 1999). Other edaphic factors such as pH, soil temperature (Rao and Tarafdar, 1999), light intensity, relative humidity, and environment aeration must also be controlled to optimize AM fungi propagation.

(ii) *In vitro* propagation on root-organ culture

Root-organ cultures consist of excised roots that proliferate under axenic conditions on a synthetic nutrient media supplemented with vitamins, minerals, and carbohydrates. Continuous cultures of vigorous root-organ cultures have been obtained through transformation of roots by the soil bacterium *Agrobacterium rhizogenes* Conn. Since 1988, several species and strains have been successfully propagated *in vitro* with various synthetic growth media and growth conditions (Fortin *et al.*, 2002), and tested with compartmentalized solid and liquid vessels. The mono-specific strains available can be used directly as starting material for large-scale inoculum production, a sole Petri dish culture being enough to generate several thousand of spores and meters of hyphae within 4 months.

In vitro bulk production of AMF inoculum is promising, offering clean, viable, contamination-free fungi (Glomeromycota *in vitro* collection, or GINCO). The cost of *in vitro* inoculum may appear prohibitive compared to the cost of a greenhouse-propagated one, but its use as starting inoculum is a warranty of purity.

Long-term preservation of AM fungi inoculum

Large-scale production of mycorrhizal inoculum requires inventory of product and the ability to provide clients with products of high and consistent quality. Although the detailed procedure for inoculum preservation is proprietary, methodologies for its preservation remain simple and inexpensive. Fungal viability and mycorrhizal efficiency can be maintained for several months at room temperature (30 to 37°C) especially when semi-dry inocula are kept in their plastic containers or packaging. The major inconvenience of such a storage period is the occurrence of spore dormancy. Long-term storage (up to 1 to 2 years) may be conducted at 5°C cold temperature storage. This method is efficient for both *in vivo* and *in vitro* propagated strains. As spore germination and mycelium potential may be stimulated by cold treatment, strain vigour can usually be recovered after long-term storage at cold temperature. Liquid inoculums should react similarly to the traditional dry ones.

More sophisticated and expensive preservation techniques are performed by research culture collections. These include the maintenance of inoculum on living plant-host grown on sterile growth substrate with regular check for mono-specificity of the cultivated strains, storage in liquid nitrogen tanks, and freeze-drying under vacuum. The last two techniques are the usual techniques used in repository culture collections.

Commercial inoculum production

Small scale AMF inoculum production began in the 1980s followed by large scale production in the 1990s. At present, several companies have officially registered and commercialized AM fungi inoculum.

(i) Methodologies

The first generation of commercial inoculum appeared in the early 1980s. Since then, basic methodologies used for *in vivo* inoculum propagation have evolved gradually. Pot-cultivation remains the preferred propagation technique, as it provides a convenient and relatively economic method to produce mycorrhizal inoculum on a large scale (Sahay *et al.*, 1998). Generally, mycorrhizal fungi propagules, such as colonized roots, spores, and hyphae, are mixed with a growing substrate, and the pots are seeded and incubated under controlled conditions. The *in vitro* propagation on root-organ culture may not change drastically the traditional procedures but will certainly facilitate the quality control of strain purity and improve the supply of massive amounts of spores as starting inoculum.

(ii) Obtaining mother inoculum for large-scale production

Both segments of colonized roots (0.08 to 0.16 inches long) containing hypha and/or vesicles and fungal spores may be used as starting fungal propagules for the production of mother inoculum. Research collections can provide such material from either *in vivo* or *in vitro* propagated fungi. Since root-organ culture technology has become available, fungal propagules may be extracted from *in vitro* cultures grown at large scales on solid or semi-liquid growing media.

(iii) Methods of inoculation

Inoculation of AM fungi can be done in different ways

- Mother inoculum added directly at seeding in large trays or pots;
- Mother inoculum incorporated at seedling transplantation of 4-to-6-week old plantlets;
- Mother inoculum added at transplantation of micro-propagated plantlets;
- Colonized seedlings produced in greenhouses and transplanted to the field.

Composition of commercial inoculum

The inoculum sold on the market are provided as granular substrates made from mixed materials such as peat, compost, vermiculite, perlite, sand, and/or expanded clay in which segments of colonized roots, spores, and filamentous networks are distributed. Most of the time these roots, spores, and hyphal networks are not detectable because of their microscopic sizes. In terms of fungal content, the tendency is to introduce a mix of several AM fungi in commercial inoculum. The most frequently used AM fungal species for commercial inoculum are *Glomus fasciculatum, G. mosseae, G. intraradices* etc. These AM fungi are sometimes in a mixture with growth-promoting bacteria and with ectomycorrhizal fungi, making a potentially better inoculum for plant protection and production.

Innovations and future developments

One innovative technique is the ready- and easy-to-use inoculum in which fungal propagules are extracted from growing media, concentrated, and mixed with carriers such as peat, sand, vermiculite, or expanded clay. Products are available in powdered form containing a specified number of active fungal propagules per volume of inoculum. Liquid inoculum dedicated to horticultural use and isolated spores are also available (Dalpe and Morice, 2004). Aeroponic inoculum production at large scale has been investigated by Souza *et al.* (1996) but has not reached commercialization. Bioreactor assays with liquid AM fungi root-organ culture propagation (Jolicoeur *et al.,* 1999) may eventually become suitable for commercialization for research needs. However, as the fungi are produced in association with *Agrobacterium*-transformed roots, it is unlikely that its use can be allowed for field inoculation.

Knowing the performance variability between AM fungal strains, the improvement of commercial inoculum quality will almost certainly come from the selection of higher performance mycorrhizal fungal strains better adapted to the plant host or crop to be colonized and to specific environmental growing conditions (Adholeya, 2003).

Constraints and Regulations

(i) Cost of inoculum versus fertilizers

The obligate biotrophic nature of AM fungi, which, unlike other fungi, requires the establishment of a plant propagation system, either under greenhouse conditions or *in vitro* laboratory propagation. These techniques result in high inoculum production costs, which still remain a serious problem since they are not competitive with production costs of phosphorus fertilizer.

(ii) Sanitary control

Another serious problem in commercializing inoculum comes from the need to control the biological composition of the product, especially from invading phytopathogenic microorganisms. When colonized roots are directly incorporated to carriers, their surface sterilization with a light solution of disinfecting product can be done without affecting the effectiveness of an inoculum. When roots and rhizosphere material are used for inoculum preparation, handling with clean apparatus is advised.

(iii) Efficiency of inoculum

In the field application of any microbial inoculum, it is essential to verify that the inoculated microorganisms possess the characteristics and the potential described by the inoculum manufacturers. With AM inoculum, such evaluations can be done using several approaches such as morphological identification of AM spores to confirm the fungus identity, and by estimating the mycorrhizal root colonization level of test plants (Dalpe, 1991). Tentative molecular techniques have been developed for the detection of AMF inoculum strains and discrimination from indigenous AMF strains naturally occurring in soils. These techniques are not yet totally reliable due to the large genetic heterogeneity in AM fungi. Reliable molecular techniques to trace the inoculated strains using rep-PCR and specific primers developments are under study (Seguin *et al.*, 2003). Such a technological breakthrough would greatly facilitate both fundamental and applied research on mycorrhizae as well as improve quality control of commercial inoculum.

Methods used in AMF Research

1) Staining roots to observe mycorrhizal colonization

Colonization of root cortex cells by arbuscular mycorrhizal fungi does not alter root morphology as in ectomycorrhizal associations. Therefore, to detect and measure mycorrhizal colonization, roots are submitted to a clearing and staining procedure. The method most extensively used by mycorrhizologists is that described by Philips and Hayman (1970).

Materials needed:

- 10% KOH solution (Potassium Hydroxide)
- 1% HCl solution (Hydrochloric Acid)
- Acidic Glycerol solution (500 ml Glycerol, 450 ml H_2O, 50 ml 1% HCl)
- 0.05% Trypan Blue in Acidic Glycerol solution (0.5 g/L)
- Alkaline H_2O_2 solution (3 ml 20% NH_4OH, 30 ml 3% H_2O_2, 567 ml tap water).

Procedure:

1. Wash roots free from soil debris and rinse in several changes of tap water,
2. Soak roots in 10% KOH at 90°C for 1 hour or at 120° C for 15 minutes,
3. Remove KOH and rinse roots with water (2-3 times) to remove KOH,
4. If roots are too pigmented, soak roots in alkaline H_2O_2 solution for 10-30 minutes. Rinse roots with water again after this step,
5. Soak roots in 1% HCl solution for 5 minutes,
6. Remove HCl. **Do not rinse** roots after this step as they must be acidified for proper staining,
7. Stain roots in acidic glycerol solution containing trypan blue at 90°C for 1 hour or at 120° C for 5 minutes,
8. Discard stain solution and keep roots on acidic glycerol (without trypan blue) or water at 4°C.

Comments: Acidic Glycerol solution can be replaced by a Lacto Glycerol solution (20 ml Lactic acid, 40 ml Glycerol and 40 ml Distilled water) if desired. Step number 4 can be omitted if roots are not highly pigmented; alkaline H_2O_2 solution must be prepared fresh just before using. For long term storage, roots can be kept in water and add some drops of 0.1% sodium azide. For steps 2 and 7, a water bath is suitable to keep root samples at 90°C.

2) Measuring mycorrhizal colonization

Measurement of root mycorrhizal colonization is frequently carried out from field collected roots (directly from the soil or from individual plants) or from experimental plants under greenhouse conditions. It estimates growth of a fungal isolate or a fungal community within the root cortex. The grid-line intersect method (Giovannetti and Mosse, 1980), which is presented below, is the most widespread method utilized that allows measurement of mycorrhizal root length and percentage of mycorrhizal colonization.

Materials needed:

- Petri Dish with a 1.1 cm x 1.1 cm grid on the bottom
- Dissecting needles

Procedure:

1. Spread out evenly stained roots in the Petri dish,
2. Under a dissecting microscope, scan vertical and horizontal lines of the grid,
3. Record: a) the total number of roots intersecting the grid lines and b) the number of intersect with mycorrhizal colonization,
4. The percent of mycorrhizal colonization is given by the formula:

$$\%\ \text{Myc Col.} = \frac{\text{Total No of intersects with mycorrhizal colonization}}{\text{Total No of intersections between root and the gridline}} \times 100$$

Comments: The data obtained at step 3 can be used to determine plant root length and mycorrhizal root length. If the entire root system is spread out in the Petri dish, the total number of roots intersecting the grid lines represents the plant root length and the total number of intersects with mycorrhizal colonization represents the mycorrhizal root length. If a small sample of roots is taken from the entire root system, then to calculate plant root length and mycorrhizal root length, appropriate calculations based on fresh and dry weights of roots are needed.

3) Spore extraction from the field

Identification of AMF species is based on the analysis of subcellular structures of assexual spores. Morphology of internal and external mycelium during mycorrhizal association is practically indistinguishable among species within the same genus and between genera. Spore subcellular structures are highly conserved and phenotypically stable regardless of environment and plant host (Morton *et al.*, 1995). Thus, spores are the only part of the fungal organism that can be used to delimit species. Spores are extracted from the soil by wet sieving (Gerdemann and Nicolson, 1963) followed by sucrose gradient centrifugation.

Materials needed:

- Set of nested sieves with different openings (or at least two with 710 µm and 45 µm opening),
- Plastic bucket or recipient,
- Beakers, Petri dishes, Watch glass,
- Centrifuge Tubes and Centrifuge,
- Sucrose solution (20% and 60%).

Procedure:

1. Place 100 g of soil sample in the bucket and suspend it with at least 500ml water,
2. Stir thoroughly to bring spores into suspension and let stand for 30 seconds,
3. Pass the suspension through two nested sieves with 710 µm and 45 µm opening,
4. Place material retained in the 710 µm sieve into a large Petri dish and observe under the dissecting microscope for large spores (*Gigaspora* and *Scutellospora* spores) and sporocarps,
5. Place material retained in the 45 µm into a becker with a small amount of water. Transfer this material to a centrifuge tube containing a 20%/60% sucrose gradient.
6. Centrifuge at 2000 rpm for 1 minute,
7. Drop the supernatant back to the 45 µm sieve and wash with tap water to remove excess of sucrose,
8. Transfer spores and debris from the 45 µm sieve to a Petri dish and under the microscope collect the spores on a clean watch glass.

Comments: There are several sieves with different openings available in the market and on step 3, other sieves can be stacked between the two recommended. This will separate spore by size but will increase the amount of work as material retained in each sieve must be centrifuge separately. The sieves recommended (with 710 μm and 45 μm opening) are suitable to recover most AMF species as spore size for the majority of species range from 40 to 600 μm. Sucrose solution is made from commercial sugar bought in the supermarket. To prepare the gradient, in a 50 ml centrifuge tube, add 15 ml of the 20% sucrose solution and then, on the bottom of it, add another 15 ml of the 60% solution.

Spores can be separated from organic debris and collected from the Petri dish by using either a fine forceps or an extrude glass Pasteur pipette. At this time, spores can be separated by morphotypes according to color and size and if the spores are in good health they can be identified up to genus and species level.

4) Species identification

In most cases, spores extracted from the field are not in large numbers to permit an accurate identification and therefore spores need to be mounted on slides and observed under a compound microscope.

Under the dissecting microscope, taxonomic important characteristics are spore size, color and shape, presence of subtending hypha, presence of suspensor cell and presence of sporiferous saccule (rarely seen from field collected spores). Under the compound microscope, some important taxonomic characteristics to be observed are presence and type of ornamentation on the spore wall, Melzer's reaction, number of germinal walls, thickness of spore wall, among others. Spores need to be mounted on permanent mounting media like PVLG (Polyvinyl-Lacto-Glycerol) and PVLG mixed with Melzer's reagent.

Materials needed:

- Glass slides, coverslips and labels,
- Dissecting needle,
- PVLG solution (100 ml distilled water, 100 ml lactic acid, 10 ml glycerol, 16.6 g polyvinyl alcohol (PVA)),
- Melzer's reagent (100 g chloral hydrate, 100 ml distilled water, 1.5 g iodine, 5 g potassium iodide). Mix Melzer's reagent with PVLG (1:1) to make a PVLG + Melzer solution.

Procedure:

1. On a glass slide, add one drop of PVLG and one drop of PVLG + Melzer solution.
2. Add spores to the center of each drop by using a fine forceps or extrude micropipette (in this case, be careful not to add too much water!),
3. With a dissecting needle, mix spores with the PVLG or PVLG + Melzer solution. Let stand for at least 5 minutes to dry slightly the surface of the drop,
4. Place gently a coverslip over the drop of PVLG and another coverslip over the drop of PVLG + Melzer.
5. Using a dissecting needle, crush each spore individually under the dissecting microscope. This step is important to expose germinal walls and their layers.
6. Label the slides including sample number, genus and species (if known) and date.

7. Keep slides at room temperature for 5 days and then place them at 40-60°C for two days to harden the mountant media.

Comments: Step 2 is crucial for mounting spores on a slide - if too much water is added, spores will slide out to the edges when the coverslip is placed on the top of each drop.

5) Most Probable Number – (MPN) to estimate number of infective propagules of AMF

The Most Probable Number method has been used to estimate the number of infective propagules (inoculum potential) of AMF in several soils. It is based on a series of soil dilution where presence or absence of mycorrhizal colonization is recorded and results given as a probability of the number of infective propagules based on a statistical table. One of the disadvantages is that the set up of the assay is time consuming and the number calculated has a high 95% confidence interval (Adelman and Morton, 1986). The advantage is that the analysis of roots to check for colonization is easy and fast to carry out if someone is familiar with mycorrhizal morphology. As the result is a single number, the method is appropriate for comparisons among land use systems if MPN assays are set up at the same time.

Materials needed:

- Plant tubes (150 cm x 2.5 cm) and racks
- Plastic bags (30 cm x 20 cm)
- Sterilized diluent (sand:soil mix 1:1)
- Onion seeds

Procedure:

1. Weigh 30 g of field soil in a plastic bag and add to it 270 g of sterilized diluent. Shake thoroughly to get 10^{-1} dilution.
2. Remove 30 g from the 10^{-1} dilution and place it into another bag containing 270 g of sterilized diluent. Shake thoroughly to get 10^{-2} dilution. Make a ten fold series dilutions up to 10^{-4} dilution (or higher dilution if needed).
3. Distribute soil from each dilution into plant tubes. Use five replicate tubes per dilution.
4. Sow seeds of onion into each tube. If onion seeds are not available, any other suitable host can be used.
5. After emergence, trim to only one plant per tube and let plants grow in a greenhouse or growth room for six weeks.
6. At harvest, wash roots away from soil and stain them with trypan blue (see below).
7. Under a dissecting microscope, determine presence or absence of mycorrhizal colonization in each replicate. Counts of positive tube (those containing mycorrhizae) in different dilutions are used to calculate MPN values. Use table of Cochran (1950).

Comments: To exemplify the calculation, let's consider that in the 5 replicates (tubes) in each of the four dilutions (10^{-1}, 10^{-2}, 10^{-3} and 10^{-4}), one obtains a combination of number of positive tubes like: 5, 5, 3, 2.

This means that all 5 replicate tubes are positive for mycorrhizal colonization at 10^{-1} and 10^{-2} dilutions, 3 replicate tubes are positive at 10^{-3} dilution and 2 replicate tubes are positive at 10^{-4} dilution. For the calculation of MPN, only three numbers of a given combination are required. The first number (N1) corresponds to the least concentrated dilution in which all (or the highest number of) tubes are positive for mycorrhizal colonization. The two other numbers (N2 and N3) are those corresponding to the next two dilutions. In the example above, the combination would be:

N1=5, N2=3 and N3=2,

Using the MPN table, the value given for this combination of positive values is 1.4. To obtain the MPN of infective propagules of AMF in the sample, this value has to be multiplied by the middle dilution (in this case 10^{-3}).

Therefore, the soil has 1.4 x 10^{-3} infective propagules/g soil.

6) Establishment of trap cultures

Trap cultures are becoming more frequently used on studies of AMF diversity because they reveal species that are not sporulating in the field at sampling time. Also, they provide fresh and healthy spores which can be used to establish single isolate culture of AMF. The method proposed is based on the protocols of Stutz and Morton (1996).

Materials needed:

- Sterile soil:sand mix (1:1)
- 1.5 Kg Plastic pots
- Plastic trays or plastic bags
- Seeds of *Sorghum sudanense* (sudangrass) and *Vigna unguiculata* (cowpea)

Procedure:

1. On a plastic tray (or plastic bag), homogenize the field soil with sterile soil:sand mix (50% of field soil + 50% sterile soil:sand mix),
2. Place this mixture in a 1.5 Kg plastic pots and seed heavily with a mixture of sudangrass and cowpea (40-50 seeds per pot). Cover seeds with sterile soil:sand mix,
3. After 3-4 months under greenhouse conditions, sample one or two 50 ml soil cores from the pot, extract spores and identify as explained above.

Comments: It is important that during sampling of field soil, plant roots are included as they also serve as propagules to start trap cultures. In step 1, sterilized pure sand can be utilized as a diluent instead of a soil:sand mix. In step 2, other plant hosts are suitable as well if sudangrass and cowpea are not available; it is recommended a C4 grass to be used as a host.

References

Abbott. L.K. and A.D. Robson. 1982. The role of vesicular-arbuscular mycorrhizal fungi and selection of fungi for inoculation. *Aust.J.Agric.Res.*. **33**: 389-408

Adelman. M.J. and J.B. Morton. 1986. Infectivity of vesicular-arbuscular mycorrhizal fungi: Influence of host-soil diluent combinations on MPN estimates and percentage colonization. *Soil Biol. Biochem.*, **18**: 7-13.

Adholeya. A. 2003. Commercial production of AMF through industrial mode and its large-scale application. 4th Int. Conf. Mycorrhizae. Montreal Quebec Canada. August 10-15. 2003. No. 240.

Almeida. R.T. and N.T. Schenck. 1990. A Revision of the Genus *Sclerocystis* (Glomaceae. Glomales). *Mycologia.* **82**: 703-714.

Atkinson. D., G. Berta and J.E. Hooker. 1994. In: Impact of arbuscular mycorrhizas on sustainable agriculture and natural ecosystems, Birkhauser. Basel. pp. 89-99.

Bagyaraj, D. J. 1984. In: VA Mycorrhiza, (ed). C.L. Powell and D.J. Bagyaraj, CRC Press, Boca Raton, USA, pp.131-154.

Bagyaraj, D.J. 1996. Mycorrhizal symbiosis in tropical trees in relation to control of root pathogens. In: Impact of diseases and insect pests in tropical forests. Kerala forest research institute, Peechi, India, pp. 239-245.

Bagyaraj. D.J., C. Tharun and A.N. Balakrishna. 2006. Utilization of arbuscular mycorrhizal fungi in Horticulture. In: Microbial Biotechnology in Horticulture (Vol 1). (ed.) R.C. Ray and P.O. Ward. Science Publishers, Enfield, UK. pp. 21-48.

Bagyaraj. D.J. and A. Varma. 1995. Interaction between arbuscular mycorrhizal fungi and plants: Their importance in sustainable agriculture and in arid and semiarid tropics. In: Advances in Microbial Ecology. Academic Press, London. pp. 119-142.

Barea, J. M. and P. Jeffries. 1995. Arbuscular mycorrhizas in sustainable soil plant systems. In: Mucorriza-Structure, function, molecular biology and biotechnology. Springer. Heidelberg. pp. 521-529.

Brundrett, M, N. Bougher, B. Dell, T. Grove and N. Malajczuk. 1996. Working with mycorrhizas in forestry and agriculture. ACIAR (*Austral. Cen. Int. Agric. Res.*) Mono. 32.

Cochran. W.G. 1950. Present status of Biometry. *Biometrics.* **6**: 75-78.

Cordier. C, S. Gianiazzi and V.Gianinazzi Pearson. 1996. Colonization patterns of root tissues by *Phytophthora nicotanae* var. *parasitica* related to reduced disease in mycorrhiizal tomato, *Plant and Soil.* **185**: 223-232.

Dalpé, Y. 1991. Vesicular-arbuscular mycorrhizae. In: Manual of Soil Sampling and Methods of Analysis. 3rd Ed. Can. Soc. Soil Sci., Lewis Pub. of CRC Press. pp. 287-301.

Dalpe, Y. and M. Morice. 2004. Arbuscular mycorrhiza inoculum to support sustainable cropping systems. Online. Crop Management doi: 10.1094/CM-2004-0301-09-RV.

Dixon, S. S. E. Smith and F. A. Smith. 1999. Characterization of two arbuscular mycorrhizal fungi in symbiosis with *Allium porrum*: Inflow and flux of phosphate across the symbiotic interface. *New Phytol.*, **144**: 173-181.

Fortin. J. A, G.Bécard, S. Declerck. Y. Dalpé. , M.. St-Arnaud. A. P.Coughlan and Y. Piché. 2002. Arbuscular mycorrhiza on root-organ cultures: A review. *Can. J. Bot.*, **80**: 1-20.

Gemma. J. N. and R. E. Koske. 1988. Seasonal variation in spore abundance and dormancy of *Gigaspora gigantea* and in mycorrhizal inoculum potential of dune soil. *Mycologia.* **80**: 211-216.

Gerdemann, J. W. 1975. In: The Development and Function of Roots, (ed.) J.E. Torrey and D.T. Clarkson. Academic Press, New York, pp. 575-591.

Gerdemann, J.W. and J.M. Trappe. 1974. Endogonaceae in the Pacific Northwest. *Mycologia Memoir.*, **5**: 1-76.

Gerdemann, J.W. and T.H. Nicolson. 1963. Spores of mycorrhizal *Endogone* species extracted from soil by wet sieving and decanting. *Trans. Br. mycol. Soc.*, **46**: 235-244.

Giovannetti, M. and B. Mosse. 1980. An evaluation of techniques for measuring vesicular arbuscular mycorrhizal infection in roots. *New Phytol.*, **84**: 489-500.

Guillemin, J.P., S. Gianinazzi, and V. Gianinazzi-Pearson. 1991. D. endomycorrhization de vitroplants *Ananas comosus:* mise en evidence d'un effect mycorrhizien. *Fruits.*, **46**: 355-358.

Jarstfer. A. G. and D. M. Sylvia. 1999. Aeroponic culture of VAM fungi.: Mycorrhiza: Structure. Function, Molecular Biology and Biotechnology. (ed.) A. Varma, and B. Hock. Springer-Verlag. Berlin. pp. 427-441

Jolicoeur, M., R. D.Williams, C. Chavarie, J.A. Fortin, and J. Archambault. 1999. Production of *Glomus intraradices* propagules, an arbuscular mycorrhizal fungus, in an airlift bioreactor. *Biotechnol. Bioeng.* **63**: 224-232.

Juge, C., J. Samson., C. Bastien., H. Vierheilig, A. Coughlan, and Y. Piché. 2002. Breaking dormancy in spores of the arbuscular mycorrhizal fungus *Glomus intraradices*: A critical cold-storage period. *Mycorrhiza*. **12**: 37-42.

Jumpponen, A, K.G. Mattson, J.M.Trappe. 1998. Mycorrhizal functioning of *Phialocephala fortinii* with *Pinus contorta* on glacier forefont soil: Interactions with soil nitrogen and organic matter. *Mycorrhiza*. **7**: 261-265.

Morton, J.B. and D. Redecker. 2001. Two new families of Glomales, Archaeosporaceae and Paraglomaceae, with two new genera *Archaeospora* and *Paraglomus*, based on concordant molecular and morphological characters. *Mycologia*. **93**: 181-195.

Morton, J.B. and G.L. Benny. 1990. Revised classification of arbuscular mycorrhizal fungi (Zygomycetes): a new order, Glomales, two new suborders, Glomineae and Gigasporineae, and two new families, Acaulosporaceae and Gigasporaceae, with an emendation of Glomaceae. *Mycotaxon*. **37**: 471-491.

Morton, J.B, S.P. Bentivenga and J.D. Bever. 1995. Discovery, measurement, and interpretation of diversity in arbuscular endomycorrhizal fungi (Glomales, Zygomycetes). *Can. J. Bot.*, **73**: S25-S32.

Oehl, F. and E. Sieverding. 2004. *Pacispora*, a new vesicular-arbuscular mycorrhizal fungal genus in the Glomeromycetes. *J. Appl. Bot.*, **78**: 72-82.

Philips, J.M. and D.S. Hayman. 1970. Improved procedures for clearing roots and staining parasitic and vesicular-arbuscular mycorrhizal fungi for rapid assessment of infection. *Trans. Br. mycol. Soc.*, **55**: 158-161.

Rao, A. V. and J. C. Tarafdar. 1999. Soil solarization for mass scale production of arbuscular mycorrhizal fungal inoculum in Indian arid zone. *Indian J. Agric. Sci.*. **69**: 271-274.

Redecker, D., J.B. Morton and T.D. Bruns. 2000. Molecular phylogeny of the arbuscular mycorrhizal fungi *Glomus sinuosum* and *Sclerocystis coremioides*. *Mycologia*. **92**: 282-285.

Rilling, M. C, S. F. Wright, M. F. Allen, and C. B. Field. 1999. Rise in carbon dioxide changes soil structure. *Nature*. **400**: 628.

Sahay, N. S. A. S. Sudha, A. Varma, and A. Singh. 1998. Trends in endomycorrhizal research. *Indian J. Exp. Biol.*, **36**: 1069-1086.

SchÜbler, A, D. Schwarzott and C. Walker. 2001. A new fungal phylum, the Glomeromycota: phylogeny and evolution. *Mycol. Res.*, **105**: 1413-1421.

Smith, S.E and D.J. Read. 1997. Mycorrhizal Symbiosis, Ed.2. Academic Press, London.

Séguin, S., N. Désaulniers, Y. Dalpé and C. A. Lévesque. 2003. Development of AMF strains specific primers for detection in field-grown colonized roots. 4th Int. Conf. Mycorrhizae. Montreal Quebec Canada, August 10-15 2003. No. 444.

Souza, de, E. S., H. A. Burity, A. C. Espirito-Santo, and M. L. R. B. Silva, da. 1996. Alternative for arbuscular mycorrhizal fungi inoculum production in aeroponic culture. *Pesqu. Agropecu. Bras.*, **31**: 153-158.

Stutz, J.C. and J.B. Morton. 1996. Successive pot cultures reveal high species richness of arbuscular endomycorrhizal fungi in arid ecosystems. *Can. J. Bot.*, **74**: 1883-1889.

Soil Microbiology and Soil Biotechnology : Vol. 2
Editor : S. Kannaiyan
Associated Publishing Co., New Delhi. 2010. pp. 136-166

7

Role of Host Genes in Symbiotic Nitrogen Fixation

S. Suyatha Lilly

Introduction

Legumes and their symbiotic bacteria (collectively referred to as rhizobia) make the largest contribution to global biological nitrogen fixation. Rhizobia elicit on their hosts, in a specific manner, the formation of specialized organs, the nodules in which they reduce molecular nitrogen to ammonia (Brewin, 1991). The *Rhizobium*-legume symbiosis, because of its agricultural importance, has ensured continuing research support world-wide and is presently the best understood of all plant microbe interactions.

The symbiosis is specific: every strain has a definable host range. Studies have revealed that in the course of infection and nodulation both partners exchange low-molecular weight signal molecules. The plant controls the expression of bacterial nodulation genes *via* phenolic compounds (Kondorosi, 1991) whereas the rhizobia signal back to the plant by secreting specific lipooligosaccharides (Nap and Bisseling, 1990).

The host, through a series of specific biochemical and morphological responses to infection by the bacterium, provides an environment and structure which supports sustained nitrogen fixation. Essential components of this response comprise a controlled microaerobic environment close to bacteroids, a continuous supply of oxidizable plant assimilates an efficient mechanism for the utilization of ammonia. The products of ammonia assimilation are nitrogenous solutes which are readily translocated and which are suitable substrates for nitrogen metabolism in the host.

Rhizobial genes controlling infection, nodulation and host range

Molecular genetics combined with the study of the symbiotic behaviour of bacterial mutant strains, has led to the identification of two major classes of rhizobial genes involved in infection and nodulation. One class includes genes in which mutations result in both

a) alteration in the synthesis, transport and assembly of surface components such as exopolysaccharides, lipopolysaccharides and β 1, 2-glucans and

b) alteration in the infection process, such as the inability to elicit the formation of infection threads, resulting in the formation of non fixing empty nodules (Nod^+

Tamil Nadu Agricultural University, Coimbatore - 641 003

Fix phenotype). These results show that these polysaccharides are involved in the infection process and a possible role of exopolysaccharides and lipopolysaccharides in the determination of host specificity has been suggested.

The second class consists of the nodulation (nod and nol) genes. The nod and nol genes are involved in infection and nodulation, and their inactivation can result in various inplanta phenotypes such as the absence of nodules (Nod$^-$) or a delayed but effective (nitrogen-fixing) nodulation (Nod d, Fix$^-$) on homologous hosts, or changes in host range. Some nod genes are common to all rhizobia whereas other are specific. The nod genes are regulated in a similar way in all rhizobia. Their regulation involves the following three elements.

a) Transcription is activated by trans-acting regulatory Nod D proteins.

b) Nod D proteins bind to promoter regions that contain extended conserved sequences, the Nod boxes.

c) Activation of nod gene transcription by the Nod D proteins requires the presence of plant signals, flavonoids, or other phenolic compounds exuded by legume roots. Nod genes reside on large symbiotic plasmids (p sym) that also carry the fix and nif (nitrogen-fixation) genes. In *Bradyrhizobium* and *Azorhizobium,* the symbiotic genes are likely to be chromosomal.

Both regulatory and structural nod genes are major determinants of host range. For example, mutations in the regulatory nod D gene of *R. trifolii* (Mclver *et al.*, 1989) and transfer of nod D1 gene of *Rhizobium* sp. NGR 234 into other *Rhizobium* species result in an extension of the host range (Horvath *et al.*, 1987). Similarly, mutations in the structural nod E gene of *R. trifolii* (Djordjevic *et al.*, 1985) and nod Q gene of *R. meliloti* (Vervantes *et al.*, 1989; Faucher *et al.*, 1989) extend the host range.

Host Bacterium specificity

Plant flavonoids as signal molecules to rhizobia

The plant-bacterium interactions begin with the diffusible chemical signals in the rhizosphere. The various flavonoid molecules released by legume roots are the natural inducers of the rhizobial nod genes and belong to subclasses of flavones and flavanones (Gottfert, 1993). In addition to flavonoids, a few non-flavonoid nod gene inducers have also been identified namely betaines-trigonelline and stachydrine (Phillips *et al.*, 1992), chalcones (Maxwell *et al.*, 1989; Recourt *et al.*, 1991), anthocyanidins (Hungria *et al.*, 1991), conjugated isoflavonoids (Dakora *et al.*, 1993) and non phenolic compounds like hydroxycinnamic acids (Kape *et al.*, 1991).

The specificity in induction of rhizobial nodulation genes by specific flavonoid signal molecules forms the first host specific step in nodule initiation. Flavones and flavanones are usually the inducers of fast growing rhizobia whereas isoflavones are often the inducers of *Bradyrhizobium* spp. (Djordjvevic *et al.*, 1987; Banfalvi *et al.*, 1988). Depending on the Nod D protein of rhizobial strain or species, the same or similar compounds can act either as inducers or inhibitors or can be effective in nod gene activation. For example, diadzein and genistein act as inducers for nod genes of *B. japonicum,* but the same compounds are nati-inducers for *R. legumionosarum* bv. *viciae* (Kosslak *et al.*, 1987). Likewise, hesperitin is the most potent inducer of *R. leguminosarum* bv. *viciae* but it is relatively ineffective with *R. meliloti* (Zaat *et al.*, 1988). The efficiency of nod gene activation also depends on the presence of different types of flavonoids in exudates because some flavonoids interfere with other inducer molecules and the effect of inducers can be additive or synergistic (Hartwig *et al*, 1989).

The nature and amount of compounds exuded are primarily controlled by the plant and its growth stage as well as influenced by the presence of combine nitrogen in the root zone. Zaat *et al.* (1989) found that *Pisum sativum, Vicia faba* and *V. hirsutum* when nodulated by the same strain of *R. leguminosarum* bv. *viciae,* a different spectrum of inducers are released. Recently, it has been reported that *Rhizobium* inoculation elicit the production and excretion of additional species of flavonoids in the respective host plant roots (Spaink *et al.*, 1991; Van Brussel *et al.*, 1990; Recourt *et al.*, 1992a; Recourt *et al.*, 1992b).

In general, nod D proteins from *R. meliloti* and *R. leguminosarum* respond to few flavonoids, while Nod Ds from the broad host range *Rhizobium* NGR 234 have a large spectrum of inducing molecules including even monocyclic aromatic compounds (Le Strange *et al.*, 1990). Some inducer molecules from heterologous hosts have also been found to induce the rhizobial nod genes. For example, luteolin, inducer for *R. meliloti* released from alfalfa roots, can also activate nod genes of *R. leguminosarum* bv. *viciae.* These studies show that plant inducers itself are not host specific and their interaction with Nod D protein initiate host specific reactions (Gyorgypal *et al.*, 1991).

Nod genes

Genes directing specific steps in nodulation of a legume by a strain of *Rhizobium* have been called as nod genes. Many nod genes from different *Rhizobium* species are highly conserved and are borne on large plasmids called sym plasmids. In addition to nod genes which direct specific nodulation events, sym plasmids contain specificity genes, which restrict a strain of *Rhizobium* to a particular host plant. Indeed, cross-inoculation group specificity can be transferred across species of *Rhizobium* by simply transferring its sym plasmid of *R. leguminosarum* (whose host is the pea) is transferred to *R. trifolii* (whose host is clover), cells of the latter species will effectively nodulate pea. Interestingly, sym plasmids can also be transferred between bacterial genera. If a sym plasmid is transferred from *Rhizobium* to a closely related bacterium, *Agrobacterium*, the latter organism forms root nodules. Although pure cultures of *Agrobacterium* will fix nitrogen, it is not clear whether *Agrobacterium* induced nodules will fix nitrogen.

In the sym plasmid of *R. leguminosarum,* nod genes are located between two clusters of genes for nitrogen fixation, the nif genes (in this species and in certain other *Rhizobium* species, nif genes are plasmid borne). Ten nod genes have been identified in this species. The entire nod region has been sequenced and the function of many nod proteins is known. The genes nod ABC direct root hair curling by production of specific polysaccharide signal molecules called NOD factors. The nod D gene encodes a regulatory protein that controls expression of other nod genes. In the presence of specific nodulation signals the nod D gene product controls transcription of other nod genes. The nod F protein is involved in the synthesis of a lipid or polysaccharide needed for effective nodulation and is thought to be involved in the transport of substances needed to begin nodulation. The nod E and nod L genes function in determining host range (cross-inoculation group), nod M encodes a protein involved in purine biosynthesis and nod I and nod J encode membrane proteins whose function is unknown.

Cloning and sequencing of nod genes in *R. leguminosarum* suggests that they are arranged in atleast three operons. As a regulatory protein, the nod D protein binds to a highly conserved 35 base pair DNA sequence which preceeds each operon. These regions, referred to as nod boxes, are presumably operator genes to which the nod D protein can bind. Following interaction with ind· ,er molecules, the conformation of the nod D protein changes, causing it to fall off the nod box DNA. This event presumably exposes a promoter site for RNA polymerase binding which initiates transcription of nod structural genes. Several inducer molecules have been identified. In most cases these are plant flavonoids, complex organic molecules that are widespread plant products. Flavonoids have many functions in plants, including growth regulation and attraction of pollinating animals. However, leguminous plants are unusual because unlike other plants, their roots secrete large amounts of flavonoids, presumably to trigger nod gene expression in nearby rhizobial cells in the soil. Interestingly, some flavonoids that are structurally very closely related to nod D inducers strongly inhibit induction of nod genes in certain rhizobial species, suggesting that part of the specificity observed between plant and bacterium in the *Rhizobium*-legume symbiosis could lie in the chemical nature of the flavonoids excreted by a particular plant.

Rhizobial nodulation genes and host range

In rhizobia, the nodulation genes are located either on large symbiotic plasmids (p sym) or chromosome which are organized into several coordinately regulated operons. So far, over 50 different nodulation (nod ore nol) genes have been characterized in different rhizobia. They are basically categorized into three main classes, namely

a) the regulatory nod D and nod V,W genes which activate the transcription of other common and host specific nod genes

b) the common nod A,B,C and M genes, which are functionally and structurally conserved among different rhizobia and

c) the host specific nod genes which are variable with bacterial species and strains, and mutations in these genes usually do not complement with cloned genes from other rhizobia (Spaink, 1994).

Regulatory nod genes

The regulatory nod genes are found in all the strains of *Rhizobium, Bradyrhizobium* and *Azorhizobium* and are expressed constitutively in cultured cells. The nod D protein of various rhizobia are functionally different in their responsiveness to different sets of flavonoids and in their ability to activate other nod genes (Spaink *et al.*, 1987; Horvath *et al.*, 1987). The structure of a given nod D protein determines which flavonoids act as nod gene inducers. In the presence of an appropriate flavonoid secreted by the appropriate host legume, nod D activates transcription of all other nod genes by binding to a conserved 47 bp long consensus nod box sequence present in the promoter of the nod genes (Kondorosi *et al.*, 1989; Fisher and Long, 1989). Therefore the flavonoid nod D interactions represent the first major host-specific step in the establishment of symbiosis.

Some *Rhizobium* species *e.g. R. leguminosarum* bv. *viciae* and bv. *trifolii, B. parasponia, A. caulinodans* and *R. loti* have a single nod D gene whereas, other rhizobia have multiple nod D genes and have a complex regulatory circuit for controlling nod gene expression (Schlaman *et al.*, 1992). In *R. meliloti,* three functional nod D genes *i.e.* nod D1, nod D2 and nod D3 and a nod D homoogy, syr M have been reported and each copy of the nod D gene is expressed and regulated differently. For example, luteolin is a nod gene inducer for nod D1 but an inhibitor for nod D2. Chalcone is an inducer both for nod D1 and nod D2 but betaines are inducers for nod D2 only.

In *B. japonicum,* the nod VW genes are required for nodulation of mung bean, cow pea and Siratro (Gottfert *et al.*, 1990). Recent results indicate that nod W is essential for isoflavone-mediated induction of the nod YABC and nod D1 operons. The mode of action of the nod V, W regulatory system, however is unknown.

Common nod genes

The nod A, nod B and nod C genes are conserved in all nodulating rhizobial strains. Mutation in any of these genes blocks nodulation completely and their products are essential for the synthesis of nod factors (Egelhoff and Long, 1985). Nod I and nod j proteins are involved in the secretion of nod factors and mutations in nod ij genes result in nodulation delay with *R. leguminosarum.* Recently, nod m and nod n genes of *R. leguminosarum* bv. *viciae* have been found to complement nod m and nod n mutations in *R. meliloti.* The nod m protein has been shown to supply precursors for the biosynthesis of nod factors.

Host specific nod genes

The host specific nodulation genes are required for correct induction of root hair curling and cell division in the compatible legume host and are species specific. Seven host specific nod genes namely nod F,E,G,H,L,P and Q have been identified in *R. meliloti* (Debelle *et al.*, 1986; Cervantes *et al.*, 1989; Scwedock and Long, 1989) while in *R. leguminosarum* bv. *viciae* five genes *i.e.*, nod F,E,L,T and O have been reported to be involved in host-specific nodulation. The nod S,U,Z and nol O genes have been identified as host specific genes in *B. japonicum* (Gottfert *et al.*, 1990; Stacey *et al.*, 1994). Mutations in the host-specific nod genes result in alteration in the host range of nodulation. For example, mutations in nod H strongly inhibit the ability of *R. meliloti* to nodulate the homologus host alfalfa and lead to delayed nodulation on sweet clover but confer the ability to nodulate the host plant vetch, while the nod Q mutant of *R. meliloti* can nodulate both alfalfa and vetch (Horvath *et al.*, 1986; Faucher *et al.*, 1989).

Nod regulators

Legume root secrete a variety of compounds, some of which regulate the expression of rhizobial nod genes. These compounds have been purified by using nod-lac Z assay system (Peters *et al.*, 1986). Most of the nod regulators are flavonoids such as anthocyanidin, chalcone, coumarin, flavanone, flavone, flavonoid or isoflavone.

Besides flavonoids, compounds such as hydroxy cinnamic acid and betaines are also known to possess nod regulatory activity. Root exudates of host plant usually contain compounds, some of which possess varying inducing activity while others antagonise the action of inducers (Firmin *et al.*, 1986; Kosslak *et al.*, 1990). In different species of rhizobia, different flavonoids are known to induce the expression of nod genes. Specificity in *Rhizobium*-legume symbiosis is also evident in the structure of nod regulators. Compounds which act as common inducers of nod genes of a specific *Rhizobium* sp. possess common structural features. Compounds acting as antagonists of nod inducers possess very strong structural similarities to the natural inducers. A compound acting as inducer of nod genes of one rhizobia may act as a antagonist of inducer of nod genes of another rhizobial species. In *Bradyrhizobium*, diadzein, act as nod gene inducer while in *R. leguminosarum* bv. *trifolii* it acts as an antagonist (Djordjevic *et al.*, 1987).

Fix genes

Those rhizobial genes which are essential for nitrogen fixation but whose homologues are not present in *K. pneumoniae* are termed as fix genes. Fix genes are organized in distinct clusters which are different in different species of rhizobia. In *R. meliloti* fix genes are organized in five operons on megaplasmids (Sorbal *et al.*, 1991). Cluster I contain fix ABCX in one operon while cluster II contains fix LJ, fix K, fix NOQP, fix GHIS organized in five operons (David *et al.*, 1987). Fix genes are localized on chromosome in *B. japonicum* and *A. caulinodans*. In *B. japonicum*, fix genes are localized on all the four clusters. Fix BCX operon is located on cluster I while on cluster II, fix R and fix A operons are localized (Kindig *et al.*, 1993). Cluster III contains Fix LJ, Fix K2 and fix NOQP operons, while the cluster IV of *B. japonicum* contains fix K1 gene. In *A. caulinodans* fix ABCX are located as a single operon on cluster II while cluster IV contains two operons for fix LJ and fix K genes. In cluster V, two operons contain gene cluster fix NOQP and fix GHIS (Pawlowski *et al.*, 1987). In *R. leguminosarum*, fix genes identified are homologous to *R. meliloti* (Lismaa *et al.*, 1989).

Plant genes

Plant genes are involved in almost all aspects of symbiosis, which include number and size of nodules, nodule morphogenesis and morphology and the rate of nitrogen fixation activity. Plant genes which take part in symbiotic nitrogen fixation can be grouped into three categories.

- Genes whose expression is induced
- Genes whose expression is repressed and
- Constitutively expressed genes

While not much is known about the genes belonging to the last two categories, products of inducible genes or nodule specific host proteins are known as nodulins. These are not found in normal root tissue. Nodulins are divided into 2 groups (c-nodulins common to all legumes and s-nodulins present in specific species. Based upon their functions, nodulins are grouped into 3 categories.

- Nodulins responsible for nodule structure
- Nodulins responsible for enzymatic activities of nodule and
- Nodulins involved in nitrogen fixation

Nodulins are numbered either according to their molecular weight (*e.g.* nodulin 35 has 35 Kd molecular weight) or by their order of size.

Among, various nodulins, leghaemoglobin is the most well characterized nodulin. At present about 30 nodulins are known and functions of most of these are not yet understood. A detailed analysis of nodulins at different stages of development of effective and ineffective nodules might provide necessary information for understanding the molecular basis of symbiosis. The nodulins appear in parallel with leghaemaglobin during the development of nodules. Their induction occur prior to, and is independent of the development of nitrogenase activity in nodule tissue. Six nodulins appear in nodules at an early stage (10-12 d) while other nodulins including leghaemoglobin are first detected at an early stage and then the amount increases until the third week. In Soybean accumulation of nodulin mRNA reaches a steady state level 10 days after infection, by which time nitrogen fixation also commences. When ineffective nodules are compared with the effective nodules, the concentration of some nodulins shows a marked decrease, while other nodulins remain unchanged and some are stimulated. The nodulins whose concentration is very low in ineffective nodules are certainly involved in nitrogen fixation.

Plant genes (Nodulins)

In leguminous plants nodule organogenesis involve several of plant genes (Verma, 1992). Various plant genes which take part in nodule structure and function include genes which are constitutively expressed, genes whose expression is induced and genes whose expression is repressed. In recent years several plant genes whose expression is induced following infection have been characterized. All such genes are collectively known as nodulin genes. Nodulins were first identified in soybean root nodules (Legock and Verma, 1979). Nodulin polypeptides are named according to their molecular weight *e.g.* nodulin-35 is 35 K Da protein.

Based on the temporal sequence of induction, the nodulins have been grouped into two categories, the early nodulins and the late nodulins. The early nodulins are expressed during infection and nodule morphogenesis (Nap and Bisseling, 1990). The late nodulin genes are expressed at the later stages of nodule formation concomittantly with or following the release of bacteria from infection threads, just before or during N_2 fixation. These genes are involved mainly in different metabolic activities necessary for the functioning of the nodule.

Early nodulin genes

The first early nodulin gene characterized was isolated for Soyabean. This gene called ENOD 2 encodes for a protein with a putative N-terminal signal peptide followed by a sequence containing repeating proline-rich pentapeptide motifs (Franssen *et al.*, 1987).

Homologous ENOD 2 genes or transcripts have reported in other legume species such as pea, alfalfa, vetch, *Sesbania,* common bean and lupine (Witty *et al.*, 1986). Since the nodule inner cortex has been suggested as a physical barrier, limiting oxygen diffusion to central nodule tissue, the presence of ENOD 2 in this tissue indicate the role of this protein in maintaining the oxygen diffusion barrier in nodules (Van de Wiel *et al.*, 1990). The early nodulin ps ENOD 12 and ps ENOD 5 are also induced in Pea root hairs after inoculation with *Rhizobium* or after treatment with culture filterate (Gloudemans *et al.*, 1989). ENOD 12 was suggested to be involved in the infection process. Other nodulins associated with nodule organogenesis are Soyabean Gm ENOD 13 (Franssen *et al.*, 1988), Gm ENOD 70 (Kouchi and Hata, 1993) and a group of related products, Nps 40 from pea (Govers *et al.*, 1986), Nvs-40 from vetch (Moerman *et al.*, 1987) and Nms-30 from alfalfa (Dickstein *et al.*, 1988).

Late nodulins

Nodulins induced at the onset of nitrogen fixation are referred to as late nodulins (Nap and Bisseling, 1990). They include enzymes involved in nitrogen assimilation, carbon metabolism and amide and ureide biogenesis. Protein present in the peribacteriod membrane and the leghaemoglobins (Lbs) also belong to this group.

Leghaemoglobins (Lbs), the most abundant nodulins, function as oxygen carrier, facilitating O_2 diffusion to bacteroids (Appleby, 1984; Wittenberg and Wittenberg, 1990). Lbs are products of a multigene family where expression is strongly activated prior to nitrogen fixation. In Soyabean, four Lbs are induced at slightly different times during nodule formation suggesting a developmental control of these genes.

Nodulin genes in legume crop improvement

Plant products directly constitute over 90 per cent of the human diet and the animal products which constitute the remaining 10 per cent. With the application of genetics and plant pathology large increase in global food production, particularly in the cereal grains have been made possible (Borlaug, 1983). About 20 per cent of the world's food comes from legumes, the cultivation of which has increased by over 30 per cent per year for the last decade, and will probably continue to do so because of their nutritional value. Legumes constitute the third largest family of higher plants (Allen and Allen, 1981). Genetic potential of both host and microsymbiont are equally responsible for the establishment of fully effective symbiosis.

Major bacterial symbiotic genes are reasonably understood, a clear conception of what constitutes a plant symbiotic gene at the molecular level is only now emerging.

Nature of plant symbiotic genes

The formation of the root nodule requires coordinate expression of essential symbiotic genes in the host plant and the microsymbiont (Verma and Nadler, 1984; Rolfe and Shine, 1984). In the plant, these genes direct the synthesis of nodule-specific host proteins, 'nodulins' (Legocki and Verma, 1980) while in bacteria they direct invasion of the plant host and subsequent differentiation into a nitrogen-fixing bacteroid form. Nodulins are involved in structure, maintenance and overall metabolism of the root nodule. Several nodulins have been characterized from soyabean (Legocki and Verma, 1981), pea (Bisseling *et al.*, 1983) and alfalfa (Langunasch and Ausubel, 1985).

Phylogeny of nodulins

On the basis of evolutionary conservation, nodulins can be subdivided into proteins common to all legume nodules ('common' or c-nodulins) and 'species specific' or s-nodulins (Verma and Nadler, 1984). C-nodulins would be required for essential nodule functions. Leghaemoglobin is clearly a c-nodulin. Lb reductase and certain pbm nodulins are also likely to be common nodulins. Some enzymes for species specific carbon nitrogen metabolism in the nodule may be s-nodulins. Since the major nitrogen assimilation pathway varies between amide and ureide plants, each group may have a subset of its own 'c' and 's' nodulins. Soybean contains 4 major leghaemoglobins, Lba, Lbc1, Lbc2 and Lbc3, which differ slightly from each other in amino acid sequence.

Nodulin gene regulation

Some nodulins are apparently synthesized early in nodule morphogenesis, but the majority (mid-phase and late nodulins) are synthesized concurrently with leghaemoglobin (Bisseling *et al.*, 1983; Fuller and Verma, 1984). Messenger RNA s for nodulins and leghaemoglobins are specifically accumulated in incipient nodule tissue 3-5 days post infection and are fully induced by 11 days. Prior to the commencement of nitrogenase activity in bacteroids nodulin gene expression is differentially influenced by sym mutations in rhizobia (Fuller and Verma, 1984), emphasizing the interdependent expression of the 2 symbiotic genomes. Not all nodulin genes are likely to possess common regulatory sites, since their products are destined for different cellular and intracellular sites or functions.

Nif genes

The DNA sequence of nif genes is highly conserved in various diazotrophs. The use of nif HDK probe in *Klebsiella pneumoniae* has enabled the localization and characterization of nif genes in different *Rhizobium* species. The information available on the structure of nif genes in different species of *Rhizobium* and *Bradyrhizobium* is very limited and it does not permit the construction of any common nif map. In most of the fast-growing rhizobia such as *R. leguminosarum, R. phaseoli, R. trifolii* and *R. meliloti* nif genes are present on a plasmid while in *Brady Rhizobium* these are present in chromosomes. In *loti,* which is a fast-growing species nif genes are present on choromosome. Based on DNA hybridization and base sequence studies, various nif genes of rhizobia and *Bradyrhizobia* show strong homology of *Klebsiella* genes. Inspite of the structural homology of nif genes of rhizobia with *K. pneumoniae* some differences exist in their organisation. In *R. meliloti* and *R. leguminosarum,* nif H, D and K genes are continguous and are part of a single operon. On the contrary, in *B. japonicum* and *R. phaseoli* nif H and nif DK genes are located on different operons.

Organisation of nif genes in different species of rhizobia show some variation. In rhizobia, only 10 nif genes have so far been identified and linkage between nif genes in rhizobia is not as tight as in *K. pneumoniae*. In *R. meliloti*, the nif HDK genes are organized in an operon along with nif E, whereas in *A. caulinodans* nif HDK and nif E form two separate operons (Denefle *et al.*, 1987). Various nif genes are organized in one cluster in *R. meliloti* while in *B. japonicum* and *A. caulinodans* nif genes are located on two and three clusters respectively. There are also difference in the number of copies of some nif genes. In *B. japonicum*, only one nif H is present, in *R. leguminosarum* bv. *phaseoli*, three copies of nif H are present, while in *A. caulinodans* a second nif H with a slight difference in nucleotides is known (Aguilar *et al.*, 1990; North *et al.*, 1993; Quinto *et al.*, 1985). Various nif genes along with the fix genes are located on two megaplasmids in *R. meliloti* while they are present on the chromosome in *B. japonicum*.

Biological nitrogen fixation (BNF) accounts for 65% of the nitrogen (N) currently utilized in agriculture, and will be increasingly important in future crop productivity, especially for sustainable systems. Among nitrogen-fixing systems, the legume *Rhizobium* symbiosis has attracted much attention because of the specific interaction of the plant and bacterial partners.

This symbiotic process begins when the bacteria stimulate the plants to induce the development of root nodules (Long, 1989). A crucial step in the signal exchange is the induction of rhizobial nodulation genes (nod) by host-specific flavonoids. Although many flavonoids are present in different leguminous plants, specific flavonoids serve as the signal molecules in specific host legumes (Denaire *et al.*, 1992, Hopper *et al.*, 1995). Plant-derived flavonoids trigger the synthesis of rhizobial signal molecules by inducing *Rhizobium* nodulation gene(s) expression through a mechanism requiring an activator gene nod D (Mulligan and Long, 1985; Peters *et al.*, 1986). These signal molecules belong to a family of lipooligosaccharides (Nod factors) that acts as species-specific morphogenes on plants (Fisher and Long, 1992; Denaire and Cullimore, 1993; Downie, 1994). The nod factors cause root hair differentiation and branching (Lerouge *et al.*, 1990), cortical cell division and formation of nodules (Truchet *et al.*, 1991), depolarization of root hair membranes (Ehrhardt *et al.*, 1992), formation of preinfection thread structures (Van Brussel *et al.*, 1992) and bacterial entry into root hairs (Ardourel *et al.*, 1994).

Several plant genes are involved in the process of *Rhizobium*-associated N2 fixation (Horvath *et al.*, 1993). Some of the plant genes that are induced during the early stage of infection include genes such as *Rhizobium*-induced peroxidase (rip 1) (Coom *et al.*, 1995), and Enod 5 and Enod 12 that are correlated with nodule infection (Scheres *et al.*, 1990). Such genes are leghaemoglobin (Appleby, 1984) are expressed in the differentiated bacteroids. Plant mutant defective in nodulation or symbiotic N_2 fixation have been generated, but little is known about the structure or function of the mutated genes (Caetano-Anolles and Gresshoff, 1991).

The advent of molecular tools has provided some insight into the plants role in this partnership. Actual exploitation of the symbiosis using molecular tools has, however, been rather limited. Two approaches have been used to enhance N_2 fixation efficiency. The first has been through chemical, physical or transposon mutagenesis where better rhizobial strains or plants have been generated which show an increased N_2 fixation ability, and in some cases, in actual yield. The second has been to enhance N_2 fixing ability by genetic engineering. This has involved either transfer of symbiotic genes or over production of key proteins that have resulted in changes in the overall efficiency of N_2 fixation.

Genetic engineering of Rhizobia to increase efficiency of N_2 fixation

Traits that have been altered using the molecular tools of genetic engineering are improvement in competition enhancement of nodulation, alteration in surface polysaccharide production and host range, energy utilization, bacteriocin production and N_2 fixation. These are essentially carried out through construction of *Rhizobium* strains that carry additional copies of genes essential for nodulation (nod), N_2 fixation (nif, fix) and transfer of essential genes, as in the case of host range determination, competition, surface polysaccharide synthesis (exo), decarboxylate substrate transport (dct), or hydrogen uptake (hup). Genes presumed to be involved in competitiveness have been isolated from *B. japonicum* USDA 438 by substractive DNA hybridization from *B. japonicum* USDA 110 (Bhagwat and Keister, 1992).

Transfer of genes that affect host restriction

One direct molecular approach in improving a strains ability to compete and nodulate its host is by the transfer of genes involved in competition to other N_2 fixing strains. Transfer of genes involved in competitiveness from *B. japonicum* USDA 438 to a less competent strain, USDA 110, enhanced the competitive nodulation ability of the strain USDA 110 (Bhagwat and Keister, 1992).

A stable genetic transfer of genes involved in trifolitoxin production (tfx), which confers competitiveness to a symbiotically effective strain of *Rhizobium,* has been achieved (Triplett, 1990). Production of antirhizobial compounds by rhizobial inoculant strains could be a useful strategy in limiting nodulation by indigenous strains. *Rhizobium leguminosarum* bv. *trifolii* T 24 produce a potent antirhizobial compound, trifolitoxin (Schwinghamer and Belkengren, 1968). Tn 5 mutant of T 24 lacking trifolitoxin production lost their competitive ability. The trifolitoxin genes have been cloned and transferred to effective strains of *Rhizobium* by conjugation of a recombinant plasmid, p TFX1, which possess the genes for trifolitoxin production and resistance (Triplett, 1988). This was the first stable genetic transfer of genes involved in nodulation competitiveness to a symbiotically effective *Rhizobium* strain. The trifolitoxin-producing transconjugant occupied significantly more nodules than its wild type indicating that functional tfx gene does not improve nodulation competitiveness.

Increase of key metabolic enzymes and regulatory proteins to improve N_2 fixation efficiency

Manipulation of rhizobia in terms of introducing key genes as additional copies has been attempted with some success. Among the earlier efforts to transfer desired traits was the mobilization of DNA fragment containing nif genes to improve N_2-fixing ability to strains. *Rhizobium* strains genetically engineered to carry additional nif A genes enhanced yield in alfalfa in field studies (Cannon *et al.*, 1988). Nif A is a regulator of other nif operons in rhizobia. Improvement in competitiveness of strains was attempted by introduction of heterologous nod D gene, which might enhance nodule occupancy by responding to specific nodulation-inducing compounds of the host legumes (Maier and Brill, 1978).

Nitrogn fixation by the *Rhizobium* legume symbiosis requires a high level of energy. The host plant supplies bacteroids with photosynthate, which is converted to ATP and reducing power. The transport of carbon substrate is one of the limiting steps. Dicarboxylic acids succinate, fumarate and malate play an important role in symbiosis, and are the most effective substrates for bacteroids (Bergerson and Turner, 1967). Dicarboxylic acid transport-defective *Rhizobium* mutant form ineffective nodules on their respective hosts (Arwas *et al.*, 1985; Glenn and Brewin, 1981; Ronson *et al.*, 1981). *Bradyrhizobium japonicum* strain carrying the *R. meliloti* dct gene enhanced the dicarboxylic acid uptake, and N_2 fixation (Birkenhead *et al.*, 1988).

Another dicarboxylic acid transport gene, dct B, that adversely affects N_2 fixation, appears to lower nod D gene expression, and therefore the induction of nodulation genes in *R. leguminosarum* bv. *viciae* (Mavridou *et al.*, 1995).

Use of inducers of nod gene induction to enhance nodulation

In 'Hairy Peruvian' (HP) alfalfa plants, which assimilate more C and N, and also have more root nodules an increase in flavone signals was observed (Kapulnik *et al.*, 1987). The superior nodulation was associated with an increase in flavone signals. Addition of 10 mM luteolin to the rhizosphere of HP plants increased nodulation, N_2 fixation, total N and total dry mass. This suggested that flavonoids, which are known to induce transcription of nod ABC genes, can be used to enhance N_2 fixation. Recently, high and, low levels of nodulation in alfalfa cultivars have been correlated with concentration of the specific flavonoids secreted by their roots (Hernandez *et al.*, 1995).

Pretreatment of inoculant strains with the specific flavonoids or host root exudate will enable the rhizobia to produce the oligosaccharide nod factor in advance and thereby make them more competitive and result in better nodule occupancy by such strains. Preconditioning of rhizobial strains with host seeds or specific flavonoids resulted in several beneficial effects including significant enhance in nodulation and biomass in chickpea, groundnut and soybean. Commercial rhizobial inoculants, pretreated with flavonoids extracted from seed, retained the enhanced nodulation activity during storage upto 12 months (Thomas *et al.*, 1997).

Plant (host) genes

Comparatively little progress has been made in understanding the contribution of plant genes in *Rhizobium*-legume symbiosis (Govers *et al.*, 1987). Classical genetic studies have documented that plant genes play an important role in the process of symbiosis right from recognition to nodule senescence, (LaRue *et al.*, 1985). Although plant genetic control of symbiosis has been made by genetic studies, the molecular, physiological and biochemical manifestations of these genes, are, however, poorly understood.

Nodulin genes and Nomenclature

Table 1: Host plants genes affecting nodulation and N_2 fixation*

Species	Number of genes	Comments
Trifolium pratense L.	7	Naturally occurring; condition non-nodulation and ineffective nodulation. Nodules vary from early senescencing to tumour-like.
Pisum sativum L.	15	Naturally occurring and EMS mutagenesis; condition non-nodulation, ineffective, and supermodulation and nodulation in presence of NO_3. Some traits temperature sensitive.
Medicago sativa L.	7	Naturally occurring; condition non-nodulation and ineffective nodulation; nodules vary from early senescencing to tumour-like.
Glycine max (L.) Merr.	8	Naturally occurring and EMS mutagenesis; condition non-nodulation, ineffective and supernodulation, and nodulation in the presence of NO_3
Trifolium incarnatum L.	1	Naturally occurring; condition ineffective nodulation
Arachis hypogeae L.	1	Naturally occurring; conditions non-nodulation
Cicer arietinum L.	5	Derived by γ-irradiation non-nodulation reduced nodulation, and ineffective nodulation. Some traits temperature sensitive.
Vicia faba L.	1	Naturally occurring, condition ineffective nodulation.

*(Vance *et al.*, 1988)

Two major types of nodulin genes have been defined by their pattern of expression and function. Nodulins expressed before the beginning of N_2 fixation are called "early nodulins" (*Enod*). They are thought to participate in the early interaction between the bacteria and the plant (curling of root hairs, formation of infection threads), as well as being involved in nodule morphogenesis (cortical cell division, formation of the primordium and establishment of a meristem). Genes expressed just before or during N_2 fixation are called "late nodulins". These genes are involved mainly in the different metabolic activities necessary for the functioning (N_2 fixation) of the nodule. The nodulin genes have also been classified as Class I and Class II (Govers *et al.*, 1987), and Class A and Class B (Gloudemans *et al.*, 1987) to account for the apparent difference in timing of expression. The term "early nodulins" is frequently used for Class I/Class A nodulins and late nodulins for Class II/Class B nodulins.

Early nodulins

Shortly after infection, in roots or root hairs, a set of plant genes known as early nodulin genes is specifically activated by the presence of symbiotic bacteria. Some of them are induced by *Nod* factor (like *Enod 12*) whereas others are activated at later stages (*e.g.*

Enod 2). The early nodulins so far identified are shown in (Table 2). Many of the early nodulins characterized seem to be structural cell wall proteins. Putative nodulins is participating in a signaling cascade are most likely of low abundance.

Table 2: Early Nodulin Genes

Plant species	Localization	No. of genes	Genes	Function	References
Medicago sativa	Empty nodules Root hairs infected/ uninfected cells	7	*Nims30* *MsEnod8* *MsEnod10*	Nodule organo genesis Infection	Deckstein *et al.*, 1993 Lobler & Hirsch, 1993
			MsEnod12		Allison *et al.*, 1993
			MsEnod12A	Infection	Bauer *et al.*, 1994
			Enod12B		
			Enod40		Schultze *et al.*, 1994
Medicago truncatula	Nodules	1	*MtEnod12*		Schultze *et al.*, 1994
Gylcine max	Root hairs infected cells	6	*GmEnod2* *GmEnod40*	Nodule morphogenesis	Rossbach *et al.*, 1989 Yang *et al.*, 1993
			GmEnod55	and Nutrition	De Blank *et al.*, 1993
			GmEnod70		Kouchi and Hata, 1993
			GmEnod93		
			GmEnod315		
Pisum sativum	Symbiotic zone infected cells Root hairs/cortical cells containing infection thread	5	*PsEnod3* *PsEnod5* *PsEnod12* *PsEnod14*	Nodule morphogenesis and infection	Scheres *et al.*, 1990

Enod 12

The *Enod12* protein sequence is composed of a putative signal peptide followed by proline-rich pentapeptide repeats that presumably represent a cell wall protein. *Enod12* was suggested to be involved in infection process (Bauer *et al.*, 1994). *MsEnod12A* and *MsEnod12B* are two early nodulin genes from alfalfa (*Medicago sativa*) and they were differentially expressed (Allison *et al.*, 1993). *MsEnod12A* RNA was detected only in nodules and not in other plant tissues. In contrast, *MsEnod12B* transcripts were found in nodules and also at low levels in roots, flowers, stems and leaves. *MsEnod12B* expression was enhanced in the root early after inoculation with the microsymbiont *Rhizobium meliloti* (Csanadi *et al.*, 1994) and after treatment with purified *Nod* factors. Alfalfa possess different mechanisms regulating *MsEnod12A* and *MsEnod12B* expression. *In situ* hybridization showed that in nodules, *MsEnod12* expression occurred in the infection zone. In empty Fix nodules, the *MsEnod12A* transcript level was much reduced and in spontaneous nodules it was not detectable. In *Pisum sativum* (Scheres *et al.*, 1990) and *M. sativa* (Allison *et al.*, 1993) two gene copies (A and B) of *Enod12A* were identified while *M. truncatula* contains only a single copy. *MsEnod12A* and *MsEnod12B* are differentially expressed whereas the spatial and temporal expression of two *Enod12* copies are identical in pea (Scheres *et al.*, 1990). *MsEnod12B* expression in roots is related to the action of *Nod* factors, whereas *MsEnod12A* expression is associated with the invasion process in nodules. Mutant plants lacking *Enod12* genes are capable of forming normal nitrogen-fixing nodules in hybrid progenies of *M. cocrulea* and *M. quasifalcata,* which indicates the *Enod12* is not essential for nodule organogenesis and functioning (Csanadi *et al.*, 1994). *Enod10,* another early nodulin gene, *MsEnod10,* highly homologous to *Enod12A* and *Enod12B* in the putative signal peptide sequence is expressed in the invasion zone of the alfalfa nodule (Lobler and Hirsch, 1993).

Enod 5

Among the pea early nodulins, *PsEnod5* and *PsEnod12* are important. *PsEnod5* encodes a 14kDa protein containing a putative N-terminal signal peptide and does not show homology to other proteins. *Enod5* is induced by *Nod* factors in pea root hairs but with different kinetics than *Enod12* (Horvath *et al.*, 1993). *PsEnod5* is expressed in the invasion zone at a low level directly adjacent to the apical meristem of nodules whereas the level of expression is high in the interzone II-III, where bacterial proliferation and enlargement of the infected cells occur. However, it is not expressed in the symbiotic zone. Contrary to *Enod12, Enod5* expression is restricted to cells containing the infection thread and is not detectable in the noninfected primordium.

Enod 40

Enod40 has been cloned from soybean (Kouchi and Hata, 1993) peas (Yang *et al.*, 1993) and alfalfa (Hirsch *et al.*, 1993). During the early stages of infection, the *GmEnod40* gene is induced in the dividing cortical cells as well as in the pericycle cells opposite to the nodule primordia. At a later stage, it is highly expressed in the pericycle of the vascular tissue of soybean nodules, and at a low level in uninfected cells. The expression pattern of *Enod40* is similar in pea, alfalfa and soybean during nodule development. *Enod40* expression in the nodule vascular bundles is infection---dependent and its expression in nodule primordia by a *BradyRhizobium* or by purified *Nod* factors is independent of infection (Franssen *et al.*, 1993). Thus the induction of *Enod40* can be triggered by two independent mechanisms, resulting in the expression of the gene at different times and locations during morphogenesis. The presence of *Rhizobium* causes an enhancement of the genes expression and also induction of the gene in the vascular tissues of developed lateral roots. *Enod40* could serve as a tool in gaining access to mechanisms involved in the determination of other developmental processess in the plant in addition to its role in nodule development (Papadopoulou *et al.*, 1996).

Enod 2

The early nodulin gene that has been studied in most detail till date is *Enod2.* Gloudemans *et al.* (1987) identified an early nodulin, Ngm-75 from soybean (*Glycine max*) by 2D-PAGE analysis of *in vitro* translation products using nodule and root RNA as templates.

This gene appears to encode two 75-kDa structural hydroxyproline rich glycoproteins (HPRG) and is required in the nodule morphogenesis. Hydroxyproline rich glycoproteins are involved in cell wall growth and development, disease-resistance root hair infection and nodule development. Cassab (1986) found high concentrations of hydroxyproline in soybean nodule cortical cell walls and in soluble arabino-galactan proteins in infected cells. The nodule specific arabino-galactan-HPRG was suggested to be involved in the initial plant-microbe interaction and its subsequent maintenance. Nodulins rich in proline are involved in early development of soybean nodules.

Enod2 was isolated from soybeans (Francine *et al.*, 1987), peas (Van de Wiel *et al.*, 1990), *Sesbania rostrata* (Strittmatter *et al.*, 1989), alfalfa (Dickstein *et al.*, 1993), and lupin (Szezyglowski *et al.*, 1989); *Enod2* is a late early nodulin expressed much later than *Enod12* or *Enod40*. It is expressed at the base of the nodule and in the parenchyma (inner cortex) cells. It also contributes to the formation of an O_2 barrier providing a microaerobic condition inside the nodule for the function of the oxygen-sensitive nitrogenase. In roots, but not in stems of *S. rostrata*, the *SrEnod2* gene is inducible by cytokinin which may be involved in the signal transduction pathway required for *Enod2* induction. *Enod2* is expressed independently of infection and N_2 fixation since it is detected in empty nodules, and in spontaneous nodules (Schultze *et al.*, 1994).

Other early nodulin genes

MsEnod8, another early nodulin gene coding for a 42kDa protein with sequence similarity to anther-specific proteins, displayed the same time course of expression as *Enod2* and was also expressed in empty nodules. The role of *Enod8* gene product is in nodule organogenesis and is expressed in a developmental pathway triggered as a result of *Rhizobium* nodulation signal transduction.

PsEnod3 and *PsEnod14* genes are members of the same gene family. They share extensive homology and are expressed in a similar way. *PsEnod3* encodes an 8kDa metal-binding protein with four cysteine residues arranged in two pairs without characteristics of DNA-binding proteins. Its expression is restricted to the symbiotic zone in cell layers where nitrogen fixation has not begun yet. Moreover, *PsEnod3* transcript is detectable only in infected cells.

GmEnod55, an early nodulin from soybean encodes a protein containing a domain rich in proline and serine flanked by hydrophobic, possibly membrane-spanning regions (De Blank *et al.*, 1993). It is expressed only in infected cells like *PsEnod5* to which it shows weak homology in the N-terminal part and is expressed later than *GmEnod2* but before maximum expression of leghaemoglobin.

GmEnod70 encodes a transmembrane protein weight a mol. weight of 53kDa containing atleast ten hydrophobic domains. It was found to be involved in cell nutrition (Kouchi and Hata, 1993). *GmEnod93* gene (Kouchi and Hata, 1993) codes for a putative 11kDa protein rich in alanine and serine residues and it does not exhibit homology to any known sequences. *GmEnod93* transcripts are first detectable in the dividing cell cluster of the outer cortex adjacent to the infected site. In emerging nodules, the expression was localized in the meristematic region, and in mature nodules, it was detected in the infected cells. Since the expression of the gene continues to increase after the onset of the nitrogen fixation, it is most likely limited to cells invaded by bacteria even in the earliest stages of infection.

GmEnod315, another early nodulin gene from soybean apart from showing homology to *PsEnod5*, it also resembles *GmEnod55* in having a domain containing proline-serine alternating residues and is expressed in infected cells (Kouchi and Hata, 1993). Three of the twelve characterized *Rhizobium*-induced, root-hair proteins of cowpea (*Vigna unguiculata*) are transiently expressed and might participate in the very early steps of infection or recognition or root-hair deformation. One of them seems to be host specific (Krause and Broughton, 1992).

Late nodulins

Genes expressed just before or during N2 fixation are called 'late nodulins'. Late nodulins include leghaemoglobin (Lb) glutamine synthetase (GS), uricase, sucrose synthase and few other enzymes such as phosphoenol pyruvate carboxylase, xanthine dehydrogenase, purine nucleosidase, aspartate aminotransferase, glutamate synthase and malate dehydrogenase. The well-known example of late nodulin is the best studied leghaemoglobin. Most late nodulins are involved in the nodule metabolism.

Leghaemoglobin

Leghaemoglobin (Lb), the protein which gives nodules their pink color, is the most well characterized nodulin. It is found in the cytoplasm of infected cells and is involved in maintaining low O_2 concentration in nodules (Morrison *et al.*, 1987). The globin portion of Lb is synthesized by plant genes and may comprise up to 30% of the total nodule soluble protein. Multiple forms of Lb, synthesized from different messages, are found in pea and soybean. In all legumes studied to date, more than one leghaemoglobin is found in the nodules and the leghaemoglobins are encoded by more than one gene.

Detailed analysis of the structure of a soybean Lb gene shows the aminoacid coding region to be interrupted by two intervening sequences at exactly the same positions found for animal globin genes. Besides the four genes responsible for the four major soybean Lb species a, C_1, C_2, C_3, there have been recognized one or more Lb pseudogenes which do not appear to be transcribed and are arranged in two independent clusters in the soybean genome. Several upstream enhancer regions necessary for "nodule-specific" expression of leghaemoglobin gene were identified (de Bruijn *et al.*, 1989).

Glutamine synthetase

Glutamine synthetase (GS) is an octomeric enzyme which catalyses the first reaction in the assimilation of ammonia into aminoacids in nodules via the GS-GOGAT pathway (Hirel *et al.*, 1987). In nodules, GS activity is increased about 14 fold relative to roots and represents about 2% of the nodule-soluble protein. Various GS isoenzymes are found in different organs and cell compartment of plants. The GS isoenzymes of leguminous plants are related immunologically with GS isoenzymes of other plants, but not with bacterial or mammalian glutamine synthetase (Cullimore and Miflin, 1984). Different GS subunits have been identified, ranging from 41 to 45kDa in molecular weight, only one of these is found primarily in the nodules. Low levels of this polypeptide is also detectable in other plant organs, but not in roots (Bennett *et al.*, 1989).

Two forms of GS are found in bean nodules, one of which is nodule-specific (GSn1) and one identical to that in roots (GSn2) (Lara *et al.*, 1983). The amount of GSn1, but not GSn2, mRNA increases following *Rhizobium* infection. In contrast to bean nodule GS, pea nodule GS is comprised of one 38kDa, three 37kDa and one 44kDa polypeptides (Tingey *et al.*, 1987). The 38kDa polypeptide is the predominant form in roots while the 37 and 44kDa polypeptides predominate in nodules and chloroplasts, respectively. Though all forms of GS can be detected in pea nodules, but only the 37kDa forms are selectively enhanced. Similarly mRNAs encoding the 37kDa GS polypeptides increased 10 to 22 fold during nodule formation.

Uricase

Another late nodulin, uricase is a key enzyme in the ureide biosynthetic pathway in exporting the reduced nitrogen from nodules in the form of ureides (Nguyen *et al.*, 1985), it was the second abundant protein in soybean nodule cytosol and found at low levels in roots and callus tissue, a nodule specific form of uricase (Uricase II) increases in soybean during nodulation while uricase I, the root form, decreases. A subunit of uricase II, initially identified as N-35, has been localized to peroxisomes of uninfected soybean nodule cells (Bergmann *et al.*, 1983). The expression of nodule-specific uricase is related to the differentiation of uninfected interestitial cells in nodules and not to the synthesis of ureides.

Sucrose synthase (SUCS)

Sucrose synthase (UDP-glucose: D-fructose-2-glucosyl transferase, EC 2.4. 1.13) is a homotetrameric enzyme, consisting of four subunits with a mol.w.t. of 92-93 kDa. In root nodules, it catalyses the cleavage of sucrose to D-fructose and UDP-glucose, leading to the formation of dicarboxylic acids. These dicarboxylic acids serve as a source of energy for the bacteroids carrying out the symbiotic process of N_2 fixation. The UDP glucose is used in cell wall and starch synthesis.

This gene is strongly expressed in nodules and 10 fold lower expression in uninfected root, hypocotyl, stem, flower tissues and very trace amounts in leaf tissues (Kuster *et al.*, 1993). The highly expressed SUCS in alfalfa nodules is antigenically related to that in roots (Miller *et al.*, 1987).

Other enzymes

There are several other late nodulins which take part in the nodule metabolism. Several of these enzymes have substantially higher activity in nodules in comparison of roots. Nodule-specific forms of enzymes that differ in the physical, kinetic and immunochemical properties from the corresponding enzymes in roots have been found for phosphoenolpyruvate carboxylase, choline kinase, xanthine dehydrogenase, purine nucleosidase, asparate amino transferase, glutamate synthase, and malate dehydrogenase. These may be due to either expression of nodule-specific genes or the nodule-specific modifications of constitutively expressed enzymes (Nap and Bisseling, 1990). As nodules senesce, cytoplasm and bacteroids lyse, nodule proteolytic enzyme activity increases substantially in concert with the lysis of cytoplasm and bacteriods which indicates that some nodulins during the final stages of nodule development may be proteolytic enzymes.

Nodulins in the peribacteroid membrane

The peribacteroid membrane separates the bacteroid from the rest of the nodule mass and is the physical and metabolic interface between the *Rhizobium* and its eukaryotic partner (Robertson *et al.*, 1984). A nodulin from soybean designated Ngm-24 is associated with PBM and does not function as a transmembrane transport molecule (Fortin *et al.*, 1987). Unlike Ngm-24, Ngm-26, another nodulin from soybean PBM acts as a transmembrane peribacteroid membrane nodulin in the exchange of metabolites between *Rhizobium* and the plant (Cheon *et al.*, 1994).

Nodulin gene expression and regulation

Immunological approach

Antibodies raised to total nodule soluble protein extracts of soybean (Legocki and Verma, 1980), pea and alfalfa (Vance *et al.*, 1988) have been used in conjunction with western blotting to identify some 25 nodulins. In soybean, 18-20 host-specific polypeptides other than leghaemoglobin are reported and all the precipitated nodulins had low mol. wt of 12-20 kDa which may be due to proteolysis during immunoprecipitation. About 30 different nodule-specific proteins in the mol. wt. range of 15-120 kDa were detected by nodule-specific antiserum in pea root nodules (Bisseling *et al.*, 1983) and 9 nodule-specific polypeptides in addition to leghaemoglobin in alfalfa by immunological approach.

The main disadvantages of immunological approach in gene expression are:

- The antisera used in these studies were generally developed against total soluble cytoplasmic proteins from nodules and subsequently made nodule-specific. They were unable to recognize nodule-specific membrane proteins.
- The inherent difficulty is that the staining intensity of a western blot, when probed with antibodies prepared to protein mixtures, may not reflect the *in vivo* concentration of a polypeptide due to differences in antigenicity between proteins.
- The total nodule protein used as the immunogen is not conducive to ascribe a function to individual proteins identified as nodulins.

***In vitro* translation studies**

The *in vitro* translation approach gives an overview of the most abundantly transcribed genes and more directly reflects the expression of plant genes. Moreover, it requires only very low amounts of RNA (1-2 μg). In alfalfa, analysis of *in vitro* translation products of nodule mRNAs showed that at least 25 plant genes encoding polypeptides ranging from 14-60kDa are expressed in mature nodules that are not detectable in roots. In addition, 9 genes expressed in roots appear to be reduced or not expressed in nodules. The expression of 17 nodulins including five leghaemoglobins is independent of nitrogen fixing ability of the nodules (Norris *et al.*, 1988).

In pea nodules, out of the 21 plant genes expressed, 4 were found to be leghaemoglobins by *in vitro* translation studies. In pea (Govers *et al.*, 1985), soybean (Gloudemans *et al.*, 1987) and alfalfa (Lullien *et al.*, 1987), nodule development is mediated by the differential expression of specific nodule and root genes.

This approach has also some disadvantages even though it is a more direct reflection of plant gene expression.

- Unstable RNAs or RNAs that are translated poorly *in vitro* are not detected.
- Isolation and translation of RNAs encoding high mol. wt. polypeptides is difficult (> 100 kDa).
- Some proteins, such as soybean Lb, lack methionine and would not be visible on fluorograms even though 35 S-methionine used *in vivo* translation is highly reactive. Therefore translations should be run with more than one radio active amino acid.
- *In vitro* translation does not allow identification of polypeptide function.

Nodulin cDNA clones

The third approach to study nodulin gene expression is by cloning the complimentary DNA of the RNA species specifically expressed in the nodules. These cloned sequences can be used in many ways *e.g.* in Northern and Southern blots, *in situ* hybridization, sequence analysis and comparison with other known genes. This approach has been used successfully with soybean (Sengupta-Gopalan *et al.*, 1986) and pea (Govers *et al.*, 1986). Nodulin cDNA clones so far used and sequenced are shown in (Table 3). In soybean, cDNA cloning resulted in the isolation and characterization of 10 nodule-specific host plant genes. Four of these genes code for Lbs. Gene N-35 codes for nodule-specific uricase and sucrose synthase is coded by N-100. The function of other soybean nodulin genes (N-23, N-24, N-27 and N-44) is unknown.

Nine cDNA clones that encode nodulins of 11-42 kDa were prepared to soybean nodule poly A^+ RNA (Sengupta-Gopalan *et al.*, 1986). Clones C51 and E27 have characteristics typical of membrane proteins and function has not been established for any of these genes. In broadbean, fourteen clones in nodule cDNA library were found to be nodule-specific. Five broadbean cross-hybridizing cDNAs showed partial homologies to the alfalfa nodulin 25 (Kuster *et al.*, 1994).

In pea, five nodule-specific cDNA clones designated PsNod6, 10, 11, 13 and 14 as well as one Lb clone pPsLb 102, coding for nodulins ranging in size from 5.8 to 19 kDa have been isolated and characterized (Govers *et al.*, 1987).

The soybean and pea cDNA clones were used to evaluate gene expression in ineffective nodules and characterize the developmental expression of nodulin genes.

The major advantage of the cDNA cloning strategy is that it can be used to evaluate directly gene structure, regulation and expression. Knowledge of the coding sequence of nodulin genes can be i) used to synthesize corresponding synthetic polypeptides which in turn can be used to produce antibodies for use in cellular localization studies. ii) the nucleotide sequence of the cloned gene can be used to compare sequence of other known genes to predict their location and function.

Table 3: Nodulin cDNA clones

Plant type	No. of cDNA	Name and type of clones	Function	Reference
Soybean (*Glycine max*)	10 nodule specific plant genes	4 Lb genes N-35 (Uricase) N-100 (Sucrose synthase) N-23 N-24 N-44 N-27a	O_2 concentration in nodules (Lb alone)	Vance *et al.*, 1988
Soybean (*Glycine max*)	9 clones	*pEnod2*	Structural nodulin	Franssen *et al.*, 1987
		Nodulins of 11-24 kDa	Nodule morphogenesis	Sengupta Gopalam *et al.*, 1986
		Clones C51 E27	Membrane proteins	
Pea (*Pisum sativum*)	5 nodule specific cDNA clones	pPsNod 6,10,11,13 and 14 (Nodulins of 5.8-19 kDa	Unknown	
	Lb clone	pPs Lb 102-leghaemoglobin	Maintenance of O_2 concentration in nodules	Govers *et al.*, 1987
Broadbean (*Vicia faba*)	28 cDNA clones 14-nodule specific	–	–	Kuster *et al.*, 1994
French bean (*Phaseolus vulgaris*)	Glutamine synthase cDNA clone 2 nGs genes	–	–	Cullimore and Miflin, 1984

Physiological and other conditions responsible for nodule specific gene expression

Oxygen, ammonium, heme and phytohormones are thought to play an important role in nodulin gene expression. Since extreme low concentration of oxygen is essential inside the nodule for the oxygen labile nitrogenase to function, it has been thought to play an important role. Auxins and cytokinins have been implicated in nodule meristem development and nodule growth. A plasmid causing constitutive synthesis of cytokinin and zeatin allowed a nod-mutant of *R. meliloti* to form primitive nodules on alfalfa suggesting the role of phytohormone balance in nodulin gene expression (Long and Cooper, 1988). *Enod2* gene from *Sesbania rostrata* is found to be induced by cytokinin (Dehio and de Bruijn, 1992).

Rhizobium signals are also very important in nodulin gene expression. Low mol. wt. factors produced by the rhizobia are important in causing initial change in plants. The nodulin genes cloned so far have provided us with the molecular tools to study the different steps of nodule organogenesis, but no functional role can be umambiguously attributed to anyone. In different legumes, the nodulins are expressed differentially even though some of these are conserved in most of the species. However, it is not clear why this symbiosis is only limited to legumes, *Parasponia* being an exception. Different groups are trying to nodulate cereals like rice and wheat (Cocking, 1995) rice, sugarcane, maize, *Sorghum* (Kannaiyan and Buvana, 1998) and oilseeds like rape and mustard (Moffat, 1990) though they have been able to develop small nodule like structure, actual nitrogen fixation in these cereal needs detailed investigation. Recently, Kannaiyan (1998) has shown the nodule like structures induced in rice, corn, *Sorghum* and sugarcane due to the inoculation of free-living *Azorhizobium caulinodans,* the microsymbiont of *Sesbania rostrata.* To be able to achieve this goal, the involvement of host genes and their expression pattern will have to be understood in detail. While molecular genetic, studies identified numerous bacterial genes and their functions, that were shown to be required for symbiotic nodule development, in the plant partner, progress on the identification of genes with functions necessary for symbiotic nodule development is still limited. Inspite of the number of nodulin genes identified recently, the function of most of the encoded gene products is not well understood which needs to be studied extensively.

References

Allen, O.N. and Allen, E.K. 1981. In: *The Leguminosae.* A Book of Characteristics, Uses and Nodulation. University of Wisconsin Press.

Allison, L.A., Kiss, G.B., Baner, P., Poiret, M., Pierre, M., Savoure, A., Kondorosi, E. and Kondorosi, A. 1993. Identification of alfalfa early nodulin genes with homology to members of the pea *Enod12* gene family. *Plant Mol. Biol.,* **11**: 375-380.

Appleby, C.A. 1984. Leghaemoglobin and *Rhizobium* respiration. *Ann. Rev. of Plant Physiol.,* **35**: 443-478.

Ardourel, M., Demont, N., Debelle, F., Maillet, F., Billy, F.D., Prome, J.C., Denaire, J. and Truchet, G. 1994. *Rhizobium meliloti* lipo-oligosaccharide nodulation factors: different structural requirements for bacterial entry into target root hair cells and induction of plant symbiotic developmental responses. *The Plant Cell,* **6**: 1357-1374.

Arwas, R., Mckay, I.A., Rowney, F.R.P., Dilworth, M.J. and Glen, A.R. 1985. Properties of organic acid utilization mutants of *Rhizobium leguminosarum* strain 300. *Journal of General Microbiology,* **131**: 2059-2066.

Baev, N., Schultze, M., Barlier, I., Ha. Cann, Ha.D., Virelizier, H., Kondorosi, E. and Kondorosi, A. 1992. *Rhizobium* nod M and nod N genes are common nod genes: nod M encodes functions for efficiency of Nod signal production and bacteroid maturation. *J. Bacteriol.,* **1174**: 7555-7565.

Banfalvi, Z., Nieuwkoop, A., Shell, M., Besl, L. and Stacey, G. 1988. Regulation of nod gene expression in *Bradyrhizobium japonicum. Mol. Gen. Genet.,* **214**: 420-424.

Bauer, P., Crespi, M.D., Szecsi, J., Allison, L.A. and Schultze, M. 1994. Alfalfa *Enod12* genes are differentially regulated during nodule development by *Nod* factors and *Rhizobium* invasion. *Plant Physiol.,* **105**: 585-592.

Bennett, M.J., Lightfoot, D.A. and Cullimore, J.V. 1989. cDNA sequences and differential expression of gene encoding the glutamine synthetase polypeptide of *Phaseolus vulgaris* L. *Plant Mol. Biol.,* **12**: 553-565.

Bergmann, H., Preddie, E. and Verma, D.P.S. 1983. Nodulin-35, a subunit of specific uricase (Uricase II) induced and localised in the uninfected cells of soybean nodules. *EMBO. J.,* **2**: 2333-2339.

Bhagwat, A.A. and Keister, D.L. 1992. Identification and cloning of *Bradyrhiozbium japonicum* mutant. *Applied and Environ. Microbiol.,* **57**: 3496-3501.

Birkenhead, K., Manian, S.S. and Gara, F.O. 1988. Dicarboxylic acid transport in *Bradyrhizobium japonicum*: Use of *Rhizobium meliloti* dct gene(s) to enhance nitrogen fixation. *Journal of Bacteriol.,* **170**: 184-189.

Bisseling, H., Preddie, E. and Verma, D.P.S. 1983. Nodulin-35: A subunit of specific uricase (uricase II) induced and localized in the uninfected cells of soybean nodules. *EMBO, J.,* **2**: 2333-2339.

Bisseling, T.C., Been, C., Klugkist, J., van Kammen, A. and Nadler, K. 1983. Nodule-specific host proteins in effective and ineffective root nodules of *Pisum sativum. EMBO, J.,* **2**: 961-966.

Borlaug, N.E. 1983. Contributions of conventional plant breeding to food production. *Science,* **219**: 589-593.

Brewin, N.J. 1991. Development of the legume root nodule. *Annu. Rev. Cell Biol.,* 7: 191-226.

Caetano-Anolles, G. and Gresshoff, P.M. 1991. Plant genetic control of nodulation. *Ann. Rev. Microbiol.,* **45**: 345-382.

Cannon, F., Beynon, J., Hankinson, T., Kwiatkowski, R., Legocki, R.P., Ratcliffe, H., Ronson, C., Szeto, W. and Williams, M. 1988. Increasing biological nitrogen fixation by genetic manipulation. In: *Nitrogen Fixation hundred years after* (Bothe, H., de Bruijn, F.J. and Newton, W.E. eds.). Stuttgart, Germany: Gustav Fisher, pp. 735-740.

Cassab, C.I. 1986. Arabinogalactan proteins during the development of soybean root nodules. *Planta,* **168**: 441-446.

Cervantes, E., Sharma, S.B., Maillet, F., Vasse, J., Truchet, G. and Rosenberg, C. 1989. The *Rhizobium meliloti* host range nod Q gene encodes a protein which share homology with translation elongation and initiation factors. *Mol. Microbiol.,* **3**: 745-755.

Cheon, C.I., Hong, Z. and Verma, D.P.S. 1994. Nodulin-24 follows a novel pathway for integration into the peribacteroid membrane in soybean root nodules. *J. Biol. Chem.,* **269**: 6598-6602.

Cocking, E.C. 1995. Interaction of rhizobia from legume stem nodules with rice and wheat for symbiotic nitrogen fixation Report. First working group meeting of the Frontier Project on Nitrogen Fixation of Rice. International Rice Reserch Institute, Mnil, Philippines, 2-4 My, 1995.

Cook, D., Dreyer, D., Bonnet, D., Howell, M., Nony, E. and Vnden Bosch, K. 1995. Trasient induction of a peroxidse gene in *Medicgo trunctul* precedes infection by *Rhizobium meliloti. The Plant Cell,* **7**: 43-55.

Csanadi, G., Szeesi, J., Kato, P., Kiss, P., Endre, G., Kondorosi, A., Kondorosi, E. and Kiss, G.B. 1994. *ENOD12,* an early nodulin gene, is not required for nodule formation and efficient nitrogen fixation in alfalfa. *The Plant Cell.,* **6**: 201-213.

Cullimore, J.V. and Miflin, B.J. 1983. Glutamine synthetase from the plant fraction of *Phaseolus* root nodules: Purification of the mRNA and *in vitro* synthesis of the enzyme. *Fed. Eur. Biochem. Soc. Lett.,* **158**: 107-112.

Cullimore, J.V. and Miflin, B.J. 1984. Immunological studies on glutamine synthetase using anti-sera raised to the two plant forms of the enzyme from *Phaseolus* root nodules. *J. Exp. Bot.,* **35**: 581-587.

Dakora, F.D., Joseph, C.M. and Phillips, D.A. 1993. Common bean root exudates contain elevated levels of diadzein and coumestrol in response to *Rhizobium* inculation. *Mol. Plant-Microbe Interac.,* **6**: 665-668.

Daora, F.D., Joseph, C.M. and Phillips, D.A. 1993. Alfalfa (*Medicago sativa* L.) root exudates contain isoflavonoids in the presence of *Rhizobium meliloti. Plant Physiol.,* **101**: 819-824.

David, M., Dommergue, O., Pognonec, P. and Kahn, D. 1987. Transcription patterns of *Rhizobium meliloti* symbiotic plasmid pSym: identification of nifA-independent fix genes. *J. Bacteriol.,* **169**: 2239-2244.

De Blank, C., Mylona, P., Yang, W.C., Katinakis, P., Bisseling, T. and Franssen, H. 1993. Characterization of the soybean early-nodulin eDNA clone *GmEnod55. Plant Mol. Biol.,* **22**: 1167-1171.

de Bruijn, F.J., Fellx, G., Grunenberg, B., Hoffmann, H.J., Metz, B., Ratet, P., Simons-Schreier, A., Szabados, L., Welters, P. and Schell, J. 1989. Regulation of plant genes specifically induced in nitrogen-fixing nodules. Role of cis-acting elements and trans-acting factors in leghaemoglobin gene expression. *Plant Mol. Biol.*, **13**: 319-325.

Debelle, F., Rosenberg, C., Vasse, J., Maillet, F., Martinez, E., Denaire, J. and Truchet, G. 1986. Assignment of symbiotic developmental phenotypes to common and specific nodulation (nod) genetic loci of *Rhizobium meliloti. J. Bacteriol.*, **168**: 1075-1086.

Dehio, C. and de Bruijn, F.J. 1992. The early nodulin gene *SrEnod2* from *Sesbania rostrata* is inducible by cytokinin. *Plant J.*, **2**: 117-128.

Denaire, J. and Cullimore, J. 1983. Lipo-oligosaccharide nodulation factors: a new class of signalling molecules recognition and morphogenesis. *Cell.*, **74**: 951-954.

Denaire, J., Debelle, F. and Rosenburg, C. 1992. Signalling and host range variation in nodulation. *Ann. Rev. of Microbiol.*, **46**: 497-531.

Denefle, P., Kush, A., Norel, F., Paquelin, A. and Elmerich, C. 1987. Biochemical and genetic analysis of the nif HDKE regiom of *Rhizobium* ORS 571. *Mol. Gen. Genet.*, **207**: 280-287.

Dickstein, R., Prusty, R., Peng, T., Ngo, W. and Smith, M.E. 1993. *Enod8*, a novel early nodule-specific gene, is expressed in empty alfalfa nodules. *Mol. Plant-Microbe Interact.*, **6**: 715-721.

Djoprdjevic, M.A., Redmond, J.W., Batley, M. and Rolfe, B.G. 1985. Clovers secrete specific phenolic compounds which either stimulate or repress nod gene expression in *Rhizobium trifolii. EMBO, J.*, **6**: 1173-1179.

Djordjevic, M.A., Schofield, P.R. and Rolfe, B.G. 1985. Tn 5 mutagenesis of *Rhizobium trifolii* host-specific nodulation genes result in mutants with altered host-range ability. *Mol. Gen. Genet.*, **200**: 463-471.

Downie, J.A. 1994. Signalling strategies for nodulation of legumes by rhizobia. *Trends in Microbiol.*, **2**: 318-324.

Egelhoff, T.T. and Long, S.R. 1985. *Rhizobium meliloti* nodulation genes: Identification of nod ABC gene products, purification of nod A protein, and expression of nod A in *Rhizobium meliloti. J. Bacteriol.*, **164**: 591-599.

Ehrhardt, D.W., Atkinson, E.M. and Long, S.R. 1992. Depolarization of alfalfa root hair membrane by *Rhizobium meliloti* nod factors. *Science*, **256**: 998-1000.

Faucher, C., Camut, S., Denaire, J. and Truchet, G. 1989. The nod H and nod Q host range genes of *Rhizobium meliloti* behave as a virulence genes in *R. leguminosarum* bv. *viciae* and determines changes in the production of plant-specific extracellular signals. *Mol. Plant-Microbe Interac.*, **2**: 291-300.

Firmin, J.L., Wilson, K.E., Rossen, L. and Johnston, A.W.B. 1986. Flavonoid activation of nodulation genes in *Rhizobium* reversed by other compounds present in plnats. *Nature*, **324**: 90-92.

Fisher, R.F. and Long, S.R. 1989. DNA foot print analysis of the transcriptional activator proteins Nod D1 and Nod D2 on inducible nod gene promoters. *J. Bacteriol.*, **171**: 5492-5502.

Fisher, R.F. and Long, S.R. 1992. *Rhizobium*-plant signal exchange. *Nature*, **357**: 655-660.

Fortin, M.G., Morrison, N.A. and Verma, D.P.S. 1987. Nodulin-26, a peribacteroid membrane nodulin is expressed independently of the development of the peribacteroid compartment. *Nul. Acids Res.*, **15**: 813-824.

Franssen, H., Nap, J.P., Gloudemans, T., Stiekema, W., van Dam, H., Govers, F., Louwerse, J., van

Kammen, A. and Bisseling, T. 1987. Characterization of cDNA for nodulin-75 of soybean: A gene product involved in early stages of root nodule development. *Proc. Natl. Acad. Sci. USA.*, **84**: 4495-4499.

Franssen, H., Yang, W.C., Katinakis, P. and Bisseling, T. 1993. Characterization of *GmEnod40,* a gene expressed in soybean nodule primordia. In: *New Horizons in Nitrogen Fixation* (eds.) R. Palacios, J. Mora and W.E. Newton. Kluwer Pub. Dordrecht, The Netherlands, pp. 275-284.

Fuller, F. and Verma, D.P.S. 1989. Appearance and accumulation of nodulin mRNAs and their relationship to the effectiveness of root nodules. *Plant Mol. Biol.,* **3**: 21-28.

Gallusci, P., Dedieu, A., Journet, E.P., Huget, T. and Baker, D.G. 1991. Synchronous expression of leghaemoglobin genes in *Medicago truncatula* during nitrogen fixing root nodules development and response to exogenously supplied nitrate. *Plant Mol. Biol.,* **17**: 335-349.

Glenn, A.R. and Brewin, J. 1981. Succinate resistant mutants of *REhizobium leguminosarum. Journal of General Microbiol.,* **126**: 237-241.

Gloudemans, T., Devries, S.C., Bussink, H.J., Malik, N.S.A., Franssen, H.J., Louwerse, J. and Bisseling, T. 1987. Nodulin gene expression during soybean (*Glycine max*) nodule development. *Plant Mol. Biol.,* **8**: 395-403.

Gottfert, M. 1993. Regulation and function of rhizobial nodulation genes. FEMS *Microbiol Rev.,* **104**: 39-64.

Gottfert, M., Grob, P. and Hennecke, H. 1990. Proposed regulatory pathway encoded by the nod V and nod W genes, determinants of host specificty in *Bradyrhizobium japonicum. Proc. Natl. Acad. Sci. USA.* **87**: 2680-2684.

Gottfert, M., S. Hitz and H. Henecke. 1990. Identification of nods and nod U, two inducible genes inserted between the *Bradyrhizobium japonicum* nod YABC and nod IJ genes. *Mol. Plant-Microbe Interac.,* **3**: 308-316.

Govers, F., Nap, J.P., van Kammen, A. and Bisseling, T. 1987. Nodulins in the developing root nodule. *Plant Physiol. Biochem.,* **25**: 309-322.

Govers, F., Nap, J.P., Moerman, M., Franssen. H.J., van Kammen, A. and Bisseling, T. 1986. cDNA cloning and developmental expression of pea nodulin genes. *Plant Mol. Biol.,* **8**: 425-435.

Govers, F., Gloudemans, T., Moerman, M., van Kammen, A. and Bisseling, T. 1985. Expression of plant genes during the development of pea root nodules: *EMBO., J.,* **4**: 861-867.

Gyorgypal, Z., Kondorosi, E. and Kondorosi, A. 1991. Diverse signal sensitivity of Nod D protein homologs from narrow and broad host range rhizobia. *Mol. Plant-Microbe Interact.,* **4**: 356-364.

Gyorgypal, Z., Kiss, G.B. and Kondorosi, A. 1991. Transduction of plant signal molecules by the *Rhizobium* Nod D proteins. *Bioessays,* **13**: 575-581.

Hartwig, U.A., Maxwell, C.A., Joseph, C.M. and Phillips, D.A. 1989. Interactions among flavonoid nod gene inducers released from alfalfa seeds and roots. *Plant Physiol.,* **91**: 1138-1142.

Hartwig, U.A., Maxwell, C.A. Joseph, C.M. and Phillips, D.A. 1990. Chrysoeriol and luteolin released from alfalfa seeds induce nod genes in *Rhizobium meliloti. Plant Physiol.,* **92**: 116-122.

Hartwig, U.A., Maxwell, C.A., Joseph, C.M. and Phillips, D.A. 1990. Effects of alfalfa nod gene-inducing flavonoids on nod ABC transcription in *Rhizobium meliloti* strains containing different nod D genes. *J. Bacteriol.,* **172**: 2769-2773.

Hernandez, G., Ramirez, M., Suarez, R. and Fuentes, I. 1995. Root exuded nod-gene inducing signals limit the nodulation capacity of different alfalfa varieties with *Rhizobium meliloti. The Plant Cell,* **14**: 626-629.

Hirel, B., Bouet, C., King, B., Layzell, D., Jacobs, F. and Verma, D.P.S. 1987. Glutamine synthetase genes are regulated by ammonia provided externally or by symbiotic nitrogen fixation. *EMBO., J.*, **6**: 1167-1171.

Hirseh, A.M., Asad, S., Fang, Y., Wycoff, K. and Lobler, M. 1993. Molecular interactions during nodule development. In: *New Horizons in Nitrogen Fixation.* (eds.) R. Palacios, J. Mora and W.E. Newton. Kluwer Pub. Dordrecht, The Netherlands. pp. 291-296.

Hopper, W., Swaminathan, R., Palinappan, S.P. and Thomas, J. 1995. Preincubation of *Rhizobium* inoculants with flavonoids enhances nodulation in chickpea and peanuts. In: *Nitrogen Fixation: fundamentals and applications* (Tikhonovich, I.A., Provorou, N.A., Romanov, V.I. and Newton, W.E. eds.). Dordrecht, Netherlands: Kluwer Academic Publishers, pp. 325.

Horvath, B., Kondorosi, E., John, M., Schmidt, J., Torok, I., Gyorgypal, Z., Barabas, I., Wieneke, U., Schell, J. and Kondorosi, A. 1986. Organization, structure and symbiotic function of *Rhizobium meliloti* nodulation genes determining host specificity for alfalfa. *Cell*, **46**: 335-343.

Horvath, B., Heidstra, R., Lados, M., Moerman, M., Spaink, H.P., Prome, J.C., van Kammen, A. and Bisseling, T. 1993. Lipo-oligosaccharides of *Rhizobium* induce infection-related early nodulin gene expression in pea root hairs. *Plant J.*, **4**: 727-733.

Horvath, B., Bachem, C.W.B. Shell, J. and Kondorosi, A. 1987. Host-specific regulation of nodulation genes of *Rhizobium meliloti* nodulation genes in *Rhizobium* is mediated by a plant-signal, interacting with the nod D gene product. *EMBO., J.*, **6**: 841-848.

Hubbell, D.H. and Elkan, G.H. 1966. Correlation of physiological characteristics with nodulating ability in *Rhizobium japonicum. Can. J. Microbiol.*, **13**: 235-241.

Hungria, M., Johnston, A.W.B. and Phillips, D.A. 1992. Effects of flavonoids released naturally from bean (*Phaseolus vulgaris*) on nodD-regulated gene transcription in *Rhizobium leguminosarum* bv. *viciae. Mol. Plant-Microne Interact.*, **5**: 199-203.

Kannaiyan, S. 1998. Induction of nodular structures in cereal crops by the N_2 fixing *Azo Rhizobium caulinodans CADRI, Newslett.*, **2**(1): 1-2.

Kannaiyan, S. and Buvana, R. 1998. Endophytic nitrogen fixation in rice. *Research Bull.* Tamil Nadu Agriculture University, Coimbatore, Tamil Nadu, India. p. 12.

Kape, R., Parniske, M. and Werner, D. 1991. Chemotaxis and nod gene activity of *Bradyrhizobium japonicum* in response to hydroxycinnamic acids and isoflavonoids. *Appl. Environ., Microbiol.*, **57**: 316-319.

Kapulnik, Y., Joseph, C.M. and Phillips, D.A. 1987. Flavone limitations to root nodulation and nitrogen fixation in flfalfa. *Plant Physiol.*, **84**: 1183-1196.

Kardailsky, I., Yang, W.C., Zalensky, A., van Kammen, A.K. and Bisseling, T. 1993. The pea late nodulin gene *PsNOD6* is homologous to the early nodulin genes *PsENOD 3/14* and is expressed after the leghaemoglobin genes. *Plant Mol. Biol.*, **23**: 1029-1037.

Kondorosi, E. and Kondorosi, A. 1986. Nodule induction on plant roots by *Rhizobium. Trends Biochem. Sci.*, pp. 298-299.

Kondorosi, E., Gyuris, J., Schmidt, J., John, M., Duda, E., Hoffmann, B., Schell, J. and Kondorosi, A. 1989. Positive and negative control of nod gene expression in *Rhizobium meliloti* is required for optimal nodulation. *EMBO, J.*, **8**: 1331-1340.

Kondorosi, E., Buire, M., Cren, M., Iyer, N., Hoffmann, B. and Kondorosi, A. 1991. Involvement of the syrM and nod D3 genes of *Rhizobium meliloti* in nod gene activation and in optimal nodulation of the plant host. *Mol. Microbiol.*, **5**: 3035-3048.

Kondorosi, E., Pierre, M., Gren, M., Haumann, U., Buire, M., Hoffmann, B., Schell, J. and Kondorosi, A. 1991. Identification of No1R, a negative transacting factor controlling the nod regulon in *Rhizobium meliloti. J. Mol. Biol.*, **222**: 885-896.

Kortt, A.A., Burns, J.E., Trinick, M.J. and Appleby, C.A. 1985. The amino acid sequence of haemoglobin I from *Parasponia andersoni* a non-leguminous plant. *FEBS. Lett.*, **180**: 55-60.

Kosslak, R.M., Joshi, R.S., Bowen, B.A., Paaren, H.E. and Appelbaum, E.R. 1990. Strain-specific inhibition of nod gene induction in *Bradyrhizobium japonicum* by flavonoid compounds. *Appl. Environ. Microbiol.,* **56**: 1333-1341.

Kosslak, R.M., Joshi, R.S., Dowen, B.A., Paaren, H.E. and Appelbaum, E.R. 1987. Induction of *Bradyrhizobium japonicum* common nod genes by isoflavones by flavonoid *Glycine max. Proc. Natl. Acad. Sci., USA,* **84**: 7428-7432.

Kouchi, H. and Hata, S. 1993. Isolation and characterization of novel nodulin cDNAs representing genes expressed at early stages of soybean nodule development. *Mol. Gen. Genet.,* **238**: 106-119.

Krause, A. and Broughton, W.J. 1992. Proteins associated with root-hair deformation and nodule initiation in *Vigna unguiculata. Mol. Plant-Microbe Interact.*, **5**: 96-103.

Kuster, H., Perlick, A.M. and Puhler, A. 1994. Members of a broadbean nodulin family with partial homologies to the alfalfa nodulin 25 are composed of two types of amino acid repeats flanked by unique amino acid sequence termini. *Plant Mol. Biol.,* **24**: 143-157.

Kuster, H., Fruhling, M., Perlick, A.M. and Puhler. A. 1993. The sucrose synthase gene is predominantly expressed in the root nodule tissue of *Vicia faba* L. Mol. *Plant-Microbe Interact.,* **6**: 507-514.

Landsmann, J., Dennis, E.S., Higgins, T.J.V., Appleby, C.A., Katt, A.A. and Peacock, W.J. 1986. Common evolutionary origin of legume and non-legume plant haemoglobins. *Nature.* **324**: 166-168.

Lang-Unnash, N. and Ausbel, F.M. 1985. Nodule specific polypeptides from effective alfalfa root nodules and from ineffective nodules lacking nitrogenase. *Plant Physiol.,* **77**: 833-839.

Laquerre, G., Mavingui, P., Allard, M.R., Charney, M.P., Louvrier, P., Mazurier, S.I., Rigottier, L. and Amarger, N. 1996. Typing of rhizobia by PCR DNA fingerprinting and PCR-restriction fragment length polymorphism analysis of chromosomal and symbiotic gene regions: application of *Rhizobium leguminosarum* and its different biovars. *Appl. and Environ. Microbiol.,* **62**: 2029-2036.

Lara, M., Cullimore, J.V., Lea, P.J., Miflin, B.J., Johnston, A.W.B. and Lamb, J.W. 1983. Appearance of a novel form of plant glutamine synthetase during nodule development in *Phaseolus vulgaris* L. *Planta,* **157**: 254-258.

LaRue, T.A., Kneen, B.E. and Gartside, E. 1985. Plant mutants defective in symbiotic nitrogen fixation. In: *Analysif of plant genes involved in the Legume-Rhizobium symbiosis* (ed.) R. Marteellin, OECD Publication, Paris, France. pp. 39-48.

Le Strange, K.K., Bender, G.L., Djordjevic, M.A., Rolfe, B.G. and Redmond, J.W. 1990. The *Rhizobium* strain NGR 234 nod D1 gene product responds to activation by the simple phenolic compounds vanillin and isovanillin present in wheat seedling extracts. *Mol. Plant-Microbe Interact.,* **3**: 214-220.

Legocki, R.P. and Verma, D.P.S. 1980. Identification of 'nodule specific' host proteins (nodulins) involved in the development of *Rhizobium*-legume symbiosis. *Cell,* **20**: 153-163.

Lerouge, P., Roche, P., Faucher, C., Maillet, F., Truchet, G., Prome, J.C. and Denaire, J. 1990. SYmbiotic host specificity of *Rhizobium meliloti* is determined by a sulphated and acylated glucosamine oligosaccharide signal. *Nature,* **344**: 781-784.

Lobler, M. and Hirsch, A.M. 1993. A gene that encodes a prolin-rich nodulin with limited homology to *PsENOD12* is expressed in the invasion zone of *Rhizobium meliloti* induced alfalfa root nodules. *Plant Physiol.*, **103**: 21-30.

Long, S.R. 1989. *Rhizobium*-legume nodulation: life together in the underground. *Cell*, **56**: 203-214.

Long, S.R. and Cooper, J. 1988. Overview of symbiosis. In: *Molecular Genetics of Plant-Microbe Interactions* (eds.) R. Palacios and D.P.S. Verma. A.P.S. Press, Minnesota, USA. pp. 163-178.

Luka, S., Sanjuan, J., Carlson, R.W. and Stacey, G. 1993. NolMNO genes of *Bradyrhizobium japonicum* are co-transcribed with nod YABCSUIJ, and nolO is involved in the synthesis of the lipooligosaccharide nodulation signals. *J. Biol. Chem.*, **268, 27**: 27053-270-59.

Lullien, V., Barker, D.G., De Lajudie, P. and Huguet, T. 1987. Plant gene expression in effective and ineffective root nodules of alfalfa (*Medicago sativa*, L.). *Plant Mol. Biol.*, **9**: 469-478.

Maier, R.J. and Brill, W.J. 1978. Mutant strains of Rhizobium japonicum *with increased ability to fix nitrogen for soybean. Science*, **201**: 448-450.

Mavridou, A., Barny, M.A., Poole, P., Plaskitt, K., Davis, A.E., JOhnston, A.W.B. and Downie, J.A. 1995. *Rhizobium leguminosarum* nodulation gene (nod) expression is lowered by an allele-apecific mutation in the dicarboxylate transport gene dctB. *Microbiology*, **141**: 103-111.

Maxwell, C.A., Hartwig, U.A. Joseph, C.M. and Phillips, D.A. 1989. A chalcone and 2 related flavonoids released from alfalfa root induce nod genes of *Rhizobium meliloti. Plant Physiol.*, **91**: 842-847.

Mclever, J., Djordjevic, M.A., Weinman, J.J., Bender, G.L. and Rolfe, B.G. 1989. Extension of host range of *Rhizobium leguminosarum* bv. *trifolii* caused by point mutations in nodD that results in alterations in regulatory function and recognition of inducer molecules. *Mol. Plant-Microbe Interact.*, **2**: 97-106.

Miller, S.S., Vance, C.P. and Barnes, D.K. 1987. The expression of plant proteins involved in C and N assimilation in effective and ineffective alfalfa nodules. In: *Proceedings of North American Rhizobium Conference.* (ed.) L. Bordeleau, Agriculture Canada, Sainte-Foy, Quebec. p. 33.

Modi, 1995. *Host gene expression in blackgram* (*Vigna mungo* (L) Hepper) nodules. Ph.D. Thesis. Tamil Nadu Agriculture University, Coimbatore, Tamil Nadu.

Moffat, A.S. 1990. Nitrogen fixing bacteria find new partners. *Science*, **250**: 910-912.

Morrison, N.A., Bisseling, T. and Verma, D.P.S. 1987. Development and differentiation of the root nodule: Involvement of plant and bacterial genes. In: *Developmental Biology* (ed.) L.W. Bowder. Plenum Press, New York.

Mulligan, J.T. and Long, S.R. 1985. Induction of *Rhizobium meliloti* nodC expression by plant exudates requires nodD. *Proceedings of the National Academy of Sciences* (*USA*), **82**: 6609-6613.

Nap, J.P. and Bisseling, T. 1990. Nodulin function and nodulin gene regulation in root-nodule development. In: *Molecular Biology of Symbiotic Nitrogen Fixation.* (ed.) P.M. Gresshoff. CRC Press. Bacoration, USA. pp. 181-229.

Nguyen, T., Zelechowska, M., Foster, V., Bergmann, H. and Verma, D.P.S. 1985. Primary structure of the soybean nodulin-35 gene encoding Uricase II localised in the peroxisomes of uninfected cells of nodules. *Proc. Natl. Acad. Sci. USA*, **82**: 5040-5044.

Norris, T.H., Macol, L.A. and Hirsch, A.M. 1988. Nodulin gene expression in effective alfalfa nodules and in nodules arrested at three different stages of development. *Plant Physiol.*, **88**: 321-328.

Olson, E., Sadowsy, M. and Verma, D.P.S. 1985. Identification of genes involved in the *Rhizobium*-legume symbiosis by Mu-dI (Kan, lac)-generated transcription fusions. *Biotechnology*, **3**: 143-149.

Papadopoulou, K., Roussis, A. and Katinakis, P. 1996. *Phaseolus ENOD40* is involved in symbiotic and non-symbiotic organogenetic processes: expression during nodule and lateral root development. *Plant Mol. Biol.*, **30**: 403-417.

Pawlowski, K., Ratet, P., Schell, J. and De Bruijn, F.J. 1987. Cloning and characterization of nifA and ntrC genes of the stem nodulating bacterium ORS571, the nitrogen fixing symbiont of *Sesbania rostrata:* regulation of nitrogen fixation (nif) genes in the free living versus symbiotic state. *Mol. Gen. Genet.*, **206**: 207-219.

Peters, N.K., Frost, J.W. and Long, S.R. 1986. A plant flavone, luteolin, induces expression of *Rhizobium meliloti* nodulation genes. *Science,* **233**: 977-980.

Phillips, D.A., Joseph, C.M. and Maxwell, C.A. 1992. Trigonelline and stachydrine released from alfalfa seeds activate NodD2 protein in *Rhizobium meliloti. Plant Physiol.,* **99**: 1526-1531.

Pillai, S.D., Josephson, K.L., Bailey, R.L. and Pepper, I.L. 1992. Specific detection of rhizobia in root nodules and soil using the polymerase chain reaction. *Soil Biol. and Biochem.,* **24**: 885-889.

Quinto, C., De La Vega, H., Flores, M., Leemans, J. and Vevallos, M.A. 1985. Nitrogenase reductase: A functional multigene family in Rhizobium phaseoli. *Proc. Natl. Acad. Sci., USA,* **82**: 1170-1174.

Recourt, K., Schripsema, J. Kijne, J.W., VCan Brussesl, A.A.N. and Lugtenberg, B.J.J. 1991. Inoculation of *Vicia sativa* subsp. *nigra* roots with *Rhizobium leguminosarum* biovar *viciae. Plant Mol. Biol.*, **19**: 411-420.

Robertson, J.G., Wells, B., Bisseling, T., Farnden, K.J.F. and Johnston, A.W.B. 1984. Immunogold localization of leghaemoglobin in cytoplasm in nitrogen fixing root nodules of pea. *Nature,* **311**: 254-256.

Rolfe, B.G. and Shine, J. 1984. *Rhizobium*-leguminosae symbiosis: The bacterial point of view. In: D.P.S. Verma and Th. Hohn (eds), Genes involved in Microbe-Plant Interactions, Springer-Verlag, Wein and New York, pp. 95-128.

Ronson, C.W., Lyttleton, P. and Robertson, J.G. 1981. C4-dicarboxylate transport mutants of *Rhizobium trifolii* from ineffective nodules on *Trifolium repens. Proc. Nat. Acad. Sci.* (*USA*), **78**: 4284-4288.

Rossbach, S., Gloudemans, T., Bisseling, T., Studer, D., Kaluza, B., Ebeling, S. and Hennecke, H. 1989. Genetic and physiologic characterization of a *BradyRhizobium japonicum* mutant defective in early bacteroid development. *Mol. Plant-Microbe Interact.,* **2**: 233-240.

Schalman, H.R.M., Lugtenberg, B.J.J. and Okker, R.J.H. 1992. The NodD protein does not bind to the promoters of inducible nodulation genes in extracts of bacteroids of *Rhizobium leguminosarum* biovar *viciae. J. Bacteriol.,* **174**: 6109-6116.

Schalman, H.R.M., Okker, R.J.H. and Lugtenberg, B.J.J. 1992. Regulation of nodulation gene expression by NodD in rhizobia. *J. Bacteriol.,* **174**: 5177-5182.

Scheres, B., Vandewiel, C., Zalensky, A., Horvath, B., Spaink, H., van Eck, H., Zwartkruis, F., Woltrs, A.M., Gloudemans, T., van Kammen, A. and Bisseling, T. 1990. The *ENOD12* gene product is involved in the infection process during the Pea-*Rhizobium* interaction. *Cell,* **60**: 281-294.

Scheres, B., Van Engelen, F., van der Knaap, E., van der Weil, C., van Kammen, A. and Bisseling, T. 1990. Sequential induction of nodulin gene expression in the developingpeanodule.*ThePlant Cell.,* **2**: 687-700.

Schultze, M., Ondorosi, E., Ratet, P., Buire, M. and Kondorosi, A. 1994. Cell and Molecular Biology of *Rhizobium*-Plant Interactions. *Inter. Rev. Cytol.,* **156**: 1-73.

Schwedock, J. and Long, S.R. 1989. Nucleotide sequence and protein products of two new nodulation genes of *Rhizobium meliloti,* nodP and nodQ. *Mole. Plant-Microbe Interact.,* **2**: 181-194.

Schwedock, J. and Long, S.R. 1992. *Rhizobium mililoti* genes involved in sulfate activation: The two copies of nodPQ and a new locus, saa. *Genetics,* **132**: 899-909.

Schwinghamer, E.A. and Belkengren, R.P. 1968. Inhibition of *Rhizobia* by a strain of *Rhizobium leguminosarum*: some properties of the antibiotic and of the strain. *Archives of Microbiology,* **64**: 130-145.

Sengupta-Gopalan, C. and Pitas, J.W. 1986. Expression of nodule-specific glutamine synthetase genes during nodule development in soybeans, *Plant Mol. Biol.,* **7**: 189-199.

Sengupta-Gopalan, C., Pitas, J.W., Thompson, D.V. and Hoffman, L.M. 1986. Expression of host genes during nodule development in soybean. *Mol. Gen. Genet.,* **203**: 410-420.

Spaink, H.P. 1995. The molecular basis of infection and nodulation by Rhizobia: The Ins and Outs of Sympathogenesis. *Ann. Rev. Phytopathol.,* **33**: 345-368.

Spaink, H.P., Wiffelman, C.A., Pees, E., Okker, R.J.H. and Lugtenberg, B.J.J. 1987. *Rhizobium* nodulation gene nodD as a determinant of host specificty. *Nature,* **328**: 337-340.

Spaink, H.P., Okker, R.J.H., Wiffelman, C.A., Pees, E. and Lugtenberg, B.J.J. 1987. Promoters in the nodulation region of the *Rhizobium leguminosarum* Sym plasmid pRL IJI. *Plant Mol. Biol.,* **9**: 27-39.

Spaink, H.P., Sheeley, D.M., Van Brussel, A.A.N., Glushka, J., York, W.S., Tak, T., Geriger, O., Kennedy, E.P., Reinhold, V.N. and Lugtenberg, B.J.J. 1991. A novel highly unsaturated fatty acid moiety of lipooligosaccharide signals determines host specificity of *Rhizobium. Nature,* **354**: 125-130.

Stacey, G., Luka, S., Sanjuna, S., Banfalvi, Z., Nieuwkoop, A.J., Chun, J.Y., Forsberg, L.S. and Carlson, R.W. 1994. Nod Z, a unique host-specific nodulation gene, is involved in the fucosylation of the lipooligosaccharide nodulation signal of *Bradyrhizobium japonicum. J. Bacteriol.,* **176**: 620-633.

Strittmatter, G., Chia, T.F., Trinh, T.H., Katagiri, F., Kuhlemeier, C. and Chua, N.H. 1989. Characterization of nodule-specific cDNA. clones from *Sesbania rostrata* and expression of th corresponding genes during the initial stages of stem nodules and root nodules formation. *Mol. Plant-Microbe Interact,* **2**: 122-127.

Szabados, L., Ratet, P., Gruenenberg, B. and de Brujin, F.J. 1990. Functional analysis of the *Sesbania rostrata* leghaemoglobin glb3 gene 5'-upstream region in transgenic *Lotus corniculatus* and *Nicotiana tabaccum* plants. *Plant Cell,* **2**: 973-986.

Szezyglowski, K., Boron, L, Szybiak-Strozycka, U. and Legocki, A.B. 1989. Characterisation of cDNA clone coding for nodulin-45 from yellow lupin (*Lupinus luteus*). *Plant Sci.,* **65**: 87-95.

Tingey, S.V., Walker, E.L. and Coruzzi, G.M. 1987. Glutamine synthetase genes of pea encode distinct polypeptides which are differentially expressed in leaves, roots and nodules. *EMBO, J.,* **6**: 1-9.

Triplett, E.W., Schink, M.J. and Nocldner, E.L. 1989. Mapping and subcloning of the trifolitoxin production and resistance genes from *Rhiozbium leguminosarum* bv. *trifolii* T24. *Mol. Pl. Microbe Interact.,* **2**: 202-208.

Triplett, E.W. 1990. The molecular genetics of nodulation competitiveness in *Rhizobium* and *Bradyrhizobium. Molecular Plant Microbe Interact.,* **3**: 199-206.

Tripplett, E.W. 1988. Isolation of genes involved in nodulation competitiveness from *Rhizobium leguminosarum* bv. *trifolii* T24. *Proc. Nat. Acad. Sci.* (*USA*), **85**: 3810-3814.

Truchet, G., Roche, P., Lerouge, P., Vasse, J., Camut, S., d Billy, F., Prome, J.C. and Denaire, J. 1991. Sulphate lipo-oligosaccharide signals of *Rhizobium meliloti* elicit root nodule organogenesis in alfalfa. *Nature,* **351**: 670-673.

Van Brussel, A.A.N., Recourt, K., Pees, E., Spaink, H.P., Tak, T., Wiffelmann, C.A> Kijne, J.W. and Lugtenberg, J.J.B. 1990. A biovar-specific signal of *Rhizobium leguminosarum* bv. *viciae* induces increased nodulation gene-inducing activity in root exudate of *Vicia sativa* subsp. *nigra. J. Bacteriol.,* **172**: 5394-5401.

Van Brussel, A.A.N., Bakhuizen, R., van Spronsen, P.C., Spaink, H.P., Tak, T., Lugtenberg, B.J.J. and Kijne, J.W. 1992. Introduction of pre-infection thread structures in the leguminous host plant by mitogenic lipooligosaccharides of *Rhizobium, Science,* **257**: 70-72.

Van de Wiel, C., Scheres, B., Franssen, H., Van Lierop, M.J., van Lammenen, A., van Kammen, A. and Bisseling, T. 1990. The early nodulin transcript *ENOD2* is located in the nodule parenchyma (inner-cortex) of pea and soybean root nodules. *EMBO, J.,* **9**: 1-7.

Vance, C.P., Egli, M.A., Griffith, S.M. and Miller, S.S. 1988. Plant regulated aspects of nodulation and N_2 fixation. *Plant Cell and Environ.,* **11**: 413-427.

Verma, D.P.S. and Nadler, K. 1984. Legume-*Rhizobium* symbiosis. Host's point of view. In: D.P.S. Verma and Th. Hohn (eds), Genes involved in Microbe-Plant Interactions, Springer-Verlag, Wein and New York, pp. 58-94.

Yang, W.C., Katinakis, P., Hendriks, P., Smolders, A. and Devries, F. 1993. Characterization of *Gm Enod40,* a gene showing novel patterns of cell-specific expression during soybean nodule development. *Plant J.,* **3**: 573-585.

Zaat, S.A.J., Wiffelman, C.A., Mulders, I.H.M., Van Brussel, A.A.N. and Lugtenberg, B.J.J. 1988. Root exudates of various host plants of *Rhizobium leguminosarum* contain different sets of inducers of *Rhizobium* nodulation genes. *Plant Physiol.,* **86**: 1298-1303.

Zaat, S.A.J., Scripsema, J., Wiffelman, C.A., Van Brussel, A.A.N. and Lugtenberg, B.J.J. 1989. Analysis of the major inducers of the *Rhizobium* nod A promoter from *Vicia sativa* root exudate and their activity with different nodD genes. *Plant Mole. Biol.,* **13**: 175-188.

Soil Microbiology and Soil Biotechnology : Vol. 2
Editor : S. Kannaiyan
Associated Publishing Co., New Delhi, 2010, pp. 167-190

8

Bioecology and Biotechnological Applications of *Trichoderma*

T. Marimuthu[1] *and S. Nakkeeran*[2]

Introduction

The success of our pest and disease management strategies through pesticides was by and large measured in terms of yield loss reduction alone. However, in field reality there are multitrophic interaction of any single approach with the host plant, pests and pathogens and the environment. We have been largely depending on pesticides and fertilizers and high yielding varieties for maximizing the yield without recognizing the interaction of these inputs with the environment and its impact on human welfare. But today food production has ceased to be a major concern, due to stable population levels and success of innovative agricultural technologies. At this juncture, the agrochemical industry that primarily recognized the farmers as their customers has failed to acknowledge the public as yet another important client. Perception of the public is that the negative aspects of pesticides seem to outnumber their benefits. The increased reflection on environmental concern over pesticide use has been instrumental in a large upsurge of biological methods of disease control. The development of pesticide resistance among insect pests and pathogens ground water and foodstuff pollution and the development of oncogenic risk has further encouraged the exploitation of beneficial microbes as an important candidate in pest and disease management. Pollution of agricultural soils with heavy metals has become an emerging problem. Industrial discharges have contaminated potential agricultural soils. Traditional engineering approaches have dominated remediation technology. Physical and chemical techniques are the most commonly used methods for remediation of pollutants. The high costs of these methods have stimulated the interest in alternative biologically-based systems, in particular, bioremediation and phytoremediation.

The decades of laboratory experiments have led to the explosion in the field of bio-control and introduced more than 50 commercial formulations comprising of fungi and bacteria.

1. Professor of Plant Pathology.

2. Associate Professor of Plant Pathology, Department of Plant Pathology, Tamil Nadu Agricultural University, Coimbatore – 641 003 ; E-mail: tha_marimuthumeha@yahoo.co.in

Among the microorganisms, *Trichoderma* plays a vital role in plant disease management and bioremediation and has opened up new avenues for commercial formulations of the fungi. To have a successful management of plant diseases and to have a pollution free environment, knowledge on the mechanism of action of bio-control agents, their ecology, mechanism behind bioremediation are highly essential to exploit the diversity of *Trichoderma* for human welfare.

Bio-ecology and diversity of *Trichoderma*

The significance of genetic variation is a key factor to evaluate biodiversity. The genetic variations in *Trichoderma* are mainly attributed to the adaptation of different species suited to different ecological conditions. The hymenomycetous genus, *Trichoderma* is a predominant cosmopolitan soil fungus, widely distributed all over the world, well known for its rapid growth and has the potential to use the varied substrates and often they are tolerant to toxic chemicals. *Trichoderma* occurs in all soils and other natural habitats that are rich in organic matter. *Trichoderma* is also found on root surfaces of various plants, decayed bark, mainly after the decay by other fungi (Elisa Esposito and Manuela da Silva, 1998). The population of *Trichoderma* decreases, when it is exposed to dry conditions for a long period of time. Some strains of *T. hamatum* and *T .psudokoningii* are adapted to excessive soil moisture and those of *T. viride* and *T. polysporum* are not found to be restricted to areas of low temperature. *T. harzianum* is predominant in warm climatic regions, while *T. hamatum* and *T. koningii* are widely distributed in areas of diverse climatic conditions (Papavizas, 1985; Grondona *et al.*, 1997). It is also distributed in different soils including agricultural, forest, prairie, salt marsh and desert soils distributed in different agro-climatic zones. The propagules of *Trichoderma* were abundant in forest soils and they proliferated well in the forest soils rich in carbon and nitrogen (Danielson and Davey, 1973). Some of the species of *Trichoderma* survive as decomposers of woody and herbaceous materials, and are also necrotrophic against the primary wood decomposers (Rossman, 1996).

The population of *Trichoderma* being very dynamic, evolved over time and have occupied the niches of higher elevations. Analysis of soil samples from higher elevations of about 2700m in the Himalayas reflected the presence of section Longibrachyatum. Based on the parsimony analysis the same was grouped under *H. audinensis*. The same was also isolated from 2300m elevation of Venezuelan Andes (Samuels *et al.*,1998). Another member of the section Longibrachyatum namely, *T. konilangbra* was isolated from locations between 1750 to 3400m in Uganda.

The diversity of the species of *Trichoderma* varies based on locations and with the fauna associated with the microsites. The soils of Taiwan was rich in *T. asperellum*, *T. sp. novum* was found in association with tree barks of Taiwan; tree bark with lichen in Chitan forest of Taiwan had *Hypocrea jecorina*, *T. viride*, *T. atroviride* and *T. harzianum*, but the tree barks that was in association with moss was found to contain *T. harzianum* alone.

Similarly the tropical forest soil of Cambodia was rich in *T. harzianum*, which indicate that it prefers warm soil temperature for its survival and multiplication. Analyses of the samples from the ruined walls of the temple indicate that they were predominant with *T.*

harzianum and to certain extent the ruined walls were also found to be associated with *T. virens*. The soils near seashore of Thailand at Coral Island and Koh Samet Island was colonized with *T.sp. novum* and *T. harzianum*. The wet soils of Kauchaburi province of Burma was colonized by *T. harzianum* and *T. ghanense*. The tropical rain forest soils of Malaysia had *T. virens, T. spirale,T. harzianum, T. asperellum* and *T.sp. novum*. The lake garden soils of Kuala Lumpur was associated with *T. virens, T. spirale,T. harzianum, T. sp. novum,T. koningii* and *T. atroviride*. The forest soil of Singapore was colonized with *T. koningii, T. asperellum, T. virens, T. spirale* and *T. harzianum*. Though there was a huge variation in the distribution pattern of the *Trichoderma* spp., the cultivated garden soils of South East Asia was predominant with *T. virens* which is being widely used as biocontrol fungi for the management of soil borne diseases. This vividly explains that the ecological requirements are different for the species colonizing tree bark, forest soils and cultivated garden soils (Kubicek *et al*., 2003).

Isolates of *T. spirale* was found to be more in South East Asia than the cosmopolitan species *T. koningii*. Similarly several isolates of *T. spirale* was observed in soils of Canada. However, the abundance of the same was more in Central America, Columbia and South Africa which suggests that this species are more abundant in tropical climates (Kubicek *et al*., 2003). *H. jecorina* a pantropical ascomycete is abundant in South–East Asia, The Pacific Isles and South America. But the same was previously reported to be restricted to the regions around equator (Samuels *et al*., 1998). Soils from Central Russia, the Urals, Siberia and the Himalayan mountains were predominant with *T. atroviride, T. hamatum, T. virens, T. asperellum, T. koningii, T.harzianum* and *T. oblongisporum* (Kullinig *et al*., 2000).

Systematic of *Trichoderma* and its significance in biodiversity

Trichoderma is a complex group of fungi and the knowledge on the systematic is highly essential to explore the biodiversity of this fungus. They are used in the production of cellulolytic and hemicellulolytic enzymes, biocontrol of plant diseases, biodegradation of chlorophenolic compounds, and in soil bioremediation. Since *Trichoderma* is being used as a very common beneficial fungus for several applications, precise identification and nomenclature of this fungus become important. For this purpose, several molecular and morphological approaches are available for identification of various isolates.

Trichoderma was introduced by Persoon in the year 1794. The taxonomy and identification of *Trichoderma* still remains inconclusive. All strains of *Trichoderma* were identified as "*T. viride*" till 1969 (*e.g.* also all the cellulase-producing strains of *T. reesei*) owing to Bisby's (1939) concept that *Trichoderma* consists of a single species. This led to the erroneous statement even in textbooks that "*T. viride* is an industrial cellulose producer". Hence, most of the taxa sampled and determined before 1969 were misidentified. Rifai (1969) adopted the concept of 'aggregate' species, and distinguished nine 'species aggregates', and admitted that some of them (particularly *T. hamatum*) are likely to contain two or more morphologically indistinguishable species. Bissett (1984; 1991a; 1991b; 1991c; 1992) revised Rifai's aggregate species by subtle recognition of continuities of morphological characters, expanded the morphological criteria to accommodate the wider range of morphological variation expressed by some anamorphs of *Hypocrea* and adopted some forms previously included in *Gliocladium*. While dissecting several of Rifai's species aggregates into several defined taxa, Bissett (1991a) also established a subdivision of the genus into five sections: *Longibrachiatum*, *Pachybasium*, *Trichoderma*, *Saturnisporum* and *Hypocreanum*. The advent of molecular tools for investigations in fungal taxonomy prompted research in the mid-nineties to re-assess the morphology-based taxonomy in *Trichoderma*. The laboratories of G.J. Samuels (Beltsville, MD, USA), T. Börner (Berlin, FRG) and C.P. Kubicek (Vienna, Austria) collaboratively pioneered a revision of Bissett's section *Longibrachiatum*. They combined the use of molecular markers (ITS1 and ITS2 sequence analysis, RAPD), physiological (isoenzyme analysis) and phenetic characters, and also for the first time included an analysis of potential teleomorphs of the *Trichoderma* spp. from this section (Samuels *et al.*, 1998). As a result, section *Longibrachiatum* was recognized to be monophyletic and to contain ten taxa, within which four pairs displayed teleomorph-anamorph relationships: *H. schweinitzii*/*T.citrinoviride*; *H. pseudokoningii*/*T. pseudokoningii*; *H.jecorina*/*T. reesei* and *H. orientalis*/*T. longibrachiatum*.

Further merged section *Saturnisporum*, which included only two species, *T. saturnisporum* and *T. ghanense* (Doi *et al.*, 1987), with section *Longibrachiatum*, and recognized the synonymy of *T.ghanense* with *T. parceramosum*. Yet, as a whole, the concept of section *Longibrachiatum* as defined by Bissett (1984) was largely confirmed, suggesting a degree of correlation between morphological and molecular approaches to taxonomy in *Trichoderma*. Section *Longibrachiatum* is a comparably small section of *Trichoderma*, and phylogenetically most distant from the other sections. The relationship between morphological characters and molecular phylogeny became more complex, however, when larger sections of *Trichoderma* were investigated. Kindermann *et al.* (1998) attempted a first phylogenetic analysis of the whole genus. Using sequence analysis of the ITS1 region of rDNA they found that the largest section, section *Pachybasium*, is actually paraphyletic. Although the use of a single gene fragment alone is insufficient by today's standards, this finding has been confirmed by phylogenetic analysis of several other genes (Kullnig-Gradinger *et al.*, 2002; Chaverri *et al.*, 2003). In this context the nomenclatural type strain of sect. *Pachybasium*, *Pachybasium hamatum* (Bonord.)=*T. hamatum*, is not a member of the major one of the two *Pachybasium* clades (clade B), but clusters together with *T. pubescens* and *T. strigosum* in a clade otherwise containing almost all taxa (*i.e.* with the exception of *T. aureoviride*) from section *Trichoderma*. Because of the lack of a distinctive morphological hiatus between clades A and B of *Pachybasium* (Kindermann *et al.*, 1998), researchers have so far maintained the name *Pachybasium* for both clades, but it is clear that this is an unsatisfactory situation from a taxonomic point of view. *Pachybasium* B contains all taxa attributed to this section by Bissett (1991b) with the exception of the three mentioned above. In addition, *Pachybasium* B also poses the problem of strong genetic variability of several of its species (*e.g.* *T. harzianum*). Since its species (both ana-as well as teleomorphs) account for the majority of taxa found in field investigations (Gherbawy *et al.*, 2004), this raises the question of how to identify species in this large heterogenous group.

Enzyme diversity of *Trichoderma*

Trichoderma is most frequently used biocontrol agent for the management of soil-borne and post harvest pathogens encountered in agriculture. The species of *Trichoderma* differ in their enzyme profile. A fundamental part of the *Trichoderma* antifungal system relies on a series of genes coding for a variety of extracellular lytic enzymes such as beta-1,3-glucanase, beta-1,6- glucanase, cellulase, chitinase and protease. Characterization of the polymorphism between five putative isoenzymatic activities [beta-1,3-glucanase (EC 3.2.1.39, EC 3.2.1.58), beta-1,6- glucanase (EC 3.2.1.75), cellulase (EC 3.2.1.4; EC 3.2.1.21, EC 3.2.1.91), chitinase (EC 3.2.1.30, EC 3.2.1.52), protease (EC 3.4.11; EC 3.4.13 - 19; EC 3.4.21 - 24, EC 3.4.99) was carried out with 18 strains from three sections of *Trichoderma*. Of these, seven strains were from T. sect. Pachybasium, nine from T. sect. *Trichoderma* and two from T. sect. Longibrachiatum. Thirty-seven different alleles in total were identified: Thirteen for beta-1,3-glucanase, four for beta-1,6- glucanase, three for cellulase, eight for chitinase and nine for protease activity (Sanz *et al*., 2004).

Diversity of Chitinase

Trichoderma spp. utilize chitinolytic enzymes, along with beta-1,3-glucanase, to degrade the pathogen's cell walls and thus reduce disease level. The chitinolytic system of *T. harzianum* was found to be more complex, consisting of six distinct enzymes. The system is composed of two beta-1,4 N-acetylglueosaminidases (CHIT102 and CHIT73), and four endochitinases (CHIT52, CHIT42, CHIT33 and CHIT31). The newly described enzymes are CHIT73 and CHIT52. All the chitinolytic enzymes were induced and excreted during growth of *Trichoderma* on chitin as the sole carbon source. Only CHIT102 was expressed intracellulary, at a low constitutive level, when *Trichoderma* was grown on glucose. Polyclonal antibodies raised against a purified 41-kDa endochitinase produced by *T. harzianumm* strain P1, reacted only with CHIT42, suggesting that serologically all the other chitinolytic enzymes are not closely related to the 41-kDa endochitinase. The complexity and diversity of the chitinolytic system of *T. harzianum* involves the complementary modes of action of six enzymes, which are required for maximum efficiency against a broad spectrum of chitin-containing plant pathogenic fungi (Haran *et al*., 1995).

Diversity of Cellulase

Trichoderma spp., produce cellulases comprising of cellobiohydrolase I, cellobiohydrolase II, and endoglucanase I to degrade cellulose. Comparison of the cellulolytic potential of the wild-type strain of *Trichoderma reesei* to members of *Trichoderma* sect, Longibrachiatum and *Hypocrea* spp, indicate a high diversity even within the same species. High cellulase activities were observed among several strains of *T. longibrachiatum* and *T. citrinoviride,* whereas *T. parceramosum* formed only low levels of activity. Among the corresponding teleomorphs, most strains of *H. schweinitzii* were comparatively poor producers, whereas the highest percentage of high producers was found among *H. jecorina* isolates. Immunoblot analysis of corresponding culture filtrates of various *H. jecorina* strains showed that the three major cellulase proteins (cellobiohydrolase I, cellobiohydrolase II, and endoglucanase I) were present in culture filtrates and was identical to that of the respective *T. reesei* proteins. With the aid of primers, corresponding to conserved sequences in the cellobiohydrolase I-encoding gene cbh1, a fragment of this gene was amplified from selected strains of *H. jecorina*, *T. reesei*, *T. longibrachiatum*, *T. citrinoviride*, and *H. schweinitzii* (Kubicek *et al*., 1996). The biochemical diversity of glycoside hydrolase family 12 (GH 12) endoglucanases revealed that *H. schweinitzii* Cell2A enzyme differed from the *T. reesei* Cell2A enzyme by only 14 amino acids (93% sequence identity), but was less thermostable (Sandgren *et al*., 2003).

Biotechnological applications

Pollution of water, both irrigation and drinking, agricultural and forest soils and the environment with pesticides, nitrites and heavy metals has become an emerging problem due to misuse of pesticides and fertilizers and industrialization throughout the globe in varying proportions and intensity. More than 300 000 ha of cultivable lands in UK near industrial areas had been contaminated. The high costs of traditional methods of remediation have stimulated interest in alternate ecofriendly biologically-based systems, in particular, bioremediation and phytoremediation. 'Bioremediation' has been defined as 'the elimination, attenuation or transformation of polluting or contaminating substances by the use of biological processes. Bioremediation consists of adding microorganisms to a polluted system, in the presence of nutrients, for degrading toxic compounds. Phytoremediation uses plants to degrade or remove toxicants from soil lacking a defined or managed root-microbial population. Combined application of both the methods is termed as phytobiation. Techniques to treat soil materials include biopiling, windrowing, land farming and bioventing, all of which depend on microbiological degradation. However, increasingly it also includes the use of vegetation to take up and/or degrade contaminants (phytoremediation) or restrict contaminant movement (phytostabilisation). Phytoremediation can be encouraged by manipulation of the rhizosphere, using selected fungal isolates in a process now defined as phytobial remediation. Among various fungi, *Trichoderma* plays a vital role in bioremediation in various fields such as:

- Bioremediation of pesticides
- Bioremediation of noxious chemicals
- Bioremediation of cellulose
- Bioremediation of heavy metals and mineralization
- Industrial application

Bioremediation of pesticides

The variety of organic pollutants released by the ever increasing number of industries is the direct cause of environmental and health-related problems that have detrimental effects on living beings. Conventional approaches (*e.g.* land-filling, recycling, pyrolysis and incineration) to the remediation of contaminated sites are inefficient and costly and can also lead to the formation of toxic intermediates. Thus, biological decontamination methods are preferable to conventional approaches because, in general, microorganisms degrade numerous environmental pollutants without producing toxic intermediates. Some environmental pollutants, however, persist in the environment.

The many different organic pollutants that are present in the environment persist for varying periods of time. Depending on their half-life, some pollutants either disappear within a short period of time or transformed into non-toxic end products. Many others, however, such as polychlorinated dibenzodioxyfurans (PCDDF/Fs), dioxins, dichlorodiphenyltrichloroethane (DDT), chlordane and hexachlorocyclohexane (HCH), have long half lives in the environment and tend to enter the food web, where they are subsequently 'biomagnified'. On entering mammalian systems, most of these pollutants are often metabolized to generate more toxic and harmful intermediates owing to the nonspecific activity of different mammalian enzymes. *Trichoderma* plays a vital role in bioremediation of heavy metals, toxic chemicals such as xenobiotic compounds, cobalt, lignite, cyanide, nitrocellulose etc.

Trichoderma are recognized as "keystone or controller organism" in forest soil due to their involvement in decomposition, nutrient cycling, and their regulation of associated mycobiota. Recent evidences have proved their potential to decompose xenobiotics. *T. harzianum* and *T. viride* have been shown to be capable of degrading organochlorides under *in vitro*. Persistence of organochloride contaminants in forest soil may be influenced by the predominant association of *T. harzianum* and *T. viride*. *T. harzianum* was capable of degrading chlorophenols and adsorbable organic halogens. It also had the capability to degrade ciliatine, glyphosate, and also phosphonic acid (Lupicka *et al.*, 1997).*T. harzianum* degraded DDT, dieldrin, endosulfan, PCNB, and pentachlorophenol. *T. harzianum* CCT-4790 degraded 60 per cent of the herbicide diuron in soil and opens a good potential for bioremediation of herbicide-contaminated soils. *T. viride* degraded pesticide, photodieldrin into water soluble and non-insecticidal compounds after 4 to 5 weeks. Thus, *Trichoderma* strains play a crucial role in bioremediation process. *T. harzianum*, a biocontrol species colonize plant rhizosphere and it metabolize several pesticides. Pentachlorophenol (PCP) was metabolized to pentachloroanisole (Rigout and Matsumura, 2002).

Bioremediation of noxious chemicals

Cultivation of cotton, beans or corn as well as ferns on a polluted site rich in heavy metals, arsenic, cyanide and metallocyanides, nitrates, etc. in association with *Trichoderma* spp. increased rapid uptake of the toxic elements by the plants. In addition there was strong reduction of toxicity caused by polycyclic aromatic hydrocarbons and phenolic compounds in water and soil by the use of *Trichoderma* (Vinale *et al.*, 2003). *T.viride* also helped in bioremediation of Co^{60} (Rashmi *et al.*, 2004).

Nitrocellulose (NC), widely used in the explosives and pharmaceutical industries is a true xenobiotic compound as it does not occur in nature. Degradation of NC in solid state fermentation was accomplished by *Trichoderma reesei*. The concentration of NC was below detection level after 28 days of colonization of the substrate (Auer *et al.*, 1999; Auer *et al.*, 2005). *Trichoderma* spp., have the potential to degrade cyanide. Biodegradation was observed in flask cultures using cyanide as the sole carbon source (Ezzi and Lynch, 2005).

iii. Bioremediation of Cellulose

Cellulose is one of the most abundant carbohydrates produced by plants. The cellulolytic microorganisms play a vital role in the biosphere by recycling cellulose. Cellulose being the simple polymer, forms insoluble, crystalline microfibrils, which are highly resistant to enzymatic hydrolysis. However, *Trichoderma* spp. play a key role in recycling the cellulose by producing battery of enzymes which degrade cellulose. In the forest ecosystem, fallen trees form the major source of lignocellulose. However, the lignocellulose are degraded by the combined action of several microbes and later the minerals are recycled in the form of cellulose ana lignin, which in turn increase the carbon pool of the soil, resulting in the improvement of soil health.

Decomposition of fallen logs of trembling aspen (*Populus tremuloides* Michx.) and white spruce (*Picea glauca* (Moench) Voss) from undisturbed and disturbed sites in northern Alberta found to have the association of 10,000 isolates, representing 292 species of filamentous microfungi, including 41 ascomycetes, 29 zygomycetes, and 222 mitosporic fungi. The most commonly isolated species were *Trichoderma viride*, *Rhinocladiella atrovirens*, *Penicillium pinophilum*, and *Mortierella ramanniana*. The tree species of the log had the greatest influence on the species composition of communities. Fungus community composition was also correlated with the stage of decomposition. Species richness was highest in logs from undisturbed sites. The most significant environmental factor was log moisture, which increased proportionately with stage of decomposition and was significantly correlated with climatic factors (Lumley *et al.*, 2001). Thus involvement of *Trichoderma* along with other fungi helped in the recycling of the polymer lignocellulose into use.

Ezell's Cave, a small limestone cave in south Texas, was investigated to establish the mycoflora associated with decomposition of organic matter within the cave. The species of *Mucor, Aspergillus, Cladosporium, Doratomyces, Fusarium, Penicillium* and *Trichoderma* were most abundant from decomposing organic material, cave sediment, bottom ground water pool, and air (Kuehn and Koehn, 1991). A mycological analysis of soil and litter pertaining to Malaysian Dipterocarp forest revealed the presence of fifty-seven taxa. Among which the major species isolated from cellulose rich soil and litter at 25°C was *T. harzianum*, *Gliocladium* and *Penicillium,* which in turn helped in addition of minerals to soil in the form of carbon and other minerals resulting in the increase of soil health (Bettuci and Roquebert, 1995).

Pleurotus ostreatus and *T. viride* degraded cellulose wastes from wine-producing industries and the agricultural waste obtained from cereal crops processing by immobilizing the fungi on radiopolymerized hydrogels. The enzymatic activity of immobilized microorganisms revealed an increasing of total reducing sugars with 20-30 mg g^{-1} (Petre *et al.*, 2001).

The solid wastes are a chronic problem that causes environmental pollution. Bioconversion using *Trichoderma* spp., offers a cheap and safe method of not only disposing these wastes, but also it has the potential to convert lignocellulosic wastes into usable forms such as reducing sugars that could be used as food. Maximum production of reducing sugars from lignocelluose waste was obtained after 6h of saccharification, while the highest concentrations of ethanol were obtained after 48 h of fermentation. Microbial bioconversion of lignocellulose component yielded up to 21% ethanol (Mtui and Nakamura, 2005).

Bioremediation of heavy metals and mineralization

Numerous microorganisms, especially those associated with roots, have the ability to increase plant growth and productivity. In a few cases, this effect has been suggested to involve solubilization of unavailable mineral nutrients. In soil, both macro- and micronutrients undergo a complex dynamic equilibrium of solubilization and non-solubilization that is greatly influenced by the soil pH and microflora and that ultimately affects their accessibility to plant roots for absorption. Phosphorus is commonly deficient in most natural soils, since it is fixed as insoluble iron and aluminium phosphates in acidic soils (especially those with pH lower than 5.0) or calcium phosphates in alkaline soils (pH above 7.0). However, insoluble calcium phosphate can be dissolved and made available to plants by soil and rhizosphere microorganisms via a mechanism involving the release of organic acids. Manganese is a microelement required for diverse physiological functions in plants and plays a major role in both plant growth and disease resistance. Manganese can occur in several oxidation states, but it is available to plants only in the reduced form (Mn^{2+}). Higher oxidation states are insoluble. Thus, microbial interactions with plant roots are known to profoundly affect plant nutrient status and for manganese at least, to affect plant resistance to pathogens.

In this connection the commonly associated plant-growth-promoting and biocontrol fungus *Trichoderma* seen in the rhizosphere and non-rhizosphere soils play a significant role in mineralization of the nutrients. It solubilizes insoluble minerals through acidification, production of chelating metabolites and through redox activity. *Trichoderma* also improves the soil health by mineralization of the toxic heavy metals. *T. harzianum* Rifai 1295-22 (T-22) solubilized MnO_2, metallic zinc, and rock phosphate (mostly calcium phosphate). Acidification was not the major mechanism of solubilization since the pH of cultures never fell below 5.0 and in cultures containing MnO_2 the pH rose from 6.8 to 7.4. Organic acids were not detected by high-performance thin-layer chromatography in the culture filtrates. Fe_2O_3, MnO_2, Zn, and rock phosphate were also solubilized by cell-free culture filtrates. In liquid culture, *T. harzianum* T-22 also produced diffusible metabolites capable of reducing Fe(III) and Cu(II). *T. harzianum* increase plant growth by chelation and reduction of metal oxides and increase solubilization of the minerals. Both of these mechanisms also play a role in biocontrol of plant pathogens, and they may be part of a multiple-component action exerted by T-22 to achieve effective biocontrol under a variety of environmental conditions (Altomare *et al.*, 1999).

Abandoned ore-washery rich in Mn and Zn was degraded by 108 taxa of soil microfungi. The most frequent soil microfungi were *Penicillium janthinellum*, *P. simplicissimum*, *Cunninghamella elegans*, *Paecilomyces lilacinus*, *Trichoderma* spp., *Mucor hiemalis f. hiemalis*, *Coniothyrium fuckelii*, *Mortierella alpina*, *Fusarium* spp., *Coemansia aciculifera*, and *Trichoderma virens* (Kubatova *et al.*, 2002).

Industrial application of *Trichoderma*

Cellulases or hemicellulases of fungal origin are being developed for textile, food, and paper industries. These applications are based on the modification of cellulose and hemicellulose by partial hydrolysis. Total hydrolysis of cellulose into glucose, which could be fermented into ethanol, isopropanol or butanol, is not yet economically feasible.

Trichoderma is an excellent producer of cellulolytic enzymes. The enzymes of *Trichoderma* are commercially used in animal food and fodder industries to improve the nutritive value of the food materials by enhancing the digestibility of the feed materials. The oat hulls are high in hydroxycinnamic acid, ferulic acid, which are inhibitory to digestion by ruminal microorganisms. Ferulic acid is produced via the phenylpropanoid biosynthetic pathway and covalently cross-linked to polysaccharides by ester bonds and to components of lignin, mainly by ether bonds. Ferulic acid also forms dimers or trimers. As a result, polysaccharides become extensively cross-linked by ferulate dimerization or trimerization and incorporation into lignin. Ferulic acid esterase of *Aspergillus* and *Trichoderma* xylanase act synergistically to release ferulic acid from feruloyl-polysaccharides in complex plant cell walls of oat hulls. This activity opens the remainder of the polysaccharides to further hydrolytic attack and facilitates the accessibility of the main polysaccharide chain to cellulase, thereby increasing the release of reducing sugars. Multi-enzyme cocktail comprising of ferulic acid esterase, xylanase, cellulase, endo-glucanase [I, II], and beta-glucanase of *Trichoderma* and *Aspergillus* origin improved ruminal fluid degradability (Yu *et al.*, 2005).

Natural and environmental-friendly disposal of wastewater sludge and textile effluents are a great concern. Recently, biological treatment has played prominent roles in bioremediation of complex hydrocarbon-rich contaminants. India is one of the leading exporters of garments. Most of the effluents drained out of the textile industries contaminate the ground water and in turn affect the agriculture productivity besides polluting the environment. Several microbes are used to degrade the effluents. Studies revealed that textile waste water polluted habitats (pools and drains) and biological textile waste water effluent treatment plant (ETP) were colonized by *Aspergillus*, *Cladosporium*, *Fusarium*, *Rhizopus*, *Trichoderma* and *Drechslera*. However, in general, *T. harzianum* was often having low spore load which was the most competitive species. Thus *T. harzianum* played a significant role in bioremediation of dyes in the polluted habitats and the effluent treatment plant (Grover *et al.*, 2004).

Trichoderma as a successful bio-control organism

Biological control of plant pathogens can be broadly defined as the reduction of inoculum density or disease producing activities of a pathogen or parasite in its active or dormant state by one or more organisms, accomplished naturally or through manipulation of environment, host or antagonist or by mass introduction of one or more antagonists (Baker and Cook, 1974). *Trichoderma* is one of the most widely used biocontrol agents. Among the various isolates of *Trichoderma*, *T. viride*, *T. harzianum*, *T. virens* and *T. hamatum* are used mostly for the management of various diseases of crop plants.

It has many attributes to call it as a potential bio-control agent. It has:

- High rhizosphere competence
- High competitive saprophytic ability
- Potentiality to enhance plant growth
- Great arsenal of inducible polysaccharide - degrading enzymes
- The amenability to mass multiplication
- Broad spectrum of action against various pathogens
- Ability to provide excellent and reliable control of diseases

Mode of action

Antagonist interacts with pathogen and host in soils either directly or indirectly. In indirect interactions plant responds to the presence of the antagonist, resulting in induced resistance or plant growth promotion. Based on the interaction, the modes of action may be direct or indirect.

Direct Mode of Action

Competition

Most of the soil-borne pathogens are controlled by the competition for space or court of infection on roots and seeds. Biological control of silver leaf disease caused by *Chondrostereum purpureum* was controlled by *T. viride* through early colonization of fresh wounds (Corke and Hunter, 1979). Competition also seems to be the most potent mechanism of *T. harzianum* T-35 in the control of *F. oxysporum*. f.sp. *vasinfectum* and *F. oxysporum* f. sp. *melonis* in cotton and melon respectively (Sivan and Chet, 1989).

Antibiotics

Antibiotics are low molecular weight (<1kDa) secondary metabolites produced during nutrient limiting conditions. Gliotoxin produced by *G. virens* G20 is a key factor in reducing damping off of Zinnia (Wilhite *et al*., 1994). *G. virens* pertaining to 'Q' group produced gliotoxin and dimethyl gliotoxin, but 'P' group produced gliovirin (diketopiperazine) and heptelidic acid. Gliovirin was inhibitory to *P. ultimum* which causes damping-off in cotton than 'Q' strains, which was effective against *R. solani* (Howell *et al*.,1993).

Trichoderma produced several kinds of antibiotics like *Trichodermain*, *dermadin*, *trichoviridin* and sesquiterpene heptalic acid. The UV mutants of *T. harzianum* produced trichozianines, the hydrophobic peptides, which interact with phospholipid membrane thus inducing membrane permeability. *T. hamatum* produced trichoviridin 3-(3-isocyano-6-oxabicyclo (3, 10) hex-2-ch-5yl) acrylic acid and 3-(3-isocyano cyclopet-2-enzylidne). In addition, it produced three isonitriles among which isonitrin A had broad spectrum activity. *T. longibrachiatum* produced *Trichodermain* and trichobrachin (Jeyarajan and Nakkeeran, 2000).

Hydrolytic Enzymes

Cellulases

Trichoderma species were potential bio-control agents and producers of powerful cellulases (Papavizas, 1985). Mandels *et al*. (1971) irradiated *T. viride* and produced new strain that secreted twice as much cellulase as its wild parent. Improved strains of *T. reesei* were isolated through mutation and selection. It produced more cellulase than did the wild strains (Montenecourt and Eveleigh, 1979). Baker (1991) found a perfect correlation between increased cellulase production and rhizosphere competence by mutants of *T. harzianum*. It is suggested that mutants use the mucigel on roots as a substrate. Mutants with rhizosphere competence were more efficient bio-control agent and induced increased growth responses from seed treatment than wild type parent. The quality and quantity of cellulases from *Trichoderma* spp. were improved through mutagenesis (Cuevas *et al*., 1994).

Proteases

Trichoderma species are well known producers of proteases. Proteolytic activity of *T. viride* was claimed to be involved in biocontrol of *Sclerotium rolfsii* (Rodriguez-Kabana, 1978). Serine proteases produced by *T. harzianum* played an important role in mycoparasitism of plant pathogens (Geremia *et al.*,1993). Autoclaved mycelium, fungal cell wall or chitin induces this enzyme and its production was repressed by glucose. Proteases are suggested to degrade cell wall, membranes and proteins released by the lysis of the pathogen, thus making the nutrients available for the mycoparasite (Goldman *et al.*, 1994).

α-glucosidases

It is an exo-carbohydrase that catalyses the splitting of α-glucosyl residues from the non-reducing terminus of substrates to liberate α-glucose. α-glucosidase is a key enzyme used in the food industries. But its role in the inactivation of phytotoxins has been recently exploited for disease management. *T. viride* produces 67kDa protein with a pI value of 5.0. *Rhizoctonia solani* causing rice sheath blight produced a glycoproteinaceous host specific RS - toxin. α-glucosidase of high molecular weight (110 kDa) from *T. viride* isolate (Tv-MNT7) was reported to degrade the RS-toxin (Sriram *et al.*, 2000).

β-1,3 glucanases

Lytic enzymes play a vital role in mycoparasitism and degrade the pathogen cell wall. The enzymes like chitinases, and β-1, 3 glucanases lyse the host wall and leads to the leakage of protoplasmic contents, which are in turn used as food material for the multiplication of antagonist (Cherif and Benhamou, 1990; Elad *et al.*, 1983; Tronsmo *et al.*, 1993). Strains of *T. harzianum* produced a mixture of lytic enzymes comprising of chitinolytic and gluconolytic enzymes, which involved in mycoparasitism (Sivan and Chet, 1989). Excretion of β-1,3 glucanase into growth medium and to the soil by *T. harzianum* was observed (Elad *et al.*, 1982). Carbon sources like laminarin, pustulan, cell walls of *R. solani* and mycelia of several fungi incorporated in the enzyme production medium induced the secretion of β-1, 3 glucanase by *T. harzianum* (Jacobs *et al.*, 1991). The purified endochitinase, chitobiosidase, N-acetyl- (-glucosaminidase and glucan, 1,3-β-glucosidase and combinations inhibited the spore germination and germ tube elongation of *Botrytis cinerea*. Combinations of the purified enzymes resulted in the synergistic increase in antifungal activity (Lorito *et al.*, 1994a, b). High synergistic activity was observed by combining 1,3-β-glucosidases and chitinolytic enzymes (Tronsmo *et al.* 1993). This corresponds to the effect observed with combination of β-1,3 glucanase and endochitinase isolated from plants (Mauch *et al.*, 1988). *T. harzianum* also produced N-acetylglucosaminidase, endochitinase, chitobiosidase and endo β-1,3 glucanases in soybean rhizosphere active against *R. solani* (Dalsoglio *et al.*, 1998).

Chitinases

Chitinase, the basic protein catalyse the hydrolysis of chitin. Chitin a linear polymer of β-1,4-N-acetyl glucosamine is often considered to be the second most abundant polysaccharide in nature (Deshpande,1986; Nicol, 1991). Relationship between mycolytic enzymes, chitinases and b-1,3 glucanases produced by mycoparasitic fungi and their significance in fungal cell wall lysis and degradation have been well established (Elad *et al.*, 1980, 1983). Assay of chitinase could be used as a basis for screening potential bio-control agents (Elad *et al.*, 1982). *Trichoderma* spp. attack several plant pathogenic fungi by excreting lytic enzymes including β-1, 3 glucanase, proteinases and chitinases enabling them to degrade host cell walls and thus reduce disease incidence (Geremia *et al.*, 1993; Chet *et al.*, 1993; Lorito *et al.*, 1993; Goldman *et al.*, 1994). *T. harzianum* strain Tm had six distinct chitinolytic enzymes. Two were identified as β-1, 4-N-acetyl glucosaminidases (Ec. 3.2.1.30). It hydrolyzes chitin to N-acetyl glucosamine (Glc NAc) monomers. Four were endochitinases (Ec 3.2.1.14) that randomly cleave at internal sites over the entire length of the chitin microfibril (CHIT 52, CHIT 42, CHIT 33 and CHIT31) (Haran *et al.*, 1995). *T. harzianum* - P1 released chimeric units from chitin and termed as chitobiosidase (40 kDa) (Harman *et al.*, 1993).

The involvement of *Trichoderma* chitinases in mycoparasitism were tested in liquid cultures supplemented with different carbon sources *e.g.* chitin, purified fungal cell walls, glucose or GlcNAc (Garcia *et al.*,1994; Limon *et al.*, 1995). Complete degradation of mycelial walls of pathogenic fungi depends to major extent on chitinolysis and hence the degrading ability to chitin by *Trichoderma* spp assumes a greater significance in antagonistic process than the other mechanisms/enzymes involved in the process (Cherif and Benhamou, 1990). Chitinases and glucanases were produced by *Trichoderma* isolates, in soil, which had live mycelium of *S. rolfsii* and *R. solani*. High level of β-1,3 glucanase and chitinase activities were also detected by dual agar cultures of *T. harzianum* with *S. rolfsii* and *R. solani* (Elad *et al.*, 1983). Direct interaction of *T. harzianum* with *R. solani*, strongly enhanced the endochitinase CHIT 42 (Carsolio *et al.*, 1994). The spore germination and hyphal elongation of *B. cinerea, F. solani, F. graminearum, Ustilago avenae* and *Uncinula necator* were inhibited substantially by the combination of chitinolytic and gluconolytic enzymes purified from *T. harzianum*. The chitinolytic enzymes from *T. harzianum* were more active than that of the enzyme from other sources like plant-bacteria and were effective against a wider range of fungi (Harman *et al.*, 1993).

Spatial distribution of glucanases and chitinases

β-1,3 glucanase degrade the substrate β-1,3 glucan, which is the major component of fungal cell wall next to chitin. Physiological concentration of these fungi effectively inhibits the growth of many potential pathogenic fungi. Though this enzyme plays a major role in cell wall lysis, it is imperative to understand the *in vitro* production and sub-cellular localization of the constitutive expression of the enzyme, β-1,3 glucanase. Expression of β-1,3 glucanase by *T. harzianum* was observed in extracellular, periplasmic, cell wall bound internal fractions during submerged fermentation in the absence of lactose and glucose as carbon source. In submerged culturing of *T. harzianum* with lactose, β-1,3 glucanase was found in all extracellular, cell wall bound, and in periplasmic fractions. The relative distribution of the enzyme in the cell is independent of the nature of carbon source and its concentration.

Chitinases are extra cellular and intracellular. Chitinases are localized either outside or inside the cell and it is localized at sub cellular level in various fractions including cellwall, periplasmic space, vacuoles, and membranes. Extracellular chitinases are more important than the intracellular chitinases, since the extra cellular chitinases provide the first line of defence against plant pathogens.

Mycoparasitism

Mycoparasitism relies on the production of lytic enzymes by the mycoparasite for the degradation of cell walls of the host fungus. Isolates of *Trichoderma* spp coiled around the hyphae of *R. solani*, which consequently lost their cell contents and collapsed (Chu and Wu, 1981). *Trichoderma* spp. attached to *S. rolfsii* or *R. solani* by hyphal coils, hooks, or appressoria. Lysed sites and penetration holes were found in hyphae of the pathogen following the removal of parasitic hyphae (Elad *et al.*, 1983). *Trichoderma* coiled over *Pythium* resulted in cell wall lysis, vacuolation and coagulation of protoplasm (Lifshitz *et al.*, 1986).

Indirect modes of action

Induced resistance

The induction of resistance in plants by *Trichoderma* spp. has been poorly studied. The first demonstration of induced resistance by *Trichoderma* was published by De Meyer *et al.* (1998). Soil treatment with *T. harzianum* strain T-39 induced resistance against foliar pathogens *B. cinerea* and *Colletotrichum lindemuthianum*, even though T-39 was present only on the roots. *T. harzianum* strain T-22 to induced systemic resistance to pathogens in maize. Therefore, the capacity to induce resistance to a range of diseases in a variety of plants seems to be widespread in *T. harzianum*.

Plant growth promotion

Growth promotion may occur either directly or indirectly. Promotion of growth may occur either through bio-control of deleterious pathogens (Baker, 1989) or through the production of plant hormones like auxin, cytokinin or gibberellins (Arshad and Franken-berger, 1991). Baker *et al.* (1984) found that *T. harzianum* can also stimulate plant growth, including floricultural and horticultural plants. Pepper seed germinated 2 days earlier in raw soils containing the fungus than in untreated controls. Flowering of periwinkle was accelerated, the number of blooms per plant on chrysanthemum was increased, and the height and weight of other plants were greater in either steamed or raw soil infested with *T. harzianum*. These responses have occurred consistently with population densities of *T. harzianum* higher than 10^8cfu g^{-1} of soil, whether the bio-control agent was applied as conidial suspension or as mycelium and spore in a peat bran medium.

Growth stimulation by *T. harzianum* could be the result of production of plant hormones by antagonist, increased uptake of nutrients by plant, through nutrient mobilization, or the control of one or more subclinical pathogens. *Pythium* spp. are capable of stunting plants and hinders the nutrient uptake (Baker and Cook, 1988). Hence suppression of *Pythium* spp. could increase the plant growth (Salt, 1978). Increased growth was also induced by a diffusable growth regulating factor produced by *Trichoderma* spp. (Windham *et al.*, 1986).

Trichoderma and plant health

Trichoderma species have been widely exploited for the management of seed and soil borne diseases of agricultural and horticultural crops (Table 1). The diseases due to *Rhizoctonia solani*, *Macrophpomina phaseolina*, and *Fusarium* spp. causes high percentage of mortality of the plants and consequent yield losses in rain-fed crops especially in pulses and oilseeds. Soil incorporation of *T. harzianum* in the field reduced the inoculum density of *R. solani* by 85 per cent and the fruit rot of tomato by 27-51 per cent. Soil application of *T. harzianum* was found effective in reducing the incidence of *Macrophomina* root rot of beans by 37 to 74 per cent. Coating of melon seeds with conidia of *T. harzianum* reduced the disease caused by *Macrophomina* to the extent of 46.3 per cent. Seed coating of green gram and cowpea @ 10g of *T. viride* isolate MG6 combined with soil application reduced dry root rot incidence and increased the yield (Nakkeeran and Sabitha Doraisamy, 2001). In addition to disease reduction, dry matter production was also enhanced which was attributed to the production of growth hormones. Treatment of sunflower seeds with *T. viride* @ 4g/kg reduced the root rot incidence to 18.15 as against 37.1 per cent in control. It also increased the yield by 21 per cent.

Table 1: Effect of *Trichoderma* spp. against plant diseases (Kannaiyan *et al.*, 2002)

Trichoderma	Pathogen/disease	Crop	Delivery system
Trichoderma sp.*C62*	*Sclerotium cepivorum*	Onion	Soil application as bran sand inoculums
Trichoderma spp.	*G. graminis var. tritici, R. solani*	Wheat Lettuce Lettuce	Spores added to soil Soil Soil
T. harzianum	*Fusarium graminearum*	Radish	Soil
T. harzianum T95I	*Pythium ultimum*	Cucumber	Soil
T. koningii	*S. rolfsii*	Tomato	Soil
T. koningii	*Pythium* spp.	Chickpea	Seed
	Pythium spp.	Pea	Seed
	P. ultimum	Pea	Soil
T. harzianum	*Pythium ultimum*	Cotton Maize Pea	Seed Seed Seed
T. harzianum + *Rhizobium*	*S. rolfsii*	Groundnut	Seed and soil
T. viride *T. harzianum* *T. longibrachiatum*	*M. phaseolina*	Sesamum	Seed
T. viride	*M. phaseolina*	Groundnut Sesamum Blackgram Greengram Pigeonpea Sunflower	Seed
T. hamatum	*R. solani*	Cotton Sugarbeet Radish	Soil-as mycelial preparations
T. viride	*M. phaseolina*	Blackgram	Seed
T. viride	*M. phaseolina*	Greengram	Furrow application

Trichoderma	Pathogen/disease	Crop	Delivery system
T. viride	*F. udum* *M. phaseolina*	Pigeon-pea	Seed and Soil
T. viride	*F. oxysporum f.* sp. *cubense* *R. solani*	Banana	*Trichoderma* capsules
	M. phaseolina	Cotton Green gram	Seed Seed
	Pythium ultimum	Cowpea Chilies, Tomato	Seed + Soil Solid matrix priming
Trichoderma spp.	*M. phaseolina*	Sesamum Blackgram	Seed Seed and soil
T. viride	*R. solani* *S. rolfsii*	Cotton Brinjal Sunflower Bhendi	Seed

Seed treatment sesame with *T. viride* @ 4g/kg reduced the root rot incidence to 12.8 per cent and increased the root, shoot length and oil content. It recorded an yield of 677 kg/ha as against 301 kg/ha in control. Results of the field demonstrations conducted in the farmer's fields for the control of root rot in groundnut, sesamum, urdbean, and chickpea against dry root rot through *T. viride* was found to be promising. *T. harzianum* was found effective in the management of *Fusarium* wilt diseases. Strain T35 checked *Fusarium* wilt of cotton and melon under natural soil conditions. Incorporation of *T. harzianum* as wheat bran-peat preparation into soil under green house conditions reduced the incidence of crown rot of tomato up to 80 per cent. Soil application increased the yield of tomato by 18.8 per cent.

Soil application of *T. koningii* @ 2g per vine of black pepper at the time of transplanting was highly effective than metalaxyl compounds against *Phytophthora* foot rot. The disease incidence was 23 to 35 per cent lower and yield was 30 to 40 per cent higher than control. Pre and post planting application of *T. viride* and *T. harzianum* was effective against foot rot disease of beetle vine. Amendment of nursery soils with *T. harzianum* prior to the colonization of damping off pathogen in nursery beds resulted in the reduction of pre and post emergence damping off of cardamom. Delivery of *Trichoderma* strains through solid matrix priming in tomato and chillies reduced the incidence of damping off under field conditions than dry seed treatments (Nakkeeran and Sabitha Doraisamy, 2001; Kannaiyan *et al.*, 2002).

Strain Improvement

Genetic engineering

A successful bio-control agent requires many characteristics. All of them may not be present in one strain. Since many bio-control agents have infrequent/unknown sexual stages, crossing and progeny testing by conventional breeding is not possible. Genetic engineering is a useful tool to:

i. Transfer the ability to produce antibiotics from one species to another,
ii. Produce location specific strains.
iii. Improve rhizosphere competence of the antagonist and,
iv. Improve its bio-control potential.

By these techniques important attributes like tolerance to pesticides, adverse environmental conditions like extremes of moisture, temperature and pH and ability to grow vigorously, capacity to sporulate in deep tank fermentation and improved shelf life can be incorporated in one organism. The strain k-1026 of *Agrobacterium radiobacter* is the first genetically manipulated antagonist, which is being exploited commercially. This genetic manipulation prevent breakdown of agrocin producing avirulent strain (Cook, 1993).

Methods of induction of new biotypes

New biotypes of bio-control organisms can be developed by:

- Exposure to fungicides,
- Exposure to chemical and physical mutagens,
- Protoplast fusion and transformation.

Exposure to fungicides

By exposing *T. harzianum* to chlorothalonil several isolates tolerant to it were developed. They differed in morphological characters, growth habit and sporulation from the wild type. Some showed enhanced biocontrol activity, which may be attributed to mutation, chromosomal aberration or mitotic recombination (Papavizas, 1987). Prolonged and repeated exposure to fungicides like chlorothalonil, iprodione, procymidone, and vinclozolin has resulted in the selection of several strains of *T. harzianum* resistant to above fungicides (Abd-El Moity *et al.*, 1982).

Exposure to mutagens

Trichoderma have one hundred and nine alleles with great genetic variability. This genetic diversity could be explored to induce improved strains from the existing wild ones. Evidence in the late 1970s showed that it was possible by physical and chemical mutagenesis to improve the yield and quality of enzymes by *Trichoderma* spp. (Montenecourt and Eveleigh, 1979). Improved strains of *T. reesii* were obtained by mutations and produced more cellulase than the wild type strains (Mandels *et al.*, 1971). But this technology was not imported to enhance the bio-control potentialities of antagonistic fungi until 1980s.

Gamma irradiation

Mutants of *T. viride* resistant to benomyl were obtained through gamma irradiation (Troutman and Matejka, 1978), but its biocontrol properties were not reported (Papavizas and Lewis, 1983). Stable mutants of *G. virens* were obtained by exposing the conidia to gamma rays. It showed variations in phenotype, growth rate, sporulation and antagonistic potential against *Rhizoctonia solani* and *Sclerotium rolfsii* from the parent strain. One mutant produced more chlamydospores but no conidia and was more aggressive than the wild type in antagonizing the pathogens *in vitro* (Mukherjee and Mukhopadhyay, 1993). But its efficacy under field conditions was not proved.

UV irradiation

Papavizas *et al.* (1982), for the first time used UV rays to induce new biotypes of *Trichoderma.* They used the inability of *Trichoderma* spp. to tolerate benomyl as a marker in a selection system following induced mutation with UV light. The biotypes of *Trichoderma* differed considerably from wild type (WT-6) in appearance, growth habit, fungitoxic metabolite production against *Sclerotium cepivorum* and ability to survive in soil. Several genetic variants from wild-type strains of *T. harzianum* and *T. viride* were developed. Several biotypes tolerated high concentrations (500 mg/ml of active ingredient) of benomyl and other fungicides like thiabendazole and carbendazim. A few benomyl tolerant biotypes of *T. viride* and *T. harzianum* differed in their growth characters, ability to sporulate, and survival in soil and in the suppression of saprophytic growth of *R. solani* in soil. Several UV-induced biotypes were consistently more effective than the wild type strains in quenching the activity of *P. ultimum* on peas, *R. solani* on cotton and radish, *S. rolfsii* on beans and *S. cepivorum* on onion. The isolate (T-1-R9) was also effective against *Fusarium* wilt of chrysanthemum (Locke *et al.*, 1985) and *Rhizoctonia* scurf of potato (Beagle-Ristaino and Papavizas, 1985). Biotype T-1-R9 was able to recolonize the soil even in the presence of benomyl. It prevented the re-colonization of soil by the pathogen and brought control similar to that of the integrated control that includes the use of benomyl, lime and nitrate nitrogen (Locke *et al.*, 1985). Papavizas *et al.* (1982) found that several UV induced mutants of *T. harzianum* produced heat-labile and heat-stable metabolite in the gliotoxin fermentation medium. It was the most effective bio-control agent against white rot of onion. The wild type strain did not produce heat labile metabolite to the assay organism (*S. cepivorum*). UV mutants of *T. pseudokoningii* inhibited the growth of *R. solani* than the parent and effectively controlled the damping-off disease of tomato (Hadwan and Khara, 1992).

Chemical mutagenesis

Exposing the conidia to N-methyl-N-nitro-N-nitrosoguanidine produced benomyl tolerant mutants of *T. harzianum* (T-95, T-12B). It produced more cellulase than the wild-type strain (Ahmad and Baker, 1987). They also found that the mutants utilized cellulose as a sole carbon source for growth and were able to colonize plant rhizosphere, whereas the wild-type strains did not. NTG induced mutants of *Trichoderma* spp. were tolerant to benomyl, but not rhizosphere competent (Ahmad and Baker, 1988). NTG induced mutants of *T. hamatum* tolerated benomyl and thiobandazole upto 100 mg a.i./ml (Katan *et al.*, 1984).

Advantages:

- Development of such fungicide resistant/tolerant strains would induce the industry to take up pilot testing and commercial production.
- Could be combined with minimal amount of fungicide for disease control.
- Such mutants can also be used in genetic recombination programme, by the introduction of specific gene encoding for increased production of antibiotics and enzymes into a recipient bio-control fungus that possess other desirable traits.

Protoplast fusion and transformation

Protoplast fusion is an excellent approach to the construction of fungal strains with novel gene combinations (Papavizas, 1987). It is possible to combine desirable traits from various parental strains to produce superior biocontrol strains. Interspecific crosses have been accomplished in *T. viride*. (Stasz *et al*., 1989). Protoplast transformation is the process by which naked DNA is introduced into a cell resulting in a heritage change. It involves two basic steps:

i. Partial or complete elimination of cell barriers to allow uptake of DNA into the cell.

ii. Stable integration, maintenance and self-replication of DNA into cell.

Isolated protoplasts are excellent materials for the introduction of foreign DNA. This can be done in three ways:

a. Fusion of whole protoplasts from two strains, species or genera

b. Transfer of mitochondrial plasmids- plasmid mediated transformation. Hybrids contain genetic material only from the naturally inbreeding microbes from which crosses strains are derived.

c. Transfer of DNA segments (autonomously replicating and integrated vectors) - recombinant DNA technique.

DNA segments are separated from donor and inserted or attached to other cells. Eukaryotic genes cloned into *E. coli*, sequence are rearranged and introduced into the new organism (Papavizas, 1987). Resistance to benomyl was constructed by transformation (Ossanna and Mischke, 1990). This was accomplished by transferring to GT-21, a wild type strain of *T. virens*, very sensitive to benomyl. The b-tubulin gene from *Neurospora crassa* that codes for resistance to benomyl was successfully expressed in *T. virens* strain GT-21. It suggests that there are tremendous possibilities for desirable transformations in bio-control fungi, provided that the genes coding for certain desirable traits have been identified.

Conclusion

The decades of laboratory research has brought the usage of *Trichoderma* spp. from lab to land and promoted the commercial application of *Trichoderma* world wide. It would be unrealistic to expect that this biocide would replace chemical fungicides in disease control if it has increased shelf life equivalent to that of chemical pesticides. Future research should be directed towards exploiting the avenues of *Trichoderma* spp. for bioremediation of pesticides, heavy metals, noxious chemicals, polymers, effluents etc. Biological agents achieve their effect through a variety of mechanisms and, thus avoid the development of resistance, which often renders fungicides obsolete. Future research should give us the insight necessary to improve the genetic potential of the organisms to have a broad spectrum of action against several plant pathogens, as well as the appropriate delivery systems to achieve consistent, effective, and environmentally sound biological control strategies in a wide range of applications.

References

Abd-El Moity, T.H., G.C. Papavizas and J.A. Lewis. 1982. Induction of new isolates of *Trichoderma harzianum* tolerant to fungicides and their experimental use for control of white rot on onion. *Phytopathology*, **72**: 396-400.

Ahmad, J. S. and R. Baker. 1988. Implications of rhizosphere competence of *Trichoderma harzianum*. *Can. J. Microbiol.*, **34**: 229-234.

Ahmad, J.S. and R. Baker. 1987. Competitive saprophytic ability and cellulolytic activity of rhizosphere-competent mutants of *Trichoderma harzianum*. *Phytopathology*, **77**: 358-362.

Altomare, C., W. A., Norvell, T. Björkman and G.E. Harman. 1999. Solubilization of Phosphates and Micronutrients by the Plant-Growth-Promoting and Biocontrol Fungus *Trichoderma harzianum* Rifai 1295-22. *Appl. Envir. Microbiol.*, **65**: 2926-2933.

Arshad, M. and W. T. Frankenberger Jr. 1991. Microbial production of plant hormones. In: The rhizosphere and plant growth (eds.) D. I. Keister and P.B. Cregan,. Kluwer Academic Publisher, Dodrecht, The Netherlands, pp. 327- 324.

Auer, N., J.N., Hedger, and C.S. Evans. 2005. Degradation of nitrocellulose by fungi. *Biodegradation*, **16**(3): 229-236.

Auer, N., R.G. Veness, and C.S. Evans. 1999. Bioremediation of nitrocellulose pollution. *Land Contamination and Reclamation*. 7: 301-4.

Backer, J.O. and R. J. Cook. 1988. The role of siderophores in suppression of *Pythium* spp. and production of increased growth response of wheat by fluorescent pseudomonads. *Phytopathology*, **78**: 778-782.

Baker, K.F. and R. J. Cook. 1974. Biological control of plant pathogens. W.H. Freeman and Co., San Francisco, C.A. p. 433.

Baker, R. 1991. Induction of rhizosphere competence in the biocontrol fungus, *Trichoderma*. In: Beltsville symposia in agricultural research. The rhizosphere and plant growth (eds.) D.L. Keister and P.B. Cregan), pp. 221-228.

Baker, R., Y. Elad and and I. Chet. 1984. The controlled experiment in the scientific method with special emphasis on biological control. *Phytopathology*, **74**: 1919-1021.

Baker, R.1989. Improved *Trichoderma* spp. for promoting crop productivity. *Trends in Biotechnology*, **7**: 34-38.

Beagle-Ristaino, J. and G. C. Papavizas. 1985. Biological control of *Rhizoctonia* stem canker and black scurf of potato. *Phytopathology*, **75**: 560-564.

Bettucci, L. and M.F. Roquebert. 1995. Microfungi from a tropical rain-forest litter and soil - A Preliminary-Study. *Nova Hedwigia*, **61**: 111-118.

Bisby, G.R. 1939. *Trichoderma viride* Pers. ex Fries, and notes on *Hypocrea*. *Trans. Brit. mycol. Soc.*, **23**: 149-168.

Bissett, J. 1984. A revision of the genus, *Trichoderma*. I. Sect. *Longibrachiatum* sect. nov. *Can. J. Bot.*, **62**: 924-931.

Bissett, J. 1991b. A revision of the genus *Trichoderma*. III. Sect. *Pachybasium*. *Can. J. Bot.*, **69**: 2373-2417.

Bissett, J. 1991c. A revision of the genus *Trichoderma*. IV. Additional notes on section *Longibrachiatum*. *Can. J. Bot.*, **69**: 2418-2420.

Bissett, J., 1992. *Trichoderma atroviride*. *Can. J. Bot.*, **70**: 639-641.

Bissett, J.1991a. A revision of the genus *Trichoderma*. II. Infrageneric classification. *Can. J. Bot.*, **69**: 2357-2372.

Carsolio, C., A. Gutierrez, B. Jimenez, M. Van Montagu and A. Herrera-Estrella. 1994. Characterization of ech-42, a *Trichoderma harzianum* endochitinase gene expressed during mycroparasitism. *Proc. natl. Acad. Sci.*, USA. **91**: 10903-10907.

Chaverri, P., F. Candoussau, G.J. Samuels. 2003. *Hypocrea phyllostachydis* and its *Trichoderma* anamorph, a new bambusicolous species from France. *Mycol. Progr.*, **3**: 29-36.

Cherif, M. and N. Benhamou. 1990. Cytochemical aspects of chitin breakdown during the parasitic action of a *Trichoderma* sp. on *Fusarium oxysporum f.sp. radicis lycopersici*. *Phytopathology*, **80**:1406-1414.

Chet, I., Z. Barak and A. Oppenheim. 1993. Genetic engineering of microorganisms for improved biocontrol activity. In: *Biotechnological prospects of plant disease control* (ed.) I. Chet). Wiley-Liss, New York. pp. 211-235.

Chu, F.F. and W. S. Wu. 1981. Biological and chemical control of *Rhizoctonia solani* by pea seed treatment. *Memoris of the college of Agriculture.* National Taiwan University. **21**: 19-28.

Cook. R.J. 1993. Making greater use of introduced micro organisms for biological control of plant pathogens. *Ann. Rev. Phytopathol.*. **31**: 53-80.

Corke, A.T.K. and T. Hunter. 1979. Biocontrol of *Nectria galligera* infection of purning wounds on apple shoots. *J. Hort. Sci.*, **54**: 47-55.

Cuevas, V. C., C. T. Culiat, R. L. Manaligod and N. E. Sajise. 1994. Induction of fungicide resistance for enhanced cellulase production of *Trichoderma harzianum* Rifai. *Philippines Journal of Science*. **123**: 177-183.

Dalsoglio, F.K., B. L. Bertagnolli, J. B. Sinclair and D. M. Eastbrn. 1998. Production of chitinolytic enzymes and endoglucanases in the soybean rhizosphere in the presence of *T. harzianum* and *R. Solani. Biol. Cont.*, **12**: 111-117.

Danielson, R. M and C. B. Davey. 1973. The abundance of *Trichoderma* propagules and the distribution of species in forest soils. *Soil Biol.Biochem.*. **5**: 486-494.

De Meyer, G., J. Bigirimana, Y. Elad and M. Höfte. 1998. Induced systemic resistance in *Trichoderma harzianum* T39 biocontrol of *Botrytis cinerea . European J. Plant Pathol.*. **104**(3): 279-286

Deshpande, M.V. 1986. Enzymatic degradation of chitin and its biological applications. *J. Sci. Indus. Res.*, **45**: 273-287.

Doi, Y., Y. Abe, J. Sugyama. 1987. *Trichoderma* sect. *Saturnisporum*, sect. nov. and *Trichoderma ghanense*, sp.nov. *Bull. Nat. Sci. Museum, Ser B (Botany)*. **12**: 1-9.

Elad, Y., I. Chet. and P. Katan. 1980. *Trichoderma harzianum*. A biocontrol agent effective against *Sclerotium rolfsii* and *Rhizoctonia solani. Phytopathology*. **70**: 119-121.

Elad, Y., I. Chet, and Y. Henis. 1982. Degradation of plant pathogenis fungi by *Trichoderma harzianum. Can J. Microbiol.*, **28**: 719-725

Elad, Y., I. Chet, P. Boyle and Y. Henis. 1983. Parasitism of *Trichodermaia* spp. on *Rhizoctonia solani* and *Sclerotium rolfsii*. Scanning electron microscopy and fluorescenct microscopy. *Phytopathology*. **73**: 85-88.

Elisa Esposito and Manuela da Silva. 1998. Systematics and environmental application of the genus *Trichoderma*. *Critical. Rev. Microbiology*.. **24**: 89-98.

Ezzi, M. I. and J. M. Lynch. 2005. Biodegradation of cyanide by *Trichoderma* spp. and *Fusarium* spp. *Enzyme Microb. Technol.*. **36**: 849-854.

Garcia, I., J. M. Lora, J. De La Cruz, T. Benitez, A. Lobell and J. A. Pintor-Toro. 1994. Cloning and characterization of chitinase CHIT 42 cDNA from the mycoparasitic fungus *Trichoderma harzianum. Curr. Genet.*, **27**: 83-89.

Geremia, R., G. H. Goldman, D. Jacobs, W. Ardiles, S. B. Vila, M. Van Montagu and E. Herera. 1993. Moelcular characterization of the proteinase-encoding gene, Prb1, related to mycoparasitism by *Trichoderma harzianum. Mol. Microbiol.*, **8**: 603-613.

Gherbawy, Y., I. Druzhinina, G. M. Shaban, M. Wuczkowsky, M. Yaser, M. A. El-Naghy, H. J. Prillinger and C. P. Kubicek. 2004. *Trichoderma* populations from alkaline agricultural soil in the Nile valley, Egypt. consist of only two species. *Mycol. Prog.*, **3**: 211-218.

Goldman, H.G., C. Hayes, and G. E. Harman. 1994. Molecular and cellular biology of biocontrol by *Trichoderma* spp. *Trends in Biotechnology*. **12**: 478-482.

Grondona.I., R. Hermosa, M. Tejada, M.D. Gomis, P.E. Mateos, P.D. Bridge, E. Monte and I. Garcia-Acha. 1997. Physiological and biochemical characterization of *Trichoderma* harzianum, a biological control agent against soilborne fungal plant pathogens. *Appl. Environ. Microbiol.*, **63**: 3189.

Grover, R., S. Kumar, P. Soni and K.P. Sharma. 2004. Ecology and decolorizing potential of fungi in the textile dye waste water polluted habitats of Sanganer (Jaipur) and biological effluent treatment plant. *Nat., Environ. Pollut. Technol.*, **3**: 225-238.

Hadwan, H.A. and H.S. Khara. 1992. Effect of UV radiation on antagonistic activity of *Trichoderma pseudokoningii* against *Rhizoctonia solani* causing damping-off of tomato. *Pl. Dis. Res.*, 7: 64-66.

Haran, S., H. Schickler, A. Oppenheim and I. Chet. 1995. New components of the chitinolytic system of *Trichoderma harzianum*. *Mycol. Res.*, **99**: 441-446.

Harman, G.E., C.K. Hayes, M. Lorito, R.M. Broadway, A. Di-Pietro, C. Petebauer and A. Tronsmo. 1993. Chitinolytic enzymes of *Trichoderma harzianum*. Purification of chitobiosidase and endochitinase. *Phytopathology*, **83**: 313-318.

Howell, C.R., R.D. Stipanovic and R.D. Lumsden. 1993. Antibiotic production by strains of *Glicaldium virens* and its relation to the biocontrol of cotton seedling diseases. *Biocontrol science and Technology*, **3**: 435 – 441.

Jacobs, D., R. A. Geremia, G. H. Goldman, O. Kamoen, M. Van Montagu and A. Herrera-Estrella. 1991. Study of the expression of â-1, 3 – glucanase of *Trichoderma harzianum*, *Petria*, **1**: 125.

Jérôme, R.M. and Fumio. 2002. Assessment of the rhizosphere competency and pentachlorophenol—metabolizing activity of a pesticide-degrading strain of *Trichoderma harzianum* introduced into the root zone of corn seedlings. *J. Environmental Science and Health*, **37**: 201-210.

Jeyarajan, R. and S. Nakkeeran. 2000. Exploitation of microorganisms and viruses as biocontrol agents for crop disease management. In: *Biocontrol potential and their exploitation in crop disease management*, (eds.) Upadhyay, *et al* Kluwer Academic/Plenum Publishers, USA , pp. 95-116.

Kannaiyan, S. Sabitha Doraiswamy, G. Ramakrishnan, R.J. Rabindra, R. Samiyappan, V. Udayasurian, K. Gunasekaran, J.S. Kennedy, S. Nakkeeran and N. Sathiah. 2002. *Recent Advances in Eco-friendly Approaches for the Management of Pests and Diseases in Crops. Tamil Nadu Agricultural University, Coimbatore*, pp. 389.

Katan, T., M. T. Dunn, and G. C. Papavizas. 1984 Genetics of fungicide resistance in *Talaromyces flavus*. *Can. J. Microbiol.*, **30**: 1079-1087.

Kindermann, J., Y. El-Ayouti, G. J. Samuels, C. P. Kubicek. 1998. Phylogeny of the genus *Trichoderma* based on sequence analysis of the internal transcribed spacer region 1 of the rDNA clade. *Fungal Genet. Biol.*, **24**: 298-309.

Kitamoto, Y., R. Kono, Y. Shimotori, N. Mori and Y. Ichikawa. 1987. Purification and some properties of an exo â-1, 3 – glucanase from *Trichoderma harzianum*. *Agric. Biol. Chem.*, **51**: 3385-3386.

Kubatova, A., K. Prasil and M. Vanova. 2002. Diversity of soil microscopic fungi on abandoned industrial deposits. *Cryptogamie Mycologie*, **23**: 205-219.

Kubicek, C.P., J. Bissett, I. Druzhinina, C.M. Kullnig-Gradinger, G. Szakacs. 2003. Genetic and metabolic diversity of *Trichoderma*: a case study on South East Asian isolates. *Fungal Genet. Biol.*, **38**: 310-319.

Kubicek, C. P., U.M. Bolzlbauer, W. Kovacs, R.L. Mach, K. Kuhls, E. Lieckfeldt, T. Borner and G. Samuels. 1996. Cellulase formation by species of *Trichoderma* sect Longibrachiatum and of Hypocrea spp with anamorphs referable to *Trichoderma* sect Longibrachiatum. *Fungal Genetics and Biology*, **20**: 105-114.

Kuehn, K. A. and R. D. Koehn. 1991. Fungal Populations Isolated from Ezell Cave in South Texas. *Stygologia*, **6**: 65-76.

Kullinig, C., G. Szakacs and C.P. Kubicek. 2000. Molecular identification of *Trichoderma* from Russia, Siberiya and the Himalya. *Mycol. Res.*, **104**(9): 1117-1125.

Kullnig-Gradinger, C.M., G. Szakacs and C. P. Kubicek. 2002. Phylogeny and evolution of the fungal genus *Trichoderma*"a multigene approach. *Mycol. Res.*, **106**: 757-767.

Lifshitz, R., M. T. Windham and R. Baker. 1986. Mechanism of biological control of pre-emergence damping-off by use of integrated chemical and biological controls. *Phytopathology*, **69**: 431-434.

Limon, M. C., J. M. Lora, I. Garcia, J. De La Cruz, A. Lobell, T. Benitez and J. A. Pintor Toro. 1995. Primary structure and expression pattern of the 33 k Da chitinase gene from the mycoparasitic fungus *Trichoderma harzianum*. *Curr.Genet.*, **28**: 478-483.

Locke, J. C., J. J. Marois and G. C. Papavizas. 1985. Biological control of *Fusarium* wilt of greenhouse grown chrysanthemums. *Plant Dis.*, **69**: 167-169.

Lorito, M., C.K. Hayes, A. Di Pietro, S.L. Woo and G.E. Harman. 1994a. Purification, characterization and synergistic activity of a glucan 1,3- â-glucosidase and an N-acetyl- â-glucosaminidase from *Trichoderma harzianum*. *Phytopathology*. **84**: 398-405.

Lorito, M., C. Peterbauer, C. K. Hayes and G. E. Harman. 1994b. Synergistic interaction between fungal cell wall degrading enzymes and different antifungal compounds enhances inhibition of spore germination. *Microbiology*, **140**: 623-629.

Lorito, M., G.E. Harman, C.K. Hayes, R.M. Broadway, A. Tronsmo, S.L. Woo and A. Di-pietro. 1993. Chitinolytic enzymes produced by *Trichoderma harzianum*: Antifungal activity of purified endochitinase and chitobiosidase. *Phytopathology*, **83**: 302-307.

Lumley, T.C., L.D. Gignac R.S. and Currah. 2001. Microfungus communities of white spruce and trembling aspen logs at different stages of decay in disturbed and undisturbed sites in the boreal mixed wood region of Alberta. *Can. J. Bot.*, **79**: 76-92.

Lupicka, K., W. Strof. K. Kubs, M. Skorupa, P. Wieczorek, B. Lejczak and P. Kafarski. 1997. The ability of soil –borne fungi to degrade organophosphanate carbon to phosphorus bounds. *Appl. Microbiol. Biotechnol.*, **48**: 549.

Mandels, M., J. Weber and R. Parizek. 1971. Enhanced cellulase production by a mutant of *Trichoderma viride*. *Appl. Microbiol.*, **21**: 152-154.

Mauch, F., B. Mauch-Mani and T. Boller. 1988. Antifungal hydrolases in pea tissue. II. Inhibition of fungal growth by combinations of chitinase and â-1,3-*glucanase*. *Plant Physiol.*, **88**: 936-942.

Montenecourt, B.S and and D. E. Eveleigh. 1979. Production and characterization of high yielding cellulase mutants of *Trichoderma reesei*. Tech. Assoc. Pulp Paper Ind. Annu. Meet. pp. 101-108.

Mtui, G. and Y. Nakamura. 2005. Bioconversion of lignocellulosic waste from selected dumping sites in Dar es Salaam, Tanzania. *Biodegradation*, **16**: 493-499.

Mukherjee, P.K. and A. N. Mukhopadhyay. 1993. Induction of stable mutants of *Gliocladium virens* by gamma – irradiation. *Indian Phytopath.*, **46**: 393-397.

Nakkeeran, S. and Sabitha Doraisamy. 2001. Genetic improvement of *Trichoderma viride* and assessment of its efficacy against soil borne pathogens. In: National Seminar on Emerging Trends in Pests and Disease Management. Center for Plant Protection Studies, TNAU, Coimbatore, p. 130.

Nicol, S. 1991. Life after death for empty shells. *New Scientist*, **129**: 46-48.

Ossamna, N. and S. Mischke. 1990. Transformation of the biocontrol fungus *Glicladium virens* to benomyl resistance. *Appl. Envir. Microbiol.*, **56**: 3052-3056

Papavizas, G.C. 1985.*Trichoderma* and Gliocladium:Biology, ecology and potential for biocontrol. *Ann.Rev. Phytopathol.*, **23**: 23.

Papavizas, G.C. 1987. Genetic manipulation to improve the effectiveness of biocontrol fungi for pland disease control. In: *Innovative approaches to plant disease control*. (ed.) I. Chet, New York, NY. John Wiley and Sons, pp. 193-212.

Papavizas, G.C. and J.A. Lewis. 1983. Physiological and biocontrol characteristics of stable mutants of *Trichoderma viride* resistant to MBC fungicides. *Phytopathology*, **73**: 407-411.

Papavizas, G.C., J.A. Lewis and T.H. Abd-El-Moity. 1982. Evaluation of new biotypes of *Trichoderma* harzianum for tolerance to benomyl and enhanced biocontrol capabilities. *Phytopathology*, **72**: 126-132.

Persoon, C.H. 1794. Disposito Methodica Fungorum in Classes, Ordines, Familias et Genera. In: Neues Magazin für Botanik. (ed.) J.J. Römer, Ziegler und Söhne, pp. 63-128.

Petre, M., M. E. Teodorescu, G. Zarnea, P. Adrian, E. Gheorghiu and V. Gheordunescu. 2001. *Meded Rijksuniv Gent Fak Landbouwkd Toegep Biol Wet*, **66**: 195-198.

Rashmi, K., Naga Sowjanya. P. Maruthi Mohan V. Balaji and G. Venkateswaran. 2004. Bioremediation of ^{60}Co from simulated spent decontamination solutions. *Science of the Total Environment.* **328**: 1-14.

Rifai. M. A. 1969. A revision of the genus *Trichoderma. Mycol. Pap.* **116**: 1-116.

Rigout. J. and F. Matsumura. 2002. Assessment of the rhizosphere competency and pentachlorophenol—metabolizing activity of a pesticide-degrading strain of *Trichoderma* harzianum introduced into the root zone of corn seedlings. *J. Env. Sci. & Health*, **37**: 201-210.

Rodriguez- Kabana. R. W. D. Kelley and E. A. Curl. 1978. Proteolytic activity of *Trichoderma viride* with mixed culture in *Sclerotium rolfsii* in soil. *Can. J. Microbiol.*, **24**: 487-490.

Rossman. A. Y. 1996. morphological and molecular perspectives on systematics of Hypocreales. *Mycologia*, **88**: 1-19.

Salt. G.H. 1978. The increasing interest in minor pathogens. In: *Soil-borne plant pathogens.* (Eds.) B. Schippers and W. Gams, Academic Press, New York. pp. 289-312.

Samuels, G. J., O. Petrini, K. Kuhls, E. Lieckfeldt and C. P. Kubicek. 1998. The *Hypocrea schweinitzii* complex and *Trichoderma* sect. *Longibrachiatum. Stud. Mycol.*. **41**: 1-54.

Sandgren. M., P.J. Gualfetti. A. Shaw, L.S. Gross, M. Saldajeno, A.G. Day, T.A. Jones and C. Mitchinson. 2003. Comparison of family 12 glycoside hydrolases and recruited substitutions important for thermal stability. *Protein Science*, **12**: 848-860.

Sanz, L., M. Montero, I. Grondona. J. Vizcaino, A. Llobell, R. Hermosa and E. Monte. 2004. Cell wall-degrading isoenzyme profiles of *Trichoderma* biocontrol strains show correlation with rDNA taxonomic species. *Current Genetics*, **46**: 277-286.

Sivan. A. and I. Chet. 1989. The possible role of competition between *Trichoderma harzianum* and *Fusarium oxysporium* on rhizosphere colonization. *Phytopathology*, **79**: 198-203.

Sriram, S., T. Raguchander, S. Babu, R. Nandakumar, V. Shanmugam, V. Vidhyasekaran, P. Balasubramanian and R. Samiyappan. 2000. Inactivation of rice sheath blight pathogen *Rhizoctonia solani. Can. J. Microbiol.*, **46**: 520-524.

Stasz. T.E., G.E. Harman and L. Gullion. 1989. Limited vegetative compatibility following intra and interspecific protoplast fusion in *Trichoderma. Experimental Mycology*, **13**: 364-371.

Troutman, J. L. and J.C. Matejka. 1978. Induced tolerance of *Trichoderma viride* to benomyl. *Phytopath News.*, **12**: 131.

Vinale. F., K. Abadi, M. Ruocco, R. Marra, F. Scala, A. Zoina. S. Woo and M. Lorito. 2003. Remediation of pollution by using biological systems based on beneficial plant-microorganisms interactions. *J. Plant Pathol.*, **85**: 301.

Wilhitte, S.E., R.D. Lumsden and D.C. Skraney. 1994. Mutational analysis of gliotoxin production by the biological fungus *Glicladium virens* in relation to suppression of Pythium damping-off. *Phytopathology*, **84**: 816-621.

Windham, M.T., Y. Elad and R. Baker. 1986. A mechanism for increased plant growth induced by *Trichoderma* spp. *Phytopathology*, **76**: 518-521.

Xu, T. 1996. Advances of molecular biology of *Trichoderma* spp. *Zhenjun Xuebao*, **15**: 143.

Yu, P., J.J. McKinnon and D.A. Christensen. 2005. Improving the nutritional value of oat hulls for ruminant animals with pretreatment of a multienzyme cocktail: *in vitro* studies. *J. Anim. Sci.*, **83**: 1133-1141.

Soil Microbiology and Soil Biotechnology : Vol. 2
Editor : S. Kannaiyan
Associated Publishing Co., New Delhi, 2010. pp. 191-202

9

Transformation Technology against Pest(s) and Disease(s)

K. Lenin, B. Powr and M.R. Davey

Introduction

Rice has been bred for many decades using traditional methods involving years of rigorous selection and backcrossing. The traditional methods involved more time for combining useful traits. Additionally, some of the genes that are found to be useful in related wild type/ other graminae species were difficult to breed into cultivated rice. With the advent of new techniques involving tissue culture and genetic engineering, transfer of gene(s) both horizontally and vertically becomes more amenable. Rice transformation has played a major role in this technology over the past decade and there are many advanced transgenic lines undergoing field trials in the International Rice Research Institute (IRRI) and other National institutes throughout the rice growing regions of the world. The introduction of traits for pest, disease, drought and saline tolerance are some examples. The breakthrough for generating rice with more useful traits became easier than ever before with the generation of breeding lines with transgenes from various sources. These transformed lines are now being used in active breeding strategies in several countries, around the world.

Rice tissue culture

Rice tissue culture is the stepping-stone to many technologies such as micro-propagation, anther culture, *in vitro* selection, embryo rescue, somaclonal variation, somatic hybridization and transformation (Braun and Thorpe, 1995; Collins and Edwards, 1998). The first cereal callus was produced from maize endosperm by La Rue in 1949. The tissue culture studies in rice began with Japanese workers trying to culture excised rice roots (Fujiwara and Ojima, 1995) and immature rice embryos (Amemiya *et al.*, 1956). The regeneration of rice plants was subsequently reported from root-derived callus (Kawata and Ishihara, 1968). These studies laid the foundation for basic cell and tissue culture studies in rice. Rice callus can be induced from different explants such as roots, shoots, leaves, ovary, pollen and endosperm (Bajaj, 1990). The success/failure of rice tissue culture systems depends on various factors

University of Nottingham, Loughborough, Leicester LE 125RD, United Kingdom like the composition of culture media (Croughan and Chu, 1991), explant source (Hartke and Lorz, 1989), carbon supply in the medium (Biswas and Zapata, 1993), age of the callus (Brar *et al.*, 1984), plant genotype influence of the plants (Hartke and Lorz, 1989) and the light and temperature conditions (Adkins *et al.*, 1993)

Rice explants used in transformation

A responsive explant is important for gene delive.y experiments and for efficient plant transformation. In early rice transformation experiments involving PEG and electroporation, protoplasts derived from embryogenic cell suspension represented the only choice of starting material (Shimamoto *et al.*, 1989; Datta *et al.*, 1990). The disadvantage of using protoplasts that their handling is time consuming and it is difficult to regenerate fertile transgenic plants. With the development of other gene delivery systems, various explants were employed in laboratories were the explants are available throughout the year. In European laboratories, mature scutellum-derived calli have been used.

Transformation of rice for agronomic characters

Production of transgenic plants has advanced so that it is now easy to transfer genes of economic importance into crop species. Since the first report of fertile transgenic rice in 1989, a variety of useful genes conferring resistance to biotic and abiotic stresses has been transferred into elite Japonica and Indica cultivars (Tyagi *et al*,. 1999). Transgenic rice conferring resistance to disease and insect pests are discussed in the review.

Resistance against viruses

Rice tungro disease caused by rice tungo spherical virus (RTSV) and rice tungro bacilliform virus (RTBV) has been a major disease in South East Asia, incurring considerable economic losses to the farming community. RTBV is a double stranded virus and RTSV in a positive sense, single stranded RNA virus. Of these, TRBV causes severe disease symptoms. These viruses are transmitted by insect vectors such as green leafhopper (GLH) *Nephotettix virescens*. To obtain resistance against RTBV, mutant viral proteins have been created which act as competitive inhibitors of viral function (Potroykus *et al.*, 1995). Transgenic Indica rice plants have been produced expressing RTBV proteins, which serve as precursors to viral coat protein and reverse transcriptase (Potrykus *et al.*, 1995).

The coat protein gene (Cp) of rice stripe virus was engineered into two Japonica rice varieties by electroporation of protoplasts. The resulting transgenic plants showed significant levels of resistance to stripe virus (Hayakawa *et al.*, 1992). To achieve resistance to rice dwarf virus, Zheng *et al.*, (1997) engineered the outer coat protein gene (S8) of the rice dwarf virus into and obtained transgenic plants. Three different coat protein genes, *Cp1*, *Cp2* and *Cp3* of RTSV, were engineered into Japonica and Indica rice (Si amani *et al.*, 1999). The transgenic plants obtained were analyzed to the R_2 generation for resistance against viral infection. The plants showed moderate resistance and the cumulative effect of the transgenes did not show any significant difference. This is contrary to popular belief that the pyramiding of genes would always help such cases.

Transgenic plants were regenerated following biolistics transformation of vectors carrying a full lengh cDNA of rice yellow mottle virus (RYMV) encoding RNA-dependent RNA polymerase of C-terminal deletions of the cDNA fused to the 35S promoter. Some of the transgenic plants obtained were highly resistant against different isolates of RYMV up to R_3 progeny (Pinto *et al.*, 1999). Rice hoja blanca virus (RHBV) is a major virus disease of economic importance affecting rice growing regions of Latin America and the Caribbean. In 1999 Mu oz and co-workers reported transformation of Costa Rican Indica rice with a coat protein gene of RHBV under the control of the rice *Actin* promoter together with AR sequences. The *bar* gene was used for the selection of transformed tissues. Molecular analysis showed the presence of the RHBV coat protein gene, in low copy number in transgenic plants.

A RHBV nucleocaspid protein gene was transferred to rice cv. Cica 8 by particle bombardment. However, nucleocaspid protein was not detected in the transgenic plants either by western blotting or by ELISA. The data indicate that resistance in these plants was conferred by RNA. When challenged with the RHBV virus, the resistant transgenic lines showed significantly increased performance for important agronomic traits including the number of grains/plant and yield compared to the wild type plants (Lentini *et al.*, 2003). Rice ragged stunt oryzavirus (RRSV) replicates in both its insect vector brown plant hopper (*Nilaparvata lugens*) and its host plant, and has complex multicomponent particles bearing spikes on its outer surface. Transgenic rice plants expressing the 39kDa spike protein showed enhanced resistance to infection by RRSV. When transgenic plants were used to feed to the insect vector, the 39kDa protein interfered with the interaction between the intact virus particles and insect cell receptors (Shao *et al.*, 2003).

Resistance against bacterial diseases

The rice Xa21 gene that confers resistance to the bacterial blight pathogen, *Xanthomonas oryzae* pv. *oryzae* was cloned by Song *et al.*, (1995). Transgenic plants harbouring the cloned gene displayed high degree of resistance with the gene being found to be effective against several isolates of the pathogen (Wang *et al.*, 1996). Subsequently, the Indica rice cultivar IR72 was transformed with the Xa21 gene (Tu *et al.*, 1998) and transgenic plants from the R1 generation were found to be resistant to bacterial blight. These workers also observed transgene silencing in some plant lines. In the same year transgenic elite Indica rice varieties IR64, IR72 and Minghui 63, a restorer line for Chinese Indica hybrid rice, were transformed with the Xa21 gene (Zhang *et al.*, 1998). Bioassays revealed that a number of lines to the R_2 generation were resistant against the Xoo isolates of the pathogen.

Resistance against fungal pathogens

A rice chitinase gene has been employed to generate transgenic plants resistant to fungal pathogens. For example, the rice chitinase gene *Chi* 11 under the control of the constitutive 35S promoter was used to transform rice protoplasts to obtain plants resistant to the sheath blight pathogen *Rizoctonia solani* (Lin *et al.*, 1995). Similarly, a basic chitinase gene from rice was employed to engineer resistance to pathogens (Xu *et al.*, 1996).

Rice chitinase genes *Cht-2* and *Cht*-3 were also engineered into Japonica rice cv. Nipponbare and Koshihikari under the control of an enhanced 35S promoter (Nishizawa *et al.*, 1999). In transgenic plants, the *Cht*-2 product was targeted extracellularly, whereas the *Cht-3* product accumulated intracellularly. Transgenic plants expressing both the genes constitutively were found to show enhanced resistance to the rice blast pathogen, *Magnaporthe grisea.* With a similar objective, Stark-Lorenzen *et al.*, (1997) employed a stilbene synthase gene from grapevine, under the control of its own promoter, to generate transgenic rice plants. The accumulation of stilbene synthase mRNA in response to inoculation with *Pyricularia oryzae* as well as wounding, elicitor treatment and UV irradiation was observed in R_2 plants. Plants expressing this gene showed resistance against *P. oryzae.*

In other experiments, transgenic rice plants expressing the maize ribosome inactivating protein gene, *Rip b-32*, were regenerated following protoplast transformation. Southern hybridization revealed that 30% of the regenerated plants had a single transgene insert. Transmission of the *Rip b-32*, as well as the selectable marker *(bar)* gene was observed into the R2 progeny, with expression levels of the protein being 0.5-1 % of the total soluble leaf protein. However, when R3 transgenic plants were challenged with *R. solani* and *M. grisea*, there was no significant reduction in disease severity as compared to the controls suggesting that normal processing of b-32 proteins may be required for *in planta* antifungal activity (Kim and Wu, 1999).

To achieve resistance against sheath blight disease, transformation with a thaumatin-like protein (pathogenesis related, PR-5) gene (t/p) under the control of the 35S promoter has also been attempted in the rice cvs. Chinsurah Boro II, IR72 and IR51500 (Datta et al., 1999). Transformation was achieved by PEG-mediated transformation of protoplasts, as well as gene delivery to immature embryos by biolistics. Southern hybridization revealed transmission of the transgenes up to the R2 progeny with 80% fertility. Western analysis confirmed the presence of 23kDa TLP-D34 proteins in the transgenic plants (Lenin, 2004). Transgenic rice (cv. Sasanishiki) plants over-expressing the wasabi defensin gene were produced by *Agrobacterium*-mediated transformation. Twenty-two T_2 homozygous lines harbouring the wasabi defensin gene were challenged with the blast fungus and two promising lines (WT 14-5 and WT 43-5) were further challenged in the T_3 generation. The average lesion size in WT 43-5 was reduced to half the size of that in non-transformed plants (Kanzaki *et al.*, 2003). Transgenic rice plants were produced and characterized with over-expression (using CaMV35S promoter) or suppression (using double stranded RNA interference [dsRNAi] of OsMAPK5 gene. In the dsRNAi plant lines, suppression of the OsMAPK5 expression resulted in the constitutive expression of PRI and PR10 genes, significantly enhancing fungal resistance against blast and bacterial (*Burkholderia glumae*) pathogens (Xiong and Yang, 2003). *Agrobacterium*-mediated transformation of Pusa Basmati 1 was carried out using a rice chitinase gene (*Chi*11) under the control of the maize ubiquitin promoter. Out of 11 lines carrying the *Chi*11 gene, 3 showed single copy genes. The progeny of these lines exhibited a 3:1 segregation ratio, reflecting a single locus insert of the transgenes (Lenin, 2004). When challenged with the rice sheath blight pathogen (*R. soalni*), plants showed enhanced resistance in comparison to non-transformed control plants (Sridevi *et al.*, 2003).

Resistance against insect pests

Insect pests are serious constraints to crop production worldwide. The cost associated with management practices and chemical control insects approaches $10 billion annually, yet global crop losses due to insects still account for 20-30% of total production (Estruch *et al.*, 1997). Conventional breeding has aided us to sustain production with increasing number of pests every season. The inability of the breeders to find new sources of resistance genes within related species has been a common concern in recent times. Scientists are now looking into alternate strategies to combat this and one such useful technology would be genetic engineering of crops. The *Bacillus thuringiensis* (Bt) δ-endotoxin, cowpea trypsin inhibitor and lectin genes have been used with success.

Use of Bt genes

The *Bacillus thuringiensis* (Bt) δ-endotoxin crystal (CRY) proteins are part of a large and still growing family of homologus proteins (Lenin *et al.*, 2001a, b). About 130 *cry* genes have been identified to date. Most of the Bt crystal proteins are synthesized in a protoxin form and are proteolytically converted into smaller toxic polypeptides in the insect midgut membranes, causing lesions. *Cry* genes are classified into four groups based on their amino acid homology as *cry* 1, *cry*2, *cry*3 and *cry*4, which are, respectively specific for insects belonging to Lepidoptera (caterpillars), and Diptera (flies and mosquitoes), Coleoptera (beetles).

A modified *cry*1 *A(b)* gene from Bt has been inserted into the Japonica rice cultivar and confers resistance to 2 major rice pests, rice leaf folder (*Cnaphalocrosis medinalis*) and stiped stem borer *(Chilo supressalis*) (Fujimoto *et al.*, 1993). A synthetic *cry 1A(b)* gene driven by CaMV35S promoter was used to engineer Indica rice cultivar IR58 by particle bombardment (Wunn *et al.*, 1996). Bioassays against important rice pests, like yellow stem borer (YSB), striped stem borer (SSB) and the leaf folder (*Cnaphlocrous medinalis*, revealed increased resistance of transgenic lines. In order to achieve high levels of expression, Nayak *et al.*, (1997) reconstructed the *cry 1A(c)* gene and used this construct to transform the Indica rice cultivar IR64. These workers found that even at low levels of expression, the transgenic lines showed resistance against damage caused by yellow stem borer (YSB). With the use of hybrid rice more prevalent in China and India, scientists at IRRI, transformed the maintainer line IR68899B, which is used for hybrid rice production with the *cry1A(b)* gene. Molecular analysis of the transgenic lines revealed stable expression and integration of the transgene into the host plant genome. Insect bioassays showed enhanced resistance against YSB of transgenic plants (Alam *et al.*, 1999). Subsequently, Indica rice Basmati 370 and M7 cultivar were transformed with a novel *cry2A* gene and showed considerable resistance to yellow stem borer and rice leaf folder (Maqbool *et al.*, 1998). There are also several other reports that describe rice transformation with *B. thuringiensis* genes including *cry1A(b)* and *cry1A(c)* (Ghareyazie *et al.*, 1997; Cheng *et al.*, 1998; Maqbool and Christou, 1999).

Use of proteinase inhibitors and lectin genes

Genes derived from plant sources have been used to attempt to combat the problem of codon usage by the Bt genes. Genes encoding proteinase inhibitors are of particular value as they are part of plants natural defence mechanisms. Transgenic rice developed using a chimeric oryza cystatin (Oc) gene and a corn cystatin (Cc) gene, through electroporation, were resistant to *Sitophilus zeamais* (Hosoyama *et al.*, 1995, Irie *et al.*, 1996). Cowpea trypsin inhibitor *(Cpti)* gene driven by *Actin1* promoter was engineered into rice that showed enhanced resistance to striped stem borer and pink dtem borer in small scale field trials (Xu *et al.*, 1999). Modified *(Cpti)* gene with ER targeting signal (KDEL) resulted in two to four times higher average proteinase inhibitor activity than that of plants transformed with *(Cpti)* gene. When the transgenic plants were challenged with larvae of *Chilo suppressalis*, the larvae showed 60-72% mortality rate (Zhen *et al.*, 1999). When a potato proteinase inhibitor II gene driven by wound inducible promoter *Pin II* was expressed in rice the transgenic plants showed stable inheritance up to 4 generations. The transgenic plants were resistant to pink stem borer (Duan *et al.*, 1996). The snowdrop lectin gene *(gna)* has been widely used to engineer transgenic rice for resistance against brown plant hopper (*Nilaparvata lugens*; BPH). It is expressed with both tissue specific (sucrose synthase) and constitutive (maize ubiquitin) promoters for achieving the desired level of protein expression to control the insect. Immunolocalization and bioassays studies confirm the tissue-specific expression and the effectiveness of controlling BPH (Rao *et al.*, 1998; Sudhakar *et al.*, 1998).

Field trials of transgenic rice

Field/performance trails involving transgenic plants are very limited and have been performed more frequently in recent times to obtain valuable data to assess the environmental risks involved in growing such plants in the midst of common crops (Schuh *et al.*, 1993; Oard *et al.*, 1996). Field trials were conducted at Maricopa Agricultural Experimental Station, Arizona, USA, to study the behaviour of *nptll* transformed into rice cv. Taipei 309 through protoplast mediated transformation. The phenotype of R_2 plants derived from 3 R1 lines was studied. Transgenic plants had shorter flag leaves, took longer to flower, and had reduced fertility when compared to non-transformed plants grown under similar conditions. The *nptll* gene was detected in all R_2 plants, but the NPTLL enzyme was not detected (Schuh *et al.*, 1993; Lynch *et al.*, 1995). Transgenic rice plants cv. Nipponbare transformed with the hygromycin resistance gene and the maize transposable activator element (Ac) were compared for major agronomic traits under field conditions. Progeny lines carrying the hygromycine resistance gene (7/8) had relatively normal agronomic traits when compared to protoplast-derived plants and wild-type plants. In contrast progeny lines carrying the maize transposable Ac element showed significant phenotypic variation in 4 out of 6 traits examined. Among 10 progeny lines produced 2 were normal resembling the protoplast-derived plants, 1 sterile and 1 albino plant were found (Li and Murai, 1995).

The Agronomic performance of 11 independently transformed lines of 2 rice cvs. '**Koshihikari**' and '**Gulfmont**' containing the bar gene were evaluated over 2 consecutive years under field conditions at Lousiana State University (LSU) Agricultural Center Rice Research Station. Seeds of the R2 and R3 transgenic lines were drilled into 2m rows in a plot measuring 3.6m x 1.42m. Normal agronomic and fertilization practices were followed. Weeds were controlled by using propanil at 3kg/ha. Glufosinate treatment was used when the plants were at the 3-4 leaf stage. Only transgenic plants survived after 7 days when glufosinate was applied at 1:12 and 2:24 kg/ha and produced normal seeds. This was the first field trial of herbicide resistance rice (Oard *et al.*, 1996). Small-scale field trials have also been conducted to evaluate transgenic rice plants expressing a *Cpti* gene. The data indicate a significant increase in resistance in transgenic plants to 2 species of rice stem borers than the wild type plants (Xu *et al.*, 1996).

Herbicide resistant plants of the rice cvs. "**Glufmont** and "**Koshihikari**" expressing the *bar* gene were evaluated for herbicide glufosinate, under field conditions for dose response studies (Sankula *et al.*, 1997). Glufosinate at concentrations of 1.1 and 2.2kg/ha inhibited the "**Koshihikari**" lines, whereas the "**Glufmont**" lines were resistant. Rice yields in transgenic "**Glufmont**" lines and 6 of 9 "**Koshihikari**" lines were not significantly different when compared to wild type plants at 2.2kg/ha glufosinate applications. A combination of post-emergence herbicides alone or in combination have been evaluated on transgenic rice carrying the *bar* gene and against the red rice under field conditions. Glufosinate at 2.2kg ai/ha alone was less phytotoxic (6%) to transgenic rice than when it was applied in combination with 0.4 kg ai/ha triclopyr (59%) or 0.6 kg ai/ha aciflurofen (22%). Rice yields were compared between treatments over a period of 2 years. Transgenic rice yields was comparable to weed free check during the first year and was 13% the second year. rice yields were reduced for glufosinate and triclopyr mixture in 1994 and 1995 by 39% and 76% when compared to glufosinate treatment alone. Propanil or aciflurofen mixed with glufosinate reduced plant height, panicle maturity and 100 seed weight in red rice when compared to glufosinate application alone (Sankula *et al.*, 1997).

Replicated field trials were carried out to evaluate agronomic traits between the transgenic line expressing the *bar* gene and weedy red rice. Eight F2 populations produced by controlled crossing of two transgenic glufosinate lines and 4 red rice biotypes was evaluated in Louisiana (LA) and Arkansas (AR) in 1997. Plant vigour and plant density were similar among populations derived from either transgenic or non-transgenic parents. Seed dormancy and seed production were not significantly different among transgenic and non-transgenic populations at either location. Dominant Mendelian segregation of glufosinate resistance was detected in 40% of the populations evaluated (Oard *et al.*, 2000). Transgenic insect resistant line KMD1 carrying a synthetic Bt *cry1Ab* gene was tested for insect resistance under field conditions for 8 lepidopteran rice pests. Under laboratory bioassays, 100% mortality was observed for the insects tested with KMD1 leaf tissues when compared to non-transgenic control. Under field infestation, KMD1 plants were not visibly damaged by striped stem borer (*C. suppressalis*), yellow stem borer (*S. incertulas*) and leaf folder (*C. medinalis*), when compared to 80, 9.3 and 88.7% damage for control plants, respectively (*Shu et al.*, 2000).

Two transgenic rice lines KMD1 and KMD2 at the R4 generation expressing synthetic Bt *cry1Ab* gene were tested for stem borer resistance in field condition during the rice growing season of 1998 in 2 areas of Zhejiang province, China. Both KMD1 and MKD2 were highly resistant to striped stem-borer and yellow stem borer and were completely undamaged during the entire rice growing season compared to untransformed parental control (Xiushui 11). Under natural infestation by striped stem borer, the damage of the control plants reached a peak of 88.7% plants with 20.1% of the tillers showing dead hearts. When the neonate larvae was introduced artificially at the vegetative and booting stage of growth,

100% of the plants were damaged, 25.6% of the tillers showed dead hearts compared to 78.9% plants and 15.6% tillers under natural infestation conditions. Damage due to yellow stem borer reached a peak of 97% plants and 22.9% of tillers damage in control plants (Ye *et al.*, 2001).

The application of breeding insect resistance from a transgenic line (KMD1 and KMD2) to conventional rice varieties were evaluated under field conditions. The inheritance of striped stem borer (SSB) resistance of the transgenic lines was investigated through their progenies, *e.g.* F_1, BC1 and F_2 populations. Under laboratory conditions, 100.0% mortality of the SSB larvae was observed after feeding on leaf tissues of F_1 and GUS positive BC1 plants after 48 hours. The area of leaf tissues consumed was similar to that of SSB resistance to susceptibility in BC1 populations was in the ratio of 1:1, which was also confirmed by laboratory experiments. In F_2 populations, the ratio was significantly smaller than a 3:1 ratio for resistance to susceptibility in all 6 Indica x Japonica crosses, though it fitted a 3:1 in all 4 Japonica x Japonica crosses. The data show that the resistance for Bt rice to SSB was controlled by a dominant gene, which was present in a homozygous condition in both KMD1 and KMD2, but the inheritance is affected by other factors (Wang *et al.*, 2002).

The inheritance and expression pattern of the synthetic *cry1A(b)* were studied in the progenies derived from different transgenic Japonica lines under field conditions. Both Mendelian and distorted segregation ratios of the *cry1A(b)* gene was seen in the selfed and crossed F_2 populations. Crosses with Japonica intra subspecies behaved normally whereas the crosses between Japonica and Indica inter subspecies showed distorted segregation pattern for *cry1A(b)*gene in the F_2 population. The *cry1Ab* protein was found to be quantitatively stable up to the F_6 generation under field conditions. Higher concentrations of *cry1Ab* protein were found in the stems, leaves and leaf sheaths than in the roots and grains. The *cry1Ab* protein content in the leaves was highest during the booting stage and during plant development (Wu *et al.*, 2002). Transgenic rice cv. Tsukinohikari plants expressing a glutein A antisense gene were tested under field conditions. Two lines H39-59 and H75-3 selected from the T_2 generation were evaluated during the T_4 generation. When grown in open paddy fields, T_4 plants of the both transgenic lines showed 20% and 40% reduction in the glutelin content of the two transgenic lines was comparable to that of wild type control of original cultivar, Tsukinohikari (Maruta *et al.*, 2002).

Transgenic elite rice lines expressing a Bt fusion gene derived from *cry1Ab* and *cry1A(c)* under the control of rice actin1 promoter were field tested in China. The lines used in the study were the Indica CMS restorer line Minghui 63 and its derived hybrid rice, Shanyou 63. The level of Bt fusion protein expressed in Minghui (T51-1) was 20ng/mg soluble proteins. The Bt Shanyou 63 showed significant protection for 2 lepidopeteran pests, leaf folder and yellow stem borer (Tu *et al.*, 2000a). Homozygous transgenic rice (cv. IR72) line T103-10, expressing the Xa21 was tested for resistance against bacterial blight (BB) pathogen under normal field conditions in Wuhan province, China in 1998 and 1999. The isolates of *Xanthomonas oryzae* pv. *oryzae* used in the experiments were PXO61, PXO79, PXO99 and PXO112 isolated from the Philippines, T2 isolated from Japan, and Zhe173 isolated from China. The results demonstrated that the transgenic homozygous line expressed the same resistance spectrum, but with a shorter lesion length to each inoculated isolates as the lesion length of the *Xa21* donor line IRBB21. The non-transformed control IR72 carrying *Xa4* was resistant to PXO61, PXO112, Zhe173 and T2, but susceptible to PXO99 and PXO79. The negative control variety IR24 was susceptible to all isolates under field conditions. The results demonstrated clearly that the *Xa21* transgene led to an excellent field performance of the introduced bacterial blight resistance trait on the recipient plants. The yield performance of this transgenic homozygous line, T103-10, is comparable with that of the control under field conditions (Tu *et al.*, 2000b).

References

Adkins, S.W., Kunanuvatchaidach, R., Gray, S.J. and Adkins, A.L. 1993. Effect of ethylene and culture environment on rice callus proliferation. *J. Exp. Bot.*, **44**: 1829-1835.

Alam, M.F., Datta, K., Abrigo, E., Oliva, N. Tu, J., Virmani, S.S. and Datta, S.K. 1999. Transgenic insect-resistant maintainer line (IR68899B) for improvement of hybrid rice. *Plant Cell Rep.*, **18**: 572-575.

Amemiya, A. and Aha, T.K. 1956. Culture conditions and growth of immature embryo in rice plant (studies on the embryo culture in rice plant 1). *Bull. Natl. Inst. Agric.*, **6**: 1-40.

Bajaj, Y.P.S. 1990. Cryopreservation of germplasm of vegetatively propagated crops. *B. Soc. Bot. France-Actualites Botaniques.*, **137**: 99-114.

Biswas, G.C. and Zapata, F.J. 1992. Plant regeneration from long-term cell suspension cultures of Indica rice (*Oryza sativa* L. cv. IR43). *J. Plant Physiol.*, **139**: 523-527.

Brar, G.S., Cohen, B.A., Vick, C.L. and Johnson, G.W. 1994. Recovery of transgenic peanut (*Arachis hypogaea* L.) Plants from elite cultivars utilizing ACCELL technology. *Plant J.*, **5**: 745-753.

Cheng, X.Y., Sardana, R., Kaplan, H., Altosaar, I. 1998. *Agrobacterium*-transformed rice plants expressing synthetic *cry1A(b)* and *cry1a(c)* genes are highly toxic to striped stem borer and yellow stem borer. *Proc. Natl. Acad. Sci. USA.* **95**: 2767-2772.

Collin, H. and Edwards, S. 1998. *Plant Cell Culture*. BIOS Scientific Publishers, Oxford, UK.

Croughan, T.P. and Chu, Q. 1990. Establishment of callus cultures and regeneration of plants. In: *Biotechnology in Agriculture and Forestry*: Y.P.S. Bajaj (Ed.) Springer-Verlag. Vol. 14. Springer-Verlag, Heidelberg, Germany.

Datta, K., Velazhahan, R., Oliva, N., Ona, I., Mew, T., Khush, G.S., Muthukrishnan, S. and Datta, S.K. 1999. Over-expression of the cloned rice thaumatin-like protein (PR-5) gene in transgenic rice plants enhances environmental friendly resistance to *Rhizotonia solani* causing sheath blight disease. *Theor. Appl. Genet.*, **98**: 1138-1145.

Datta, S.K., Peterhans, A., Datta, K. and Potrykus, I. 1990. Genetically engineered fertile Indica-rice recovered from protoplasts. *Bio/Technol.* **8**: 736-740.

Duan, X.L., Li, X.G., Xue, Q.Z., Abo-El-Saad, M., Xu, D. and Wu, R. 1996. Transgenic rice plants harboring an introduced potato proteinase inhibitor II gene are insect resistant. *Nature Biotechnol.* **14**: 494-498.

Estruch, J.J., Carozzi, N.B., Desai, N., Duck, N.B., Warren, G.W. and Koziel, M.G. 1997. Transgenic plants: an emerging approach to pest control. *Nature Biotechnol.* **15**: 137-141.

Fujimoto, H., Itoh, K., Yamamoto, M., Kyozuka, J. and Shimamoto, K. 1993. Insect resistant rice generated by introduction of a modified delta-endotoxin gene of *Bacillus thuringiensis*. *Bio/Technol.*, **11**: 1151-1155.

Fujiwara, A. and O'Jima, K. 1955. Physiological studies of plant roots (Part 1). Influence of some environmental conditions on the growth of isolated roots of rice and wheat. *J. Sci. SOil Manure. Japan.* **28**: 9-12.

Ghareyazie, B., Alinia, F., Menguito, C.A., Rubia, L.G., Palma, J.M., de Liwang, E.A., Cohen, M.B., Khush, G.S. and Bennett, J. 1997. Enhanced resistance to two stem borers in an aromatic rice containing a synthetic *cry1A(b)* gene. *Mol. Breed.*, **3**: 401-414.

Hartke, S. and Lorz, H. 1989. Somatic embryogenesis and plant regeneration from various Indica rice (*Oryza sativa* L.) genotypes. *J. Genet. Breed.*, **43**: 205-214.

Hayakawa, T., Zhu, Y., Itoh, K., Kimura, Y., Izawa, T., Shimamoto, K. and Toriyama, S. 1992. Genetically engineered rice resistant to rice stripe virus, an insect-transmitted virus. *Proc. Natl. Acad. Sci. USA*. **89**: 9865-9869.

Hosoyama, H., Irie, K., Abe, K. and Arai, S. 1995. Introduction of a chimeric gene encoding an oryzacystatin-beta-glucuronidase fusion protein into rice protoplasts and regeneration of transformed plants. *Plant Cell Rep.*, **15**: 174-177.

Irie, K., Hosoyama, H., Takeuchi, T., Iwabuchi, K., Watanabe, H., Abe, M., Abe, K. and Arai, S. 1996. Transgenic rice established to express corn cystatin exhibits strong inhibitory activity against insect gut proteinases. *Plant Mol. Biol.*, **30**: 149-157.

Kawahigashi, H., Hirose, S., Hayashi, E. Ohkawa, H. and Ohkawa, Y. 2002. Phytotoxicity and metabolism of ethofumesate in transgenic rice plants expressing the human CYP2B6 gene. *Pestic. Biochem. Phys.*, **74**: 139-147.

Kawata, S. and Ishihara, A. 1968. The regeneration of rice plant in the callus derived from the seminal root. *Proc. Jpn. Acad.*, **44**: 549-553.

Kim, J.K., Duan, X.L., Wu, R., Seok, S.J., Boston, R.S., Jang, I.C., Eun, M.Y. and Nahm, B.H. 1999. Molecular and genetic analysis of transgenic rice plants expressing the maize ribosome-inactivating protein b-32 gene and the herbicide resistance bar gene. *Mol. Breed.*, **5**: 85-94.

La Rue, C. 1949. Cultures on the endosperm of maize. *Amer. J. Bot.*, **34**: 585-586.

Lenin, K., Udayasuriyan, V. and Kannaiyan, S. 2001. *Bacillus thuringiensis* and its use in agriculture. *Madras Agri. J.*, **88**: 187-197.

Lenin, K. 2004. *Evaluation of approaches for rice Transformation.* Ph.D. Thesis. University of Nottingham, United Kingdom, p. 311.

Li, Z.J. and Murai, N. 1995. Agronomic trait evaluation of field-grown transgenic rice plants containing the hygromycin resistance gene and the maize activator element. *Plant Sci.*, **108**: 219-227.

Lin, W., Anuratha, C.S., Datta, K., Potrykus, I., Muthukrishnan, S. and Datta, S.K. 1995. Genetic engineering of rice for resistance to sheath blight. *Bio/Technol.*, **13**: 686-691.

Lynch, P.T., Finch, R.P., Davey, M.R. and Cocking, E.C. 1991. Rice tissue culture and its application. In: *Rice biotechnology*. G.S. Kush, G.H. Toenniessen, (Eds.) CAB International, Wallingford, UK. pp. 135-155.

Maqbool, S.B. and Christou, P. 1999. Multiple traits of agronomic importance in transgenic Indica rice plants: analysis of transgene integration patterns, expression levels and stability. *Mol. Breed.*, **5**: 471-480.

Maqbool, S.B., Husnain, T., Riazuddin, S., Masson, L. and Christou, P. 1998. Effective control of yellow stem borer and rice leaf folder in transgenic rice Indica varieties Basmati 370 and M 7 using the novel delta-endotoxin *cry2A Bacillus thuringiensis* gene. *Mol. Breed.* **4**: 501-507.

Maruta, Y., Ueki, J., Saito, H., Nitta, N. and Imaseki, H. 2002. Transgenic rice with reduced glutelin content by transformation with glutelin A antisense gene. *Mol. Breed.*, **8**: 273-284.

Nayak, P., Basu, D., Das, S., Basu, A., Ghosh, D., Ramakrishnan, N.A. and Sen, S.K. 1997. Transgenic elite Indica rice plants expressing *Cry1Ac* delta-endotoxin of *Bacillus thuringiensis* are resistant against yellow stem borer (*Scirpophaga incertulas*). *Proc. Natl. Acad. USA.*, **94**: 2111-2116.

Nishizawa, Y., Nishio, Z., Nakazono, K., Soma, M., Nakajima, E., Ugaki, M. and Hibi, T. 1999. Enhanced resistance to blast (*Magnaporthe grisea*) in transgenic Japonica rice by constitutive expression of rice chitinase. *Theor. Appl. Genet.*, **99**: 383-390.

Oard, J.H., Linscombe, C.D., Braverman, M.P., Jodan, F., Brouin, D.C., Leech, M., Kohli, A., Vain, P., Cooley, J.C. and Christou, P. 1996. Development, field evaluation, and agronomic performance of transgenic herbicide resistant rice. *Mol. Breed.*, **2**: 359-368.

Pinto, Y.M., Kok, R.A. and Baulcombe, D.C. 1999. Resistance to rice yellow mottle virus (RYMV) in cultivated African rice varieties containing PYMV transgenes. *Nature Biotechnol.*, **17**: 702-707.

Potrykus, I., Burkhardt, P.K., Datta, S.K., Futterer, J., Biswas, G.C., Kloti, A., Spangenberg, G. and Wunn, J. 1995. Genetic engineering of Indica rice in support of sustained production of affordable and high quality food in developing countries. *Euphytica.*, **85**: 441-449.

Rao, K.V., Rathore, K.S., Hodges, T.K., Fu, X., Stoger, E., Sudhakar, D., Williams, S., Christou, P., BHarathi, M., Bown, D.P., Powell, K.S., Spence, J., Gatehouse, A.M.R. and Gatehouse, J.A. 1998. Expression of snowdrop lectin (GNA) in transgenic rice plants confers resistance to rice brown planthopper. *Plant J.*, **15**: 469-477.

Sankula, S., Braverman, M.P., Jodari, F., Linscombe, S.D. and Oard, J.H. 1997. Evaluation of glufosinate on rice (*Oryza sativa*) transformed with the bar gene and red rice (*Oryza sativa*). *Weed Technol.*, **11**: 70-75.

Sankula, S., Braverman, M.P. and Linscombe, S.D. 1997. Glufosinate-resistant, bar-transformed rice (*Oryza sativa*) and red rice (*Oryza sativa*) response to glufosinate alone and in mixtures. *Weed Technol.* **11**: 662-666.

Shao, C.G., Wu, J.H., Zhou, G.Y., Sun, G., Peng, B.Z., Lei, J.L., Jin, D.D., Chen, S.X., Upadhyaya, N.M., Waterhouse, P. and Gong, Z.X. 2003. Ectopic expression of the spike protein of Rice Ragged Stunt Oryzavirus in transgenic rice plants inhibits transmission of the virus to insects. *Mol. Breed.*, **11**: 295-301.

Shimamoto, K., Terada, R., Izawa, T. and Fujimoto, H. 1989. Fertile transgenic rice plants regenerated from transormed protoplasts. *Nature*. **338**: 274-276.

Shu, Q.Y., Cui, H.R., Ye, G.Y., Wu, D.X., Xia, Y.W., Gao, M.W. and Altosaar, I. 2002. Agronomic and morphological characterization of *Agrobacterium*-transformed Bt rice plants. *Euphytica*. **127**: 345-352.

Shu, Q.Y., Ye, G.Y., Cui, H.R., Cheng, X.Y., Xiang, Y.B., Wu, D.X., Gao, M.W., Xia, Y.W., Hu, C., Sardana, R. and Altosaar, I. 2000. Transgenic rice plants with a synthetic *cry1Ab* gene from *Bacillus thuringiensis* were highly resistant to eight lepidopteran rice pest species. *Mol. Breed.*, **6**: 433-439.

Sivamani, E., Huet, H., Ping, S., Ong, C.A., Kochko, A., de Fauquet, C. and Beachy, R.N. 1999. Rice plant (*Oryza sativa* L.) containing rice tungro spherical virus (RTSV) coat protein transgenes are resistant to virus infection. *Mol. Breed.*, **5**: 177-185.

Song, W.Y., Wang, G.L., Chen, L., Kim, H., Pi, L., Holsten, T., Gardner, J., Wang, B., Zhai, W., Zhu, L., Fauquet, C. and Ronald, P. 1995. A receptor kinase-like protein encoded by the rice disease resistance gene, *Xa21*. *Science*. **270**: 1804-1806.

Sridevi, G., Sabapathi, N., Meena, P., Nandakumar, R., Samiyappan, R., Muthukrishnan, S. and Veluthambi, K. 2003. Transgenic Indica rice variety Pusa Basmati 1 constitutively expressing a rice chitinase gene exhibits enhanced resistance to *Rhizoctonia solani*. *J. Plant Biochem. Biotechnol.* **12**: 93-101.

Stark-Lorenzen, P., Nelke, B., Hanssler, G., Muhlbach, H.P. and Thomzik, J.E. 1997. Transfer of a gerapevine stilbene synthase gene to rice (*Oryza sativa* L.) *Plant Cell Rep.* **16**: 668-673.

Sudhakar, D., Fu, X.D., Stoger, E., Williams, S., Spence, J., Brown, D.P., Bharathi, M., Gatehouse, J.A. and Christou, P. 1998. Expression and immunolocalisation of the snowdrop lectin, GNA in transgenic rice plants. *Transgenic Res.*, **7**: 371-378.

Tu, J., Datta, K., Khush, G.S., Zhang, Q. and Datta, S.K. 2000a. Field performance of *Xa21* transgenic Indica rice (*Oryza sativa* L.) IR72. *Theor. Appl. Genet.*, **101**: 15-20.

Tu, J., Ona, I., Zhang, Q., Mew, T.W., Khush, G.S. and Datta, S.K. Transgenic rice variety IR72 with *Xa21* is resistant to bacterial blight. *Theor. Appl. Genet.*, **97**: 31-36.

Tu, J.M., Zhang, G.A., Datta, K., Xu, C.G., He, Y.Q., Zhang, Q.F., Khush, G.S. and Datta, S.K. 200b. Field performance of transgenic elite commercial hybrid rice expressing *Bacillus thuringiensis* delta-endotoxin. *Nature Biotechnol.*, **18**: 1101-1104.

Tyagi, A.K., Mohanty, A., Bajaj, S., Chaudhury, A. and Maheshwari, S.C. 1999. Transgenic rice, a valuable monocot system for crop improvement and gene research. *Crit. Rev. Biotech.*, **19**: 41-79.

Wang, G.L., Song, W.Y., Ruan, D.L., Sideris, S. and Ronald, P.C. 1996. The cloned gene *Xa21*, confers resistance to multiple *Xanthomonas oryzae* pv. *Oryzae* isolates in transgenic plants. *Mol. Plant-Microbe In.*, **9**: 850-855.

Wang, Z.H., Shu, Q.Y., Ye, G.Y., Cui, H.R., Wu, D.X., Altosar, I. and Shu, Q.Y. 2002. Genetic analysis of resistance of Bt rice to stripe stem borer (*Chilo suppressalis*). *Euphytica.*, **123**: 379-386.

Wu, L.Y., Nandi, S., Chen, L.F., Rodriguez, R.L. and Huang, N. 2002. Expression and inheritance of nine transgenes in rice. *Transgenic Res.*, **11**: 533-541.

Wunn, J., Kloti, A., Burkhardt, P.K., Biswas Ghosh, C.G., Launis, K., Iglesias, V.A. and Potrykus, I. 1996. Transgenic Indica rice breeding line IR58 expressing a synthetic *cry1A(b)* gene from *Bacillus thuringiensis* provides effective insect pest control. *Bio/Technol.*, **14**: 171-176.

Xiong, L.Z. and Yang, Y.N. 2003. Disease resistance and abiotic stress tolerance in rice are inversely modulated by an abscisic acid-inducible mitogen-activated protein kinase. *Plant Cell.*, **15**: 745-759.

Xu, Y., Zhu, Q., Panbangred, W., Shirasu, K. and Lamb, C. 1996. Regulation, expression and function of a new basic chitinase gene in rice (*Oryza sativa* L.) *Plant Mol. Biol.*, **30**: 387-401.

Ye, G.Y., Shu, Q.Y., Yao, H.W., Cui, H.R., Cheng, X.Y., Hu, C., Xia, Y.W., Gao, M.W. and Altosaar, I. 2001. Field evaluation of resistance of transgenic rice containing a synthetic *cry1Ab* gene from *Bacillus thuringiensis* Berliner to two stem borers. *J. Econ. Entomol.*, **94**: 271-276.

Zhang, H.M., Yang, H., Rech, E.L., Golds, T.J., Mulligan, A.S., Cocking, E.C. and Davey, M.R. 1988. Transgenic rice plants produced by electroporation-mediated plasmid uptake into protoplasts. *Plant Cell Rep.*, 7: 379-384.

Zhen, Z., Wu, Q., Xu, H., Zeng, Q., Deng, Q., Deng, Z., Xiao, G., Wang, F. and Xie, H. 1999. Targeting CPTI into endoplasmic reticulum (ER) results in high level accumulation in transgenic rice. In: *Abstracts, General Meeting of the International Programme on Rice Biotechnology*, Phuket, Thailand, pp. 31.

Soil Microbiology and Soil Biotechnology : Vol. 2
Editor : S. Kannaiyan
Associated Publishing Co., New Delhi, 2010, pp. 203-211

10

Commercial Production of Mushrooms and Constraints

R.P. Singh[1] *and K.K. Mishra*[2]

Introduction

Cultivated mushrooms have now become popular all over the world. In 1997, the world production of cultivated edible mushrooms was estimated to be greater than 7 miilion tons. The combined total market value for medicinal and edible mushrooms for that year is conservatively estimated at more than US $30 billion. In addition, the contribution of mycorrhizal mushrooms collected in forest eco-systems annually adds serveral billion US dollar. The world production of mushrooms increased steadily during the last two decades as follows: 1.2 million tons in 1981, 4.9 million tons in 1994, and 6.2 million tons in 1997 (Table 1); and it was estimated to be more than 7 million tons in 1999. China has now become the biggest mushroom producer, consumer and export country. For the first time production reached more than 1 million tons in 1990, about 4 million tons in 1997 and 6.6 million tons in 2000 (Table 2). Mushroom cultivation and the processing of mushroom products have been beneficial to millions of people in China, India and other developing countries in terms of financial, social and health improvement. In addition, cultivation and development of mushroom industries positively generate economic growth and have already had an impact at national and regional levels. This impact is expected to continue to increase and to expend in the future with sustainable research and development of mushroom production and mushroom products generating a non-green revolution (Chang, 1999).

Table 1: World production of cultivated edible mushrooms during the period 1981-1997.

Year	Production (MT)	Average annual increase(%)
1981	1237000	--
1986	2182000	14.7
1990	3763000	18.1
1994	4909000	7.6
1997	6158000	8.5

Source: *Int. J. Med Mush., 1999*

1. Director, Centre of Advanced Studies in Plant Pathology. GBPUA&T. Pantnagar-263145 (UA)
2. Senior Research Fellow, Mushroom Research and Training Centre. GBPUA&T. Pantnagar-263145 (UA)

Table 2: The farmed production mushrooms in China during the period 1978-2000.

Year	Production	Average annual increase (%)
1978	60,000	--
1986	5,86,000	109.6
1990	10,00,000	17.6
1994	26,00,000	40.0
1996	35,00,000	17.3
1997	40,00,000	14.3
1998	43,50,000	8.8
2000	66,30,000	26.2

Source: Int. J. Med. Mush., 1999; The Market of Edible Fungi, 2002.

Commercial cultivation of mushroom in India is hardly 3 decades old and with initial lag-phase it has started showing upward trend from a meagre 4000 tons production during 1985, India's total mushroom production reached 30,000 tons in 1995, which has further increased to 40,000 tons in 1996-97 and 70,000 tons in 2003-04. Similarly its productivity has increased from 6kg/q compost in the year 1970 to 22kg/q compost during 2004. States-wise estimates of production of mushroom during 1996, 2000 and 2003 (Table-3) although indicate that most of the states have ventured into this non-traditional farming activity, no effort to develop the domestic marketing consumption of this highly perishable commodity has been made in any of these states either by the private or public sector.

Table 3: State wise production of cultivated mushroom in India.

State	Production (MT)			Type of mushrooms
	1996	2000	2003	
Jammu & Kashmir	400	400	500	*Agaricus*
Punjab	4000	13,000	22500	*Agaricus*
Himanchal	500	600	2500	*Agaricus*
Haryana	4000	5000	6000	*Agaricus*
Uttar Pradesh	4000	2000	3000	*Agaricus/Pleurotus*
Uttaranchal	---	2640	4050	*Agaricus*
Madhya Pradesh	2000	2500	2800	*Agaricus/Pleurotus*
Rajasthan	50	80	150	*Pleurotus/Calocybe*
Maharashtra	7000	7000	7500	*Agaricus/Pleurotus*
Goa	1000	1000	1000	*Agaricus/Pleurotus*
Karnataka	600	700	750	*Agaricus/Pleurotus/Volvariella*
Kerala	100	300	500	*Pleurotus/Volvariella*
Tamil Nadu	10,000	11000	11500	*Agaricus/Pleurotus/Volvariella/Calocybe*
Andhra Pradesh	2500	3000	3500	*Agaricus/Pleurotus*
Orissa	700	700	800	*Pleurotus/Volvariella*
West Bengal	250	300	400	*Agaricus/Pleurotus*
North Eastern States (Including Sikkim)	1000	1200	1250	*Agaricus/Pleurotus/Volvariella/Lentinula*
Delhi/Pondichery/ Chandigarh/Other	1000	1200	1300	*Agaricus/Pleurotus*
Total	**40,000***	**52,620****	**70,000*****	

Source: Mushroom Conference Solan, 2000, Coimbatore, 2002** and AICMIP Workshop 2004****

Species cultivated commercially

A. *Agaricus bisporus*

The cultivated mushroom produced in greatest amounts, about 2 million metric tons (MT) in terms of fresh equivalent weight in 1997, is *A. bisporus*. It is cultivated in more than 100 countries located in every continent. In 2000, the largest amount of 637,304 MT were grown in China; followed by 391,000 MT in the United States; 263,000 MT in the Netherlands; 180,000 MT in France; 105,000 MT in Poland; and 102,000 MT in Italy.

B. *Lentinula edodes*

The mushroom that is second in total worldwide production is *L.edodes*. In 1997, the worldwide production was more than 1,564,000 MT with almost 88.8% (1,397,000 MT) produced in China. As late as 1983, Japan accounted for more than 82% of the total world production of this mushroom. At that time, China represented only 9.4%. Only 8 years later, in 1991, China produced 57% and in 1997 its share had risen to 88.8%. On the other hand, the precentage of *Lentinula* production of this mushroom, which was initially limited to Asian countries, has now spread to the United States, Australia, Canada, Brazil, and a few European countries. It is expected that more countries will grow this mushroom, and it could surpass *A. bisporus* and rank in first place for cultivated edible mushrooms worldwide in the near future.

C. Volvariella volvacea

Volvariella volvacea (the straw mushroom) is a high-temperature mushroom that is grown in largest amounts in tropical and subtropical regions of Asia, although it can also be grown seasonally in temperate regions. In 1997, 180,000 MT were produced, with Vietnam, China, Taiwan, Thailand, and Indonesia the largest producers. Although Southeast Asia is the stronghold for growing *V.volvacea*, it is also grown in Madagascar, Africa, and India. In recent years, this mushroom has become the fastest expanding mushroom in Vietnam, where 20,000 MT were produced 1996, 41,000 Mt in 1997, and more than 100,000 Mt in 1998.

D. Flammulina velutipes

Another mushroom of great popularity in Asia is *F. velutipes,* the winter mushroom, which, as the name implies, is a low-temperature mushroom. The worldwide production in 1997 totalled 284,000 MT with China, Japan, Korea and Taiwan the leading producers. In the period from 1990 (143.0 thousand MT) to 1997 (284.7 thousand MT), production of *Flammulina* in the world has doubled. *Flammulina* has been as an experimental organism in laboratories all over the world, and the requirement of low temperature for fruiting is the only significant factor that limits increased cultivation of this wood-rotting mushroom in temperate regions this need not be a major drawback.

***E. Auricularia* spp.**

Long popular in China as a food and for its medicinal properties, it is not surprising that *Auricularia* is reported to be the first mushroom cultivated by humans (Chang & Miles, 1987). This jelly fungus is commonly known as *Mu Erh* (wood ear, wooden ear, or tree ear) in Chinese. It is also known as Jew's ear (a contraction of Judas' ear). It is worldwide in its distribution in temperate and tropical regions. *Auricularia polytricha* is a widely distributed tropical and subtropical species. while *A. auricula* is a temperate species that occurs occasionally in the subtropics. In 1997, 285.3 thousand MT of *Auricularia* were cultivated. Extensive cultivation takes place in China, Taiwan, Thailand and the Philippines, and, because the mushroom can be produced on basically a sawdust substrate of a variety of woods in plastic bag culture, cultivation of *Auricularia* need not be seriously restricted by geography. Its production in 1997 ranked it as the fourth most important edible mushroom in the world.

E. *Pleurotus* spp.

Cultivated members of the genus *Pleurotus*, the oyster mushroom, is rapidly increasing, and 875.6 thousand MT were produced in 1997. Various species of this wood-rotting fungus are found the world over, and its edibility is especially appreciated in Europe and Asia. In the past decade its popularity has increased, and it is cultivated in many countries in Europe and Asia, including Italy, Germany, the Netherlands, Belgium, China, Japan, Taiwan, India, Singapore, Thailand, Pakistan and Indonesia. To illustrate the increasing popularity of *Pleurotus* during the period from 1986 to 1997, the production in the world has increased by a factor of almost 5.2. Cultivation has begun in Nigeria, Mexico, Brazil, Colombia, Canada, the United States, Namibia, Tanzania, Zambia and Malawi. One reason for the great interest in species of *Pleurotus*, besides their taste and nutritional appeal, is that they secrete a wide range of enzymes, which can degrade all the three key categories of polysaccharides found in the biomass of forest and agricultural crop residues: lignin, cellulose, and hemi-cellulose. They are therefore capable of growing on a wide range of substrates. Furthermore, there are some species that will grow and fruit at a relatively high temperature, a feature that make for lower production costs in tropical or subtropical areas, or even in temperate regions during the summer season. It is probable that production of *Pleurotus* in geographic regions of diverse environmental conditions on a variety of available substrates.

G. *Pholiota nameko*

A wood-rotting fungus whose popularity and production are greatest in Japan, *P. nameko* (the viscid mushroom) has been produced in relatively stable amounts since 1975, with an overall annual increase amounting to about 5 or 6% in Japan between 1975 and 1985. In 1997, the worldwide total production of *Pholiota* was 55.5 thousand MT, ranking it ninth in total production. With a low-temperature requirement for fruiting, it is unlikely that this species will be extensively produced outside temperate regions, unless the mushroom breeder develops some high-temperature fruiting stocks of *P.nameko*. This may be a goal with reasonable expectation of success in light of Arita's (Arita, 1978) report that his earlier studies of temerature requirements for fruiting in this fungus revealed that 45 stocks separated into two groups on the basis of temperature requirement. The higher-temperature group requires temperature in the range from 8 to 20°C, not very high but sufficiently different from the lower temperature group (5 to 15°C) to indicate some promise for breeding for fruiting at more elevated temperatures. One feature of fruiting at the higher temperatures that would have to be overcome, however, is the tendency for small pilei and slender stipes.

H. Tremella fuciformis

Tremella fuciformis is known as the white jelly fungus or silver ear mushroom. It is a long-time favourite in China, especially for medicinal purposes, and the dried fruiting bodies of *Tremella* are conspicuous in most pharmacies specializing in Chinese medicines. It is also consumed as a food and usually as a special dessert. *Tremella* is being produced in greater amounts now than in some years past, because new techniques have been developed for its stable production using sawdust mixed with cottonseed hulls, which have removed some of the hazards in *Tremella* cultivation and the requirement for wood logs. In 1997, 130,000 MT (fresh weight) of *Tremella* were produced, mostly in China. The dried mushroom is shipped to countries in Southeast Asia, Japan, and to Europe, North America, and Australia. It will certainly be welcomed by the people of Asian origin in those areas. From the standpoint of increasing its use as a food in Western countries, the probabilities of *Tremella* being widely accepted rank below those of the other mushrooms that have been considered. Although Chen and Hou consider the fact that it has no strong flavor an advantage in making it acceptable to Westerners, it is our experience that most Westerners find it tasteless although it has an interesting texture. There is room for marketing an ample increase in production in the countries where it is produced, in Hong Kong, and in various countries of southeast Asia, however. It is interesting that in nature the distribution of this fungus is wide, but what effect this may have on local acceptance as a food is uncertain.

I. Mushroom species commercially cultivated recently

Among the 14,000 or more species of mushrooms, more than 40 have been commercially cultivated (Table 4). During the last decade, the following 15 species have been considered as rare and delicious species with great potential for further development: *Pleurotus nebrodensis, P. eryngii, P. citrinopileatus, Coprinus comatus, Agaricus blazei, Agrocybe aegerita, A. chaxinggu, Hypsizygus marmoreus, Grifola frondosa, Clitocybe maxima, Dictyophora indusiata, Pholiota adiposa, Lentinula giganteus, Oudemansiella radicata, and Tricholoma giganteum*. They are all commercially cultivated on various scales in China, Taiwan and Japan.

J. Mycorrhizal fungi: tuber, *Tricholoma*

The most difficult mushrooms to cultivate are those associated with another organism in a symbiotic relationship, such as mycorrhizal fungi with the roots of plants. Because some of these mushrooms are among the most highly prized, *Tuber melanosporm* and *Tricholoma matsutake*, for example, much study has been made on methods to encourage mushroom development on acreage in which a suitable ground is prepared and planted with seedlings of host species that have been inoculated with the proper fungal mycelium. Extensive ecological studies of the Perigord truffle, *Tuber melanosporum*, and the matsutake mushroom, *Tricholoma matsutake*, have been responsible for the development of techniques leading to better and more consistent yields of both of these important mushrooms. While the amounts of both of these species that become items of commerce are not great, each brings a high price. Korea exports its total production of *Tricholoma matsutake* to Japan, and in 1984 the value from Korea of all other mushrooms combined. In 1985, Japan produced matsutake mushroom valued at approximately $82,667,000, which was twice the value of the production in 1975. Production of truffles in France in 1975 was estimated at 60 to 120 MT, with a worldwide production of 150 to 300 MT. It is interesting to note that a hundred years ago, France produced 1000 to 1500 MT of truffles per year, and Italy produced 100 MT (Delmas, 1978). As stated above, with modern growing techniques, a consistent increase in production of *Tuber* by semi-cultivation methods is anticipated. The semi cultivation methods of farming these mycorrhizal mushrooms will have application to other mycorrhizal fungi whose mushrooms are esteemed for their taste appeal, such as *Boletus edulis*, *Cantherellus cibarius*, and *Amanita caesarea*.

K. *Termitomyces*

Another group of delicious mushrooms, termite mushrooms, has not yet been successfully cultivated. The termite mushroom species, which can, however, be "grown" or "cultivated" by termites, are grouped in the genus *Termitomyces*. Most termite species make their nests in the soil. Sometimes the nests are practically invisible; sometimes their mounds are 10m high. Pseudorhiza of *Termitomyces* is found in the chambers of nests or mounds, where primordial fruiting bodies are developed.

Table 4: Species of commercially cultivated edible mushrooms

*A. garicus bisporus**	*Clitocybe maxima*
*A. bitorquis**	*Dictyophora duplicate*
A. blazei	*Flammulina velutipes**
Agrocybe aegerita	*Grifola frondosa*
Auricularia auricula	*Hericium erinaceus*
A. fusccosucinea	*Hypsizygous marmoreus**
Aricula delicate	*Lentinula edodes**
A. polytricha	*L. giganteus*
Coprinus comatus	*Lepista nuda*
Lyophyllum ulmarium	*Tremella aurantialba*
*Pleurotus ostreatus**	*Pleurotus ferulae*
P. florida	*P. djamor*
P. sajor-caju	*T. lobayense*
P. eryngii	*Volvariella volvacea**
*Tremella fuciformis**	*V. diplasia*
Tricholoma giganteum	*V. esculenta*

* Species whose production has reached an industrial scale.

Constraints

Cultivation of *Agaricus* in developed countries has become a high-technology industry, and in fundamental research *Agaricus* has become the leader in mushroom science. Millions of dollars have been spent in Western countries to develop the industry. The mushroom farms are usually supplied with advanced equipment. The support from industry and governments for *Agaricus* mushroom research is most impressive in some developed countries. Such institutions as the Mushroom Research Center at the Pennsylvania State University in the United States; the Glasshouse Crops Research Institute in Littlehampton, England (now the Horticulture Research International, Wellesbourne, Warwick, England); the Mushroom Science Unit at Aston University, Birmingham, England; and the Mushroom Experimental Station and Mushroom Growers Training Center, Horst, the Netherlands, are the result of this type of support.

By contrast, the cultivation of mushrooms, particularly tropical mushrooms, in tropical and subtropical countries is still primitive. Several reasons, which may be directly or indirectly related to the slow development of mushroom cultivation in tropical regions, are presented.

1. Social concept

Mushrooms are usually eaten for their gastronomic properties, providing a flavoring and garnish for other foods. They are cultivated with special techniques and are usually consumed by the more affluent prople, as the price of mushrooms is usually much higher than that of most common vegetables. Because mushrooms are rich in protein and contain several vitamins and mineral salts, they should be considered high protein vegetables to enrich human diets, and they should be produced in quantities sufficient to lower the price so that they can be available to all.

2. Lack of support from Government and industry

There is a lack of governmental and industrial support for research, even though research on tropical mushrooms is relatively inexpensive, requiring neither huge research establishment nor massive, highly complex equipment.

3. Lack of interest of academia

There is a lack of interest of academic scientists in the fundamental biological studies of edible mushrooms. This is responsible for delaying the massive production of edible mushrooms in the tropical regions, and fundamental knowledge of the biological nature of mushrooms cultivated in tropical regions is still meagre. Without such basic knowledge, the development of mushroom industry is difficult, because cultivation of mushrooms requires very strong regional and local adaptation. The greatest needs are for creation of national laboratories or centers of fundamental biological and applied technical research on edible mushrooms in tropical or subtropical countries, and for the training of personnel to work in these laboratories or centers.

4. Additional concerns

The three problems mentioned above are the most basic and important problems. There are, to be sure, several other problems, such as a shortage of technical expertise, a lack of regional conferences and cooperative activities and a lack of regulated marketing practices, etc.

In practice, because of the hot and humid climate, contaminatin occurs more easily; it is difficult to prolong the shelf-life of mushrooms; and the manufacture and preservation of pure culture mushroom spawn are also problems for mushroom growers who lack a basic knowledge of microbiology. On the whole, the yield of mushrooms produced in these regions is generally low and unpredictable.

The techniques used for cultivation of mushrooms are, as mentioned, mainly based on the principles of microbiology, fermentation, and environmental engineering; and continuous production of successful crops requires both practical experience and scientific knowledge.

Last, but not the least, is the problem of keeping the techniques secret. Mushroom growers in general do not trust their competitors and, unfortunately, some mushroom researchers also adopt the same attitude toward their colleagues. They should share their experience and knowledge in order to enlarge and promote mushroom cultivation. By sharing their results, they will share the benefits.

Prospects

The prospects for cultivation of edible mushrooms in tropical regions are basically very good. Let us consider the following points:

1. Mushrooms can convert waste materials into human food. Tropical regions generally have a wet and warm climate and an abundant supply of agricultural wastes. These materials are resistant to degradation because they contain mainly cellulose, hemicellulose, and lignin. The mycelium of mushrooms excretes extensive enzyme complexes, which can directly attack and degrade these components. Therefore, mushrooms can use these wastes as nutrients for their growth, and the mushrooms produced become food for human consumption.
2. Mushrooms are relatively fast-growing organisms. Some tropical mushrooms can be harvested and consumed within 10 days after spawning. By susing different varieties, mushrooms can be produced all the year. They can be cultivated by using primitive farming techniques in rural areas or by using high technology in populated regions.
3. Mushroom cultivation is labour intensive; however, labour is plentiful in many tropical regions.
4. While land availability is usually a limiting factor in most types of primary production, mushroom culture can be concentrated within a relatively small space.
5. Mushrooms have been accepted as human food from time immemorial and can immediately supply additional protein to the human diet. Other sophisticated and unconventional sources of food protein, such as yeast, algal cultures, and single-cell proteins, have relatively more complicated requirements and need to be processed before they can be consumed.
6. Mushrooms should be used as a type of vegetable. After improving the culture and cultivation techniques, it should be possible to grow them as widely and cheaply as other common vegetables, which would be greatly beneficial to the public.
7. In view of their pleasing flavour, adequate protein, and health values, mushrooms unquestionably represent one of the world's greatest relatively untapped sources of nutritious and palatable food for the future.

In spite of the many problems that exist in the cultivation of mushrooms in developing countries in tropical regions, there is definitely a possibility of using mushrooms in a more important role as a source of protein to enrich human diets in these regions where the shortage of protein is most marked. There is no reason why both rural and urban areas cannot share in this new possibility and prospect?

The continuing challenge remains today to introduce new technology, not necessarily involving expensive equipment, to maximize mushroom production and mushroom products per unit area with minimum costs to provide a cheap source of food protein, mushroom dietary supplements, and nutriceuticals for people in the developing countries of the world. Furthermore, the spent composts or substrates can be utilized for organic fertilizers or soil conditioners. Based on an analysis by the China Agricultural University (Zheng *et al.*, 2002) the content of organic matter remaining after the production of edible mushrooms is more than 10%. Other indices reached or surpassed those of human waste, pig manure, and cow manure. Mushroom cultivation provides (1) edible mushroom production for food protein, (2) medicinal mushroom products for nutriceuticals, (3) diminution of unused lingo-cellulosic wastes from activities of agriculture, forestry, and food processing and minimization of environmental pollution arising from burning, and (4) production of organic fertilizer and the promotion of organic agriculture. This list is also indicative of their ranking in terms of their importance in relation to production volumes and market values worldwide.

Conclusion

The number of cultivated species is increasing, and this will have the effect of augmenting the total world production in the future because new species are unlikely to be substituted for others by the consumers, but they will be added to the ones that are already popular. Production of the existing cultivated species has shown a steady increase for decades, and more countries are engaging in mushroom cultivation as an agricultural technology. These factors argue against total mushroom production reaching a plateau in the near future.

References

Arita, I. 1978. *Pholiota nameko, In:* The Biology and Cultivation of Edible Mushrooms. (Chang. S.T. and Hayer. W.A. Eds.) Academic Press, New York. pp. 475-496.

Chang. S.T. 1999. Global impact of edible and medicinal mushrooms on human welfare in the 21st century: nongreen revolution. *Int. J. Med. Mushrooms.* **1**: 1-7.

Chang. S.T. 1999. World production of cultivated edible and medicinal mushrooms in 1997 with emphasis on *Lentinula edodes* (Berk.) in China. *Int. J. Med. Mushroom.* **1**: 291-300.

Chang, S.T. and Miles, P.G. 2004. Mushrooms. II edn. CRC Press *LLC.* pp. 406-408.

Chen. P.C. and Hou. H.H. 1978. *Tremella fuciformis, In:* The Biology and Cultivation of Edible Mushrooms, (Chang, S.T. and Hayer, W.A. Eds.) Academic Press, New York. pp. 629-643.

Delmas. J. 1978. *Tuber* spp. *In:* The Biology and Cultivation of Edible Mushrooms (Chang. S.T. and Hayer. W.A. Eds.) Academic Press. New York. pp. 645-681.

Lin. C.M. 2002. Special report of China edible fungi association. *The marker of edible fungi.* (In Chinese). **8**: 5.

Zheng. S., Liu. Q., Wang. H. and Ng., T.B. 2002. Can edible mushrooms promote sustainability in Beijing? *Mycol. Res.* **106**: 754-756.

Soil Microbiology and Soil Biotechnology : Vol. 2
Editor : S. Kannaiyan
Associated Publishing Co., New Delhi, 2010, pp. 212-221

11

Status and Recent Developments in the Production Technology of Paddy Straw Mushroom

*S. Kannaiyan***

Introduction

Volvariella volvacea (Bull. ex Fr.) Sing. is known as paddy straw mushroom or Chinese mushroom. It is also called "fukurotake" in Japan and "choku" in China. Chinese were the first to cultivate this mushroom almost 300 years ago. Migrating overseas Chinese introduced this mushroom to other South East Asian countries including Philippines and Malaysia (Chang 1977). In India, although Su and Seth (1940) had first cultivated *Volvariella displasia* (Bull. ex Fr.) Sing., Thomas *et al.* (1943) demonstrated the successful scientific cultivation by using the spawn. Well suited to the tropical and sub-tropical climate of Asia, *Volvariella* prefers to grow at high temperature and high humidity. Optimum temperature for the production of paddy straw mushroom is 30-35°C. Among all the cultivated mushrooms it is the fastest growing mushroom. The fruiting bodies are formed within 10-12 days of spawning.

The genus *Volvariella*

Volvariella has more than one hundred species, sub species and varieties. Previously the genus was known as *Volvaria*. Shaffer (1957) published the only monograph of this species and reported that more than 100 species, sub species and varieties of *Volvariella* are distributed in tropical, subtropical and temperate regions of both eastern and western hemisphere. In India, 13 species of *volvariella* have been recorded. But, only three species *viz.*, *V. volvacea*, *V. diplasia* and *V. esculenta* are under cultivation. The difference between the species is apparently colour. *V. diplasia* is white while, *V. volvacea* is dark grey. *V. bombycina* is white to yellow. Tu and Cheng (1979) proved that an isolate of *Volvariella* was white at low light intensity as compared to dark brown mushrooms at high light intensity. Hence, colour is not a constant character to distinguish the species. *V. bombycina* and *V. volvaea* can be differentiated based on the habitat of occurrence. *V. bombycina* is found on woods while, *V. volvacea* is found on soil or compost.

** Former Chairman, National Biodiversity Authority, Chennai-600041. Chairman, QRT-Mushroom.

World production as compared to other edible mushrooms

World production of *Volvariella* is approximately 1,81,000 MT. China is the leading producer of paddy straw mushroom. Although the mushroom has greater potential in tropical and subtropical climate, because of its poor productivity and shelf life, paddy straw mushroom has not attained a remarkable growth rate of *Pleurotus* and *Lentinula* (Table 1).

Table 1: World production of important edible mushrooms

Species	Fresh weight (x 1,000t)				Increase
	1986	%	1977	%	%
Agaricus bisporus	1,227	56.2	1,956	31.8	59.4
Lentinula edodes	314	14.4	1,564	25.4	398.1
Pleurotus spp.	169	7.7	876	14.2	418.3
Volvariella volvacea	178	8.2	181	3.0	1.7
Total	2,182	100.0	6,158	100.0	182.2

Source: USDA Bulletin Apr. 1998.

Paddy straw mushroom is also grown in other countries like Mexico, Puerto Rico, Mauritius and Madagascar.

Spawn base

A number of substrates *viz.*, wheat or sorghum grains, paddy straw, paddy husk, rice bran and saw dust have been tried by several workers to raise the spawn (Kannaiyan and Prasad, 1978; Ramaswamy and Kandaswamy, 1976). Chang (1982) observed enchanced mycelial growth in cotton waste + used tea leaves as compared to any other spawn base. Although grain spawn especially sorghum grain spawn is more popular in many places, paddy straw bits mixed with horse gram powder (2% supplementation) seems to be the best base for paddy straw mushroom spawn production (Krishnamohan, 1975).

Quimio (1993) has recommended the use of sawdust compost mixture for bed spawn preparation. Cotton wastes, oil palm wastes, tea wastes, coffee hulls and tobacco midrib are also used for spawn making in different countries. Generally *V. volvacea* lack phenol-oxidizing enzymes. Hence, if the substrates contain more phenolic compounds and tannin; the growth of the mushroom fungus will be highly inhibited. So, the spawn base needs to be selected carefully.

Paddy straw spawn

Polypropylene bags (45x30cm) are used for spawn preparation. Comparatively well-dried paddy straw is cut into smaller bits of approximately 3 cm. They are soaked or drenched with running tap water for one hour. The straw bits are then shade dried until excess moisture dries up. When the moisture content is approximately 65%, they are thoroughly mixed with coarsely powdered and autoclaved (at 1.42 kg/cm^2 for 90 min) horse gram powder @ 20g per kg of substrate (dry weight basis). Then the straw bits are filled in polypropylene bags up to three fourth height. The open end of the bag is inserted with a PVC ring of 1" diameter to make a mouth. The mouth is carefully plugged with non-absorbent cotton and covered with a piece of paper and autoclaved at 1.42 kg/cm^2 pressure for 90 min. The bags are then inoculated with the stock cultures of *V. volvaceae* maintained on PDA medium under aseptic conditions. After that the bags are incubated at 32-36°C over racks in a dust free room. The spawn run will be very quick and the fungus covers the entire bag with in 7-10 days. At this stage pink chlamydospores will appear in groups in the inner margins of the bag. Chlamydospore formation is characteristic of *V. volvacea* and its appearance is indicative of culture quality and maturity and hence mushroom spawn showing pink or brown chlamydospores is mostly preferred. The spawn bags will be ready for mushroom bed preparation within 10-12 days. The virulence of the spawn mainly depends upon the quantity of chlamydospores formed. Sangeetha *et al.*, (2005) have identified an efficient strain of *V. volvacea* PS 1 based on the chlamydospore production (Table 2). Kannaiyan and Tara (1977) reported a new medium for the spawn production of straw mushroom by supplementing rice bran which stimulated the pink coloured chlamydospores.

Table 2: Spawn running and chlamydospore production of different strains of *V. volvacea*

Strain	Days taken for complete spawn run	Days taken for the appearance of chlamydospores	Chlamydospores density	Yield (kg/bed containing 3 kg of paddy straw)	Bio-efficiency (%)
PS 1	10.2	12.2	++++	1.17[a]	39.1
PS 2	11.8	13.5	++	1.00[b]	33.5
PS 3	11.8	14.2	+	1.02[b]	34.0
PS 4	10.3	12.2	+++	1.15[a]	38.2

Substrates

Although a variety of substrates have been tried for the production of *Volvariella* till date partially fermented paddy straw and cotton waste have been reported to be the best substrates. Yield on cotton waste composts under controlled condition are much higher than other substrates (Table 3).

Table 3: Yield of paddy straw mushroom on different substrates

Substrates	Yield (%)	Source
Paddy straw	6.0	Thomas *et al.*, 1943
Dried banana leaves	10.0	Wade, 1970
Cotton waste compost	25-45	Hu *et al.*, 1973; Chang, 1974
Oil palm pericarp	4.0	Goh and Graham, 1975
Oat, Barley and Wheat straw	5.6	Garcha & Kalra, 1979
Sugarcane bagasse + cotton waste (1:1)	19	Khan *et al.*, 1991
Water hyacinth, Water hyacinth + Cotton waste (1:1)	28-76	Saeed *et al.*, 1994
Kallar grass (*Diplachine fusca*)	63	Abdul *et al.*, 1995
Oil palm bunch waste	31	Pramod *et al.*, 2004

Although paddy straw and cotton waste compost have been reported to be the best substrate by several workers, Saeed *et al.* (1994) and Abdul *et al.* (1995) obtained remarkable yield with water hyacinth with cotton waste and kallar grass.

Supplement

Addition of organic nitrogen to paddy straw substrate or cotton waste compost always helped to increase the yield (Table 4). Enzymes especially cellulases play a major role in substrate utilization by *Volvariella* spp. Ramasamy and Kandaswamy (1976) indicated that *V. volvacea, V. esculenta* and *V. diplasia* had responded well to cellulose supplementation through tapioca starch. They also established a positive correlation between supplementation, enzyme production and yield.

Table 4: Effect of supplementation on yield of paddy straw mushroom

Supplements	Yield %	Source
Paddy straw + Red gram powder	6.6	Ramaswamy and Kandaswamy, 1976
Paddy straw + Cooked apple extract	10.6	Ramaswamy and Kandaswamy, 1976
Paddy straw + cooked apple extract + Red gram powder	12.6	Ramaswamy and Kandaswamy, 1976
Paddy straw + Red gram powder	10.5-19.0	Ramakrishnan *et al.*, 1968; Kaul, 1968; Munjal, 1975
Paddy straw + Rice bran	22.0-27.1	Kaur *et al.*, 2004
Paddy straw + Red gram/Horse gram/Bengal gram powder	10.7-20.8	Sangeetha *et al.*, 2004

Red gram powder as supplement to paddy straw has been reported to increase the sporophore yield (Kannaiyan and Ramasamy, 1980).

Cultivation technology

As early in 18th centuary some one in the Nanhua temple in China accidentally discovered the primitive method of straw mushroom cultivation. The method is that a heap of straw mixed with cow manure and split beans was built in the centre of the field and surrounded with a ditch that could be flooded and used as a source of water for the heap. After a few days, the straw was made into small bundles and laid in a straight line five or six layer thick on the ground and finally covered up with other straw so as to be protected from direct sun and rain. It also had helped to maintain perfect bed moisture. In a fortnight fruit bodies would grow in the shape of umbrella. These mushrooms were split and dried on a tire before storage. Such mushrooms were regularly offered as tribute to the Emperor from the Nanhua temple (Chang, 1977). Several techniques for indoor and outdoor cultivation of paddy straw mushroom have been practiced in different countries according to the local conditions available. Kannaiyan and Prasad (1976) reviewed the cultivation of Paddy Straw Mushroom.

Indoor cultivation - industrial approach

In China an indoor cultivation method was introduced in early 1970s to enable year round cultivation of paddy straw mushroom with different agricultural wastes. More particularly, the use of cotton waste was first introduced in Hong Kong, which increased and stabilized the yields considerably (up to 50%). Indoor cultivation is now recommended for industrial scale production in China, Indonesia and Malaysia. Quimio (1993) has given a detailed account on the indoor production technology. The process of composting for indoor cultivation includes Phase I outdoor composting and Phase II indoor composting. Phase I process basically involves mixing the ingredients, pre wetting and forming piles for the initial breakdown of the substrates to provide ideal environment for the microbial growth. Phase II involves pasteurization and conditioning. Rice or wheat straw, cotton waste, cotton seed hull, sugarcane bagasse are supplemented with substrates like soybean powder, FYM and rice bran at definite proportions. Calcium hydroxide or limestone is also mixed with the ingredients, pre-wet overnight and partially composted for four days with one turning.

After 4 days the fermented substrate is filled in shelves or rack to 6" thickness. Surface spawning is done at 1.5% to the compost used and the racks are covered with a transparent polythene sheet for 4 days. The temperature inside the room will be maintained at 32-389°C and the relative humidity should be more than 80%. On 5th day, the spawn run will be completed and small pinheads appear on the surface. More fresh air is needed at pinning stage. Subsequently, within a day or two egg shaped fruiting bodies can he harvested.

Bundle method

The method was suggested by Thomas *et al.* (1943). Paddy straw bundles normally 8-10" thickness and 3-4 ft. length are prepared and soaked in water for 24 h. Four to five such bundles are laid side by side to form the base for the bed. Over this an equal number of bundles are placed with the bases on the opposite side. Both these layers together form the first layer of the bed. Paddy straw spawn bits are placed on this layer all around the bed 4-5" from the edge. The spawned layer is then sprinkled with a handful of dhal powder. The second layer of the bundles is placed over this by putting the bundles across the first layer. Following the same procedure upto five layers are made and on each layer spawn bits are placed and dhal powder (red gram – Cajanus cajan) is sprinkled. The entire bed size will be approximately 4 cubic feet. The bed is regularly sprinkled with water and maintained in a shaded place. Mushrooms will appear on the beds in 15-20 days after spawning. Instead of bundles paddy straw twists are also used in this method.

Twist method - circular compact bed

Well-dried good quality paddy straw is made into small twists of 2.5 m long and 5 to 8 cm diameter weighing about one kg (dry weight). Totally five twisted bundles will form one bed (45 cm dia x 30 cm height). The twists are immersed in clean water for about 24 h. After draining the excess water the twists are shade dried until the substrate moisture is 65-75 per cent. One twist is in clockwise in a circular fashion as compact as possible (measuring to about 45 cm in diameter) and placed over a small wooden plank to form the first layer. All over this layer 8-10 spawn bits are placed leaving equal space between them. While placing the spawn care should be taken so that they are not placed at the edge of the paddy straw twist. Then coarsely powdered and autoclaved horse gram powder is sprinkled over the spawn bits. Second layer of twist is placed in the same way sandwiching the spawn bits. Similarly, third and fourth layers are formed. Completed beds are tightly pressed and incubated in a poly house at 32-35°C temperature and 80-85 per cent relative humidity. The beds are regularly sprayed with water. Spawn running will be completed within 8-10 days.

The hyphae will be seen as silk threads. Some times large number of chlamydospores are seen on the paddy straw bits. As the mycelium is highly delicate beds should not be disturbed at this stage. At the end of spawn run small pinhead like fructifications occur all over the bed surface. They will grow fast through peanut shaped structures to egg shaped undifferentiated fruiting bodies. This is the best stage for harvest. If net harvested, the fruiting bodies will become flattened with well-differentiated volva, stipe and pileus. Flattened mushrooms possess poor shelf life and taste in addition to weight loss.

Cylindrical hollow bed

In this method, straw twists are wound around a cylindrical PVC pipe (15 cm diameter) and finally the container is removed to make a hollow at the center of the bed. The size of the bed will be 60 cm in diameter with 30 cm height and a central hollow space of about 15 cm. Spawning is done both in the inner and outer periphery of each layer.

Among the different bed methods tried, circular compact bed gave significantly higher yield of 820.0 g per 5.0 kg of paddy straw (Sangeetha, 2004). Comparatively lesser yield was obtained from bundle and hollow round beds. The variability in the yield is attributed to certain physical factors particularly, temperature, aeration, wetness and compactness of the beds. Circular compact bed always help to maintain homogenous growing conditions that may favour better proliferation of mycelia and subsequent fruiting. In square compact bundle method, the straw became loose especially in the corners of the bed. At this region, proliferation of the mycelium will be rather weak. So also the hollow bed receives more aeration coupled with lower temperature in all the layers resulting in retarded mycelial growth and poor yield. However, Krishnamohan (1975) reported increased yield in hollow bed method. Garcha (1975) reported an improvement in the yield of *V. volvacea* with hollow interior bed system on square type orientation with a maximum bio-efficiency of 13.6 per cent. In a series of experiments, Garcha and Kalra (1979) found small beds to be more economical than large sized beds.

Basket method

Vela and Carrera (1989) cultivated *V. volvacea* and *V. bakeri* in partially fermented barley straw taken in plastic baskets. Barley straw was mixed with lime (3%) and piled up in a pyramidal heap for fermentation for 5 days. The substrate was turned on 3rd day in order to avoid anaerobic fermentation in the inner layers. The fermented straw was immersed in hot water at about 70-75°C for 45 min. The excess water was drained and approximately 76% moisture was retained in the substrate. Plastic baskets of 32x26x8 cm size were filled with 3 kg of fresh spawned barley straw and perfectly covered with a polythene sheet to prevent dehydration. The baskets were arranged in the cropping room for fruiting.

Cage cultivation

Thakur *et al.* (2003) suggested a cage method of cultivation. They used 1.00 m x 0.50 m x 0.25 m size wooden cages covered with polythene sheets for the cultivation of paddy straw mushroom. The cages were filled with pretreated paddy straw bundles and the spawn. Chickpea flour is sprinkled on the spawn to enhance the growth. In this method up to 17.4% bio-efficiency could be achieved. Similarly, Alicbusan (1975) recorded up to 25% yield in a box method of cultivation than the conventional bed system. This method showed greater promise in Philippines on a commercial scale.

Out door cultivation

In Hong Kong and Philippines, outdoor cultivation is practiced in shaded places where special care is needed to shelter and protect the beds. This method is also practiced in some parts of Chattisgarh, where rural people spread one or two year old paddy straw in their home garden as a thick layer. Whenever, fruiting bodies under optimum conditions appear during rainy days, they crush the fruiting bodies with water and make a homogeneous suspension. They then sprinkle the entire suspension over the straw substrate especially at the edges of the paddy straw layers. The heap moisture is maintained by regular sprinkling. Mushrooms appear over the heaps when conducive conditions occur. The farmers also practice spraying 2% common salt or urea over the paddy straw heaps to enhance fruit body production. Spent straw either mixed with fresh straw or thrown as such in the manure pit along with decayed paddy straw also produced very good yield (Thakur *el al.*, 2003).

Inter cropping paddy straw mushroom—a new approach

The possibility of introducing paddy straw mushroom as an inter row crop in maize was experimented by Krishnamoorthy *et al.* (2005). Paddy straw bundles weighing one kg each were soaked in water in a tank for 48 h. Using the partially fermented bundles, compact square beds (5 bundles x 4 layers) were laid in the inter row spaces of 30 days old maize crop (var. CO MH 4). Total weight of each bed was 20 kg. Layer spawning was done with paddy straw spawn @ 2%. Steamed horse gram powder (1.46 kg/cm^2 pressure) was sprinkled on the spawn bits every time after placing. The entire bed was placed in the inter row space without disturbing the routine agronomic practice. In this method an average yield of 1.73 kg per bed (BE 8.75%) was harvested in 15 days. This type of cropping system helps in establishing pennaculture, soil amelioration, addition of organic nitrogen, weed control and all the most ecorestoration and *in situ* organic recycling.

Future research

Although the genus *Volvariella* is known for its peculiar taste and medicinal properties, till today the mushroom has not attained a commercial impetus in India. The main reason for the set back is its colour, less productivity and poor shelf life. People are also not made aware of its real taste. Though the fungus is an aggressive colonizer of any cellulosic substrate, its competitive saprophytic ability is much poor. *Volvariella* spp. lack phenol-oxidizing enzymes. Hence, if the substrates contain more phenolic compounds and tannin, the growth of the mushroom fungus will be highly inhibited. Hence, suitable substrate treatment methods or improved composting processes need to be elaborately investigated. A strong positive correlation has been established between the chlamydospore production and yield of *Volvariella*. Some studies indicate that sugars and sugar alcohols like mannose sucrose mannitol and sorbitol induce more chlamydospores. Attempts to formulate certain boosters containing such sugars or sugar alcohols may have greater advantage in spawn production or substrate supplementation for increasing the productivity. Bioinoculants including the live or dead cells of certain bacteria, yeast and actinomycetes either individually or as consortia may be developed to increase the yield. Research on biotechnological approaches to regulate the functional genes like mannitol dehydrogenase and tyrosinase may be given priority to regulate the morphogenesis and shelf life of this praise worthy mushroom. Developing simple and sustainable indoor and outdoor cultivation methods based on local needs will have greater scope and impact on the farming system as a whole. Paddy straw mushroom best suited to introduce as an integral part of farming system both in wetland as well as dry land so as to have constant income to the growers and also for the nutritional security of the individual family. All the more, constant efforts should be made to create public awareness about the nutraceutical and pharmaceutical importance of paddy straw mushroom.

References

Abdul. R., Khan. S., Mohamd. M. and Inaram ul-Haq. 1995. Cultivation of paddy straw mushroom *Volvariella volvacea. Pakistan J. Phytopathol.* 7: 5.

Aliebusan. R.V. 1975. Commercial production of paddy straw mushroom in wooden boxes. *The Philippine Biota*, 9: 120.

Chang. S.T. 1974. Production of the straw mushroom (*Volvariella volvacea*) from cotton wastes. *The Mushroom J.* **21**: 348-353.

Chang. S.T. 1977. The origin and early development of straw mushroom cultivation. *Economic Bot.*, **31**: 374-376.

Chang. S.T. 1982. Cultivation of *Volvariella* mushroom in South East Asia. In *Tropical mushrooms - Biological nature and cultivation methods* (Chang. S.T. and Quimio. T. H. ed.). The Chinese University Press. Hong Kong. pp. 182.

Garcha. H.S. 1975. Mushroom Culture. Communication Centre. PAU. Ludhiana. Panjab. Garcha. H.S. and Kalra, K.L. 1979. Spawn production and its viability *Mush. Sci.* **10**: 645-652.

Goh. S.C. and Graham. K.M. 1975. Growing paddy straw mushroom in oil palm pericarp. *Malany. Agric. Res.*, **4**: 109-116.

Hu. K.J., Song. S.F. and Xiu. P. 1973. Experiments on Chinese Mushroom Cultivation 1. The comparison of cultivating materials. *J. Taiwan. Agric. Res.*, **22**: 145-148.

Khan. S.M., Haq. R., Dongar. M.A. and Maher. M.J. 1991. Sugarcane bagasse and cotton waste compost for growing paddy straw mushroom-Musfcraom *Science XJ1.2* Proceeding of the 13th International Congress on the Science and Cultivation of Edible Fungi held at Dublin. pp. 579-584.

Kannaivan. S. and Prasad. N.N. 1976. Production of paddy straw mushroom—A brief review. Proc. First Symp. On Survey and Cultivation of Edible Mushrooms in India. **2**: 31 36.

Kannaiyan, S. and Tara, K. 1997. A new medium for spawn production of straw mushroom. *Farm Sci.*, **3**(10): 19-20.

Kannaiyan, S. and Prasad, N.N. 1978. Production of paddy straw mushroom in India. Review. In: Indian Mushroom Science (Atal, C.K.. Bhan, B.K. and Kaul, T.N. ed.), Indo-American Literature House, New Delhi, pp. 305-310.

Kannaiyan, S. and Ramasamy, K. 1980. A Hand Book on Edible Mushrooms. Today and Tomorrow's Printers and Publishers, New Delhi.

Kaul, T.N. 1968. Prospects of large-scale cultivation of paddy straw mushroom in UP. In: Indian Mushroom Science (Atal, C.K.. B.K. Bhat and T.N. Kaul ed.). Indo American Literature House, New Delhi, pp. 310-315.

Kaur, K., Kapoor, S. and Phutela, R.P. 2004. Suitability of organic supplements in paddy straw and cotton waste compost for the production of *Volvariella volvacea*. *Mushroom Research*, **13**: 31-34.

Krishnamohan, G. 1975. Studies on Paddy Straw Mushroom. M.Sc. (Ag.) Thesis submitted to Tamil Nadu Agricultural University, Coimbatore. p. 72.

Krishnamoorthy, A.S., Thiribhuvanamala, G., Shanthi, K. and Marimuthu, T. 2005. Outdoor cultivation of paddy straw mushroom as inter crop in maize field. *Mushroom Research*, **14**: 9-12.

Munjal, R.L. 1975. Cultivation ol paddy straw mushroom. *Indian J. Mushroom*. **3**: 32-38

Pramod, R., Balakrishnan, B. and Lulu Das. 2004. Evaluation of different substrates for the cultivation of paddy straw mushroom *Volvariella volvacea*. *Mushroom Research*. **13**: 29-30.

Quimio, T.H. 1993. Indoor cultivation of the straw mushroom *Voivariella vohacea*. *Mushroom Research*. **2**: 87-90.

Ramaswamy, K. and Kandaswamy, T.K. 1976. Effect of certain amendments on cellulase (s) and yield of straw mushroom. *Indian J. Mushroom*. **2**: 4-7.

Ramakrishnan, K., Lalithakumari, D., Shanmugam, N. and Krishnamoorthy, C.S. 1968. A simple technique for increasing the yield of straw mushroom (*Volvariella diplasia*). *Madras. Agr. J.*, **55**: 194-195.

Samajpati, N. 1982. Cultivation of *Volvariella volvacea*. In: Tropical Mushrooms - Biological Nature and Cultivation Methods (Chang, S.T. and Quimio, T.H. ed.) The Chinese University Press, Hong Kong. pp. 267-74.

Sangeetha, G., Prakasam, V. and Krishnamoorthy, A.S. 2004. Enhancing yield in paddy straw mushroom by supplementation. *J. Mycol. Pl. Pathol.*, **34**: 863-866.

Sangeetha, G., Prakasam, V. and Krishnamoorthy, A.S. 2005. Selection of the best commercial strain of paddy straw mushroom (*Volvariella volvacea* (Bull.ex.Fr.) Sing.) *Smith Indian Nort.*, **52**: 117-121.

Saeed, M., Khan, S.M. and Imam-ul-haq, M. 1994. Growing paddy straw mushroom on water hyacinth . *Pakistan. J. Phytopathol.* **6**: 130-134.

Shaffer, R.L. 1957. *Volvariella* in North America. *Mycologia*, **49**: 545-579.

Su, U.T. and Seth, L.N. 1940. Cultivation of Straw Mushroom. *Indian Farm.* **1**: 332-33

Thakur, M.P., Godara, D.R., Shukla, C.S. and Sharma, R.L. 2003. Recent Advances in the Production Technology of Paddy straw Mushroom. In: Current Vistas in Mushroom Biology and Production (Upadhayay, R.C., Singh, S.K. and Rai, R.D. eds). Mushroom Society of India, Chambaghat, Solan (H.P.).

Tu, C.C. and Cheng, S. 1979. A new white variety of *Volvariella volvacea* (Bull. ex.Fr.) Sing. *Mushroom Sci.*. **10**(2): 185-90.

Thomas, K.M., Ramakrishan, T.S. and Narsimhalu, I.L. 1943. Paddy straw Mushroom. *Madras Agric. J.*. **31**: 57-59.

Vela. R.M. and Carrera, M. 1989. Cultivation of *Volvariella bakeri* and *V.volvacea* in Mexico: A comparative study. *MUSH. J.Tropics.* **9**: 99-108.

Wade. T. 1970. Top banana-mushroom. *Singapore Trade Ind.*, Apr. pp.70-71.

Soil Microbiology and Soil Biotechnology : Vol. 2
Editor : S. Kannaiyan
Associated Publishing Co., New Delhi. 2010. pp. 222-232

12

Production Technologies and Constraints in Oyster Mushrooms

B. Balakrishnan[1] *and Lulu Das*[2]

Introduction

Oyster mushrooms (*Pleurotus* spp.) have become an integral part of world mushroom industry. The desirable attributes like rapid mycelial growth, high ability for saprophytic colonization, simple cultivation techniques and wide choice of species available for cultivation under varying agroclimatic conditions have made it popular among the mushroom growers.

India is a country of varied topography, climate and agricultural resources. The vast amount of agriculture wastes coupled with climatological diversity can be successfully exploited for the cultivation of oyster mushrooms. The commercial cultivation of this mushroom in India is of recent origin. More than twenty five species have been described under this genus from different parts of the country and about a dozen out of this are under cultivation. The popular cultivated species are *Pleurotus ostreatus* (Jacq.) Fr., *P. sajor-caju* (Fr.) Singer, *P. eryngii* (DC.ex.Pers.) Quel., *P.tuberregium* (Fr.) Singer, *P. cornucopiae* (Paul. ex. Fr.) Sing; *P. fossulatus* (Fr.) Sing., *P. florida* Eger., *P.flabellatus* (Berk. & Br.) Sacc., *P. opuntiae* (Dur. & Lev.) Sacc., *P. citrinopileatus* Singer, *P. membranaceus* Massee and *P. platypus* (Cook & Massee) Sacc.

The cultivation of *Pleurotus* spp. started first, by the begininning of 19th century on tree stumps. The technique was further refined by using sawdust as the substrate. *Pleurotus* cultivation was first standardized by (Bano and Srivastava, 1962) utilizing *P. flabellatus*. The first domesticated species under *Pleurotus* was *P. ostreatus*. But later *P. sajor - caju* has gained much importance as a commercial species in major part of the country and its first cultivation was reported on banana pseudostem and chopped paddy straw (Jandaik and Kapoor, 1974). Followed by this the successful cultivation of *P. sajor-caju* was reported

1. Associate Professor. Department of Plant Pathology. College of Agriculture. Vellayani. Thiruvananthapuram - 695522, Kerala. E.mail: balakrishnanbhanu@yahoo.co.in
2. Associate Professor (S.Gr.). Department of Plant Pathology. College of Agriculture. Vellayani. Thiruvananthapuram - 695522. Kerala. E.mail: luludaspavi@yahoo.co.uk

from Tamil Nadu using different waste materials like paddy straw, saw dust and wood shaving (Rangaswamy *et al.*, 1975).

The different species of *Pleurotus* are suited to growing within a temperature range of 15 to 30°C. *P. sajor - caju* is tolerant of a tropical temperature at 28 to 30°C although it fruits faster and produces larger mushrooms at 25°C during the cooler months of the year or in the highlands of the tropics. *P. ostreatus* is the so called low temperature *Pleurotus*, fruiting mostly at 12-20°C (Quimio *et al.*, 1974).

A lot of technological advancements have been achieved over the years in the production of this mushroom including the development of improved strains. Now efficient strains of *Pleurotus* spp. suited to different climatic zones, namely, temperate, tropic and sub-tropic, have been in use which are either identified and domesticated from the wild natural flora or developed by various strain improvement programmes. Even though researches on various aspects of production technology of oyster mushrooms have been advanced a great deal during the last one decade in India, its production at the commercial level has not yet been achieved in the country, probably may be due to the great attention given to the production of white button mushrooms.

Production technologies

Different *Pleurotus* spp. and their growth requirements

The growth requirements vary for different species of this mushroom. Under natural conditions, *Pleurotus* spp. mostly grow on dead and rarely on living parts of plants, which are generally poor in nutrients and vitamins. But it has been established by various workers that for both mycelial growth and fruit body development on lignin - cellulose materials a C:N ratio of 1:50 is essential. Materials namely, maize cobs, straw of all cereals, paper, wood shavings, saw dust, vegetable wastes as well as food industry wastes are sufficient to meet the growth requirements of most of the species under cultivation (Bano and Srivastava, 1962; Block, 1965).

The environmental factors are of much importance in the production of *Pleurotus* spp. The spread of mycelial growth of *Pleurotus* is greatly dependent on temperature and humidity, which also varies from species to species (Zadrazil and Schneidereit, 1972). A temperature range of 20-26°C (24±1°C being optimum) and relative humidity of 70-90 per cent is ideal for *Pleurotus sajor-caju*. A fairly good crop can be obtained up to 30°C. *P. eous* prefers a temperature of 21-35°C and a humidity of 65-100 per cent. Proper ventilation is essential for fruit body production (Purkayastha and Jana, 1983). *P. fossulatus* prefers 20±1°C and light is very essential for fruiting (Puri *et al.*, 1981). Maximum growth of *P. ostreatus* was recorded at 25°C (Rangad and Janadaik, 1977). Whereas *P. florida* gave highest yield at 30°C. *P. flabellatus* also have a similar temperature requirement (Srivastava and Bano, 1970). It has been pointed out that though the rate of growth is low at low or at very high temperatures the tropical and sub tropical species of *Pleurotus* continue to grow at 10 or 12°C and upto 35°C (Sharma, 1983). It has also been suggested that different species of *Pleurotus* need variable temperature conditions and hence they are ideally suited for cultivation at different times of the year in various regions (Sohi, 1988). For most of the *Pleurotus* spp., the optimum pH was found to be 5.6 (Srivastava and Bano, 1970; Janadaik and Kapoor, 1975).

Variation of *Pleurotus* spp. with respect to climatical differences have also been reported. An observation from Mysore (Bano and Rajarathnam, 1982) have stated that maximum yield could be obtained during rainy seasons, when the temperature was 20-26°C and relative humidity 70-90 per cent. Similar observations were reported form Tamil Nadu also. The yield of *Pleurotus sajor-caju* was found to be increased three fold during the north-east monsoon than summer (Zadrazil, 1978). Seasonal variations in the yield of certain *Pleurotus* spp. have been recorded in Kerala also. (Balakrishnan, 1994). It was observed that during monsoon period, *P. sajor-caju* gave significantly higher yield which was followed by *P. sapidus*, *P. petaloides* and *P. flabellatus*. During summer all the species tested were recorded lower yield compared to the monsoon yield. The cropping duration was also little bit higher for all the species during summer (Table 1).

Table 1: Seasonal variation in the yield of various *Pleurotus* spp. under Kerala conditions

Sl. No.	Name of species	Seasons			
		Monsoon (N:E)		Summer	
		Avg. no. of days from spawning to harvest	Avg. yield (g) per kg substrate	Avg. No. of days from spawning to harvest	Avg. yield (g) per kg substrate
1.	*P. sajor-caju*	16	1043.33	29	973.33
2.	*P. citrinopileatus*	23	643.333	29	508.33
3.	*P. ostreatus*	18	286.67	23	236.67
4.	*P. flabellatus*28	28	740.00	36	516.67
5.	*P. opuntiae*	19	188.67	27	116.67
6.	*P. florida*	18	356.67	23	220.00
7.	*P. membranaefeus*	16	680.00	26	566.62
8.	*P. platypus*	16	766.67	16	606.67
9.	*P. petaloides*	16	940.00	16	886.00
10.	*P. sapidus*	13	960.00	15	823.30
	C.D (0.05)	—	206.60	—	206.60

Pleurotus fruiting is directly related to the composition of the air mixture in the fruiting atmosphere (Zadrazil, 1978). It is also emphasised that air factor is more important than light factor for development of primordia.

Effect of CO_2 on the growth of different species of *Pleurotus* has also been worked out. It was reported that the mycelial growth of *P. ostreatus* is stimulated by CO_2 concentration upto 28 per cent (volume) in air (Shanel, 1970). The same phenomenon was also reported in the case of *P. florida*, *P. osteratus*, and *P. eryngii* (Zadrazil, 1975). Inhibition of growth becomes noticeable only at CO_2 level of 36 per cent and above. This has a great bearing in commercial cultivation that higher the CO_2 concentration in the substrate the other competitors will be reduced.

Development of improved strains

In *Pleurotus* spp. sufficient information is available on sexual life cycle. Intermating and fruictification is quiet easy. Genetic variability is available for the genus, only thing is to improve and preserve it properly. Efficient strains of oyster mushrooms suited to different climatic zones namely, temperate, tropic and sub-tropic have been in use, which are either identified and domesticated from the wild natural flora or developed by various strain improvement programmes. Like any other mushrooms the techniques used for strain improvement in this group of mushroom are introduction, selection, mutation, hybridization and genetic manipulations. A lot of achievements have been reported in this aspect from various parts of the world. Development of heterokaryons of *Pleurotus ostreatus* was succeeded. By hybridizing a temperature sensitive American strain of *P. ostreatus* a recombined variety with changed temperature sensitivity has been developed (Eger *et al.*, 1976). They also suggesed that temperature sensitive and insensitive populations will help to a gene drift on both sides. Two improved varieties of *P. florida* were developed by crossing two different strains of *P. florida*, (Martonffy, 1987). The introduction and selection strategy has improved a large number of *Pleurotus* spp. namely, *P. ostreatus* (Dhar, 1976); *P. eryngii* (Sharma and Jandaik, 1985), *P. sapidus* (Mehta and Jandaik, 1989), *P. cystidiosus* (Beig and Jandaik, 1989) etc. A sporeless mutant of *P. florida* has been developed and studied at IIHR, Bangalore which has good commercial prospects (Pandey and Tewari, 1989). Selective dikaryotisation exert some exciting changes over the traditional culture of *P. sajor-caju*. Improvement in quality character govern fast colonizing ability, crop earliness, reduced percentage of bud mortality, size, shape, colour of pileus, and also protein content of the fruit bodies (Ghosh and Chakravarty, 1991). Increase of yield may be attributed to number of factors among which genetic potential of the strains is more important (Bahukhandi and Munjal, 1990). The strategies for breeding potentials of edible mushrooms with special reference to hyphal anatomosis was shown in *P. sajor-caju* (Jandaik, 1997; Pahil, 1997). An interstock hybrid of *Pleurotus petaloides* has been developed in Kerala by crossing monokaryon germlings of its wild and a stress induced mutant which was found to out yield the parents, with milky white large sporcarps in bunches, short crop duration and tolerant to the most dangerous weed mould, *Trichoderma* spp. (Balakrishnan and Nair, 1997).

Kinds and availability of substrates for cultivation

In India, the total quantity of agricultural by products or wastes which are cellulosic in nature account for nearly 25 million tonnes in a year (Ghose and Ghose, 1978). These materials assume the perennial source of raw material for mushroom cultivation. The important ones among these include different cereal straw, sugarcane bagassee, cotton waste, jute, coirpith (coir waste), cocoa pods, mango seed kernels, different oil cakes including rubber seed cake, sunflower straw, ground nut pod shell, tapioca starch waste, water hyacinth and certain other dry and wet land weeds, banana leaves, stem and its market wastes, rubber wood wastes, mango wood chips, oil palm mesocarp waste, fermented coffee pulp, soft wood saw dust etc. (Balakrishnan and Nair, 1997).

A lot of work has been done in India on the suitability of various substrates namely rice straw, wheat straw, ragi straw, hulled maize cob, waste cotton, banana pseudostem and waste paper (Jandaik, 1974, Bano *et al.*, 1978; Sivaprakasan *et al.*, 1979). In a study, maximum yield of *P. sajor-caju* was obtained from banana pseudostem (Thilagavathi *et al.*, 1991). A comparative study with pseudostem and rhizome of different banana varieties in Kerala have shown that *P. sajor-caju* has performed well in both the pseudostem and rhizome of Red banana variety which has given 103 per cent Biological Efficiency (B:E) in its pseudostem (Balakrishnan, 1994), (Table 2).

Table 2: Comparitive efficiency of pseudostem and rhizome of three common banana varieties of Kerala as substrates for cultivation of *P. sajor-caju*

Sl. No.	Variety of banana and type of substrate	Avg. no. of days from spawning to harvest	Mean yield (g) per kg substrate	Avg. Per cent of B:E
A	**Pseudostem**			
1.	Nendran	23	390.0	39.0
2.	Palayanthodan	23	364.0	36.0
3.	Red Banana	23	1034.4	103.4
B.	**Rhizome**			
1.	Nendran	28	382.0	38.2
2.	Palayanthodan	25	216.0	21.6
3.	Red banana	26	784.0	78.4
C.	**Control (paddy straw)**	21	892.0	89.0
	C.D. (0.05)		257.7	

Similarly, various workers have suggested different substrates for *Pleurotus* cultivation. Some of them are Jowar straw and groundnut pod for *P. florida* (Khandar *et al.*, 1991), oil palm mesocarp waste (Babu and Nair, 1991); *Eupatorium odoratum* and water hyacinth (Das *et al.*, 1988); fermented coffee pulp (Upadhyay and Sohi, 1988); mango wood chips, coconut petiole chips and match work wastes (Balakrishnan, 1994). Sugarcane by-products pretreated with sodium hydroxide (4%) increased the (B:E) considerably (Chandrsekhar *et al.*, 2001).

Processing of substrates

Various methods of substrate preparation have been developed for *Pleurotus* cultivation. Steam pasteurization, hot water treatment or chemical pasteurization are the widely adopted and effective methods. Some of the important recommendations in this area are hot water dip of the substrate at 65±5°C for one hour (Bano *et al.*, 1978); 60 to 120 minutes of hot water dip at 80°C (Singh and Dwivedi, 1991); chemical pasteurization method using formalin (500 ppm) and carbendazim (75 ppm) in substrate soaking water for 18 hours (Vijay and Sohi, 1987). Increasing the pH of the substrate soaking water between 7 to 8.5 by adding lime had shown minimum infestation by the weed mould, *Trichoderma* without any reduction in yield (Balakrishnan, 1994). Solarisation and UV radiation of the substrates were found equally effective as chemical pasteurization of substrates for oyster mushrooms (Ram &

Thakur, 2005). An efficient method of composting of saw dust has also been recommended for tropics for better yield (Zadrazil and Kurtzman, 1982). In their long method of processing for commercial level cultivation, saw dust combined with 1% urea and 1% lime are used. After sufficient wetting the material is piled into a heap of about 1.5m and during the fermentation period of seven days time sufficient turning is also given and the compost will be ready by 30 days. At this stage the compost will soften without producing any unpleasant smell. If the compost is not to be used immediately it should be dried before storage. Composting may not be necessary when using older and weathered saw dust.

In short composting rice bran (20%) sugar (1%) and lime (1%) are mixed with the sawdust and allowed to ferment for 6 to 7 days. When the mixture is ready, 10-20% rice bran is added and the compost is packed into 15 × 3 cm (or larger) polypropylene bags. The compost is punched down to make it compact. Alternatively, blocks may be prepared with the compost. For this, compost is steamed first in bulk and is inoculated after cooling by thoroughly mixing with the spawn. Blocks are then prepared by the spawned compost in 5 × 10 kg capacity plastic bags and flattering them to form blocks. The use of composted or uncomposted substrates varies according to personal preference (Quimio *et al.*, 1995).

The use of autoclave to sterilize *Pleurotus* substrates is not recommended since sterilization kills beneficial microorganisms present in the substrates and favours the harmful ones. Heating at 100°C (pasteurization) is more acceptable. When using lower temperature (60-70°C) as in the case of room or bulk pasteurization, steaming is done for 6 to 8 hours (Quimio *et al.*, 1995).

Spawn production

The success of the production technology of mushroom is greatly dependant on the availability of a quality and productive spawn. The main problem of the spawn production is the isolation of a stable strain which will fulfill the expectation of the growers. A poor strain will be ultimately unsatisfactory. No matter, how ideal are the conditions of the cropping atmosphere (Chang and Miles, 1985). Therefore selection of a suitable and stable strain is the prime and important aspect in spawn preparation. The second aspect of a quality spawn is selection of a suitable spawn substrate. Even though lot of spawn substrates have been suggested by various workers, grain spawn is mostly suited and widely adopted for oyster mushrooms. Commonly used grains are wheat, sorghum, maize, jowar and paddy. The choice of the grain depends on the availability of the same in a particular locality. Use of bacterial fortified (bioinoculant) spawn has been reported recently to restrict the undesirable growth of moulds and to get higher yield in *Pleurotus* spp. (Ahlawat and Vijay, 2003).

Spawning and containers for incubation

Different types of containers are used for spawning and incubation. Earthen pots, wooden trays, bamboo basket, gunny bags, perforated plastic trays and polythene bags are common. Among these, polythene bags (Bano and Nagarajan, 1976) or compact polybag method (Bhaskaran *et al.*, 1978; Easwaramoorthy *et al.*, 1983) were found to be the best ones.

Layer spawning over the entire surface recorded increased yield over thorough spawning and spawning on the periphery only (Sivaprakasam and Ramraj, 1991). The rate of spawn varies from 3-15 per cent by dry weight of the substrate, but the optimum is reported to be 10 per cent (Shanmugam, 1986). But it has also been reported that increase in inoculation level could increase the yield (Sivaprakasam and Kandaswamy, 1982; Sivaprakasam and Ramraj, 1991).

The highly industrialized method involves bulk pasteurization and bulk spawning. The system is labour saving but requires more complex equipments and has the risk of contamination (Quimio *et al.*, 1995).

Supplementation for better yield

Supplementation of the substrate with various materials is recommended prior to spawning for enhancement of yield of oyster mushrooms. Various oil seed meals and cakes, powdered pulses, wheat and rice bran, etc., are added as supplements (Bahukhandi, 1990). Use of sterilized chicken manure has been suggested to supplement *P. sajor-caju* and *P. flabellatus* (Vijay and Upadhyay, 1989). Jandaik (1974) recommended addition of oatmeal or arhar dal powder for better yield of *Pleurotus* spp. Supplementation of mushroom beds with horse gram (75 g/kg straw) after the spawn-run increased the yield significantly (Bano *et al.*, 1978). Starch at 10 g/kg substrate improved the B:E from 28.2 to 40.00 per cent (Eswaramoorthy *et al.*, 1983). It has also been reported that use of urea or orchid fertilizer dissolved in water (1 g in 1 litre water) as spray immediately before fruiting will increase the yield of *Pleurotus* spp. (Quimio *et al.*, 1995). Inorganic food supplements namely, urea and KH_2PO_4 (one per cent each) sprays at primodial initiation on beds alternating with each water spray upto half mature stage of sporocarp gave maximum sporocarp weight (Balakrishnan, 1994).

Incubation of spawned beds and cropping

The spawned compost bags are kept in a dark room until the mycelium has fully penetrated from the bottom to top of the substrate. After completion of the spawn-run in the substrate (varying incubation periods for different *Pleurotus* spp. usually 10-25 days), the polythene/gunny bags are cut open and the coverings are removed from the trays (for tray method). The trays are stacked in well ventilated growing rooms where provisions for proper light and humidity are made. Maintaining proper humidity (80-90 per cent) and congenial temperature (25±5°C) within the growing room are two vital factors for getting a good crop. Too much variations in the temperature and humidity of the cropping room may delay the cropping and also reduce the yield. The trays are stacked one above the other keeping a space of about 15-20 cm in between the trays to a height convenient for workers. Whereas, in other cases, the compact mass of the spawn-run substrate (bed) is taken out of the polythene/gunny bags and placed either on the shelves made in the walls or stacked on wooden shelves. The shelves can also be prepared in rows, one above the other so that maximum beds can be accommodated in a room.

Cross but not very heavy ventilation is required at this stage. Light is also very essential for the fruit body formation. The doors of the cropping room are kept open for a few hours in the day time to allow natural light to filter in. It is advisable to have few small glass windows in the cropping room so as to facilitate free entry of light and fresh air as and when required. During night, fluorescent light may be provided for few hours. Watering should be done 2-3 times a day during summer. The beds should not be over watered, which may cause rotting of the substrate and hindrance to the fruit body formation. Small fruit bodies begin to appear within a week of providing aeration and light in the room. The crop cycle continues for 50-70 days (Pal and Thapa, 1979).

Harvesting

Approximately 3-4 days after opening the bags, mushroom primordial will start to emerge. Mature mushrooms will be ready for harvesting in another 2-3 days. It has been suggested that the optimum time for picking is before starting the inward rolling of the margins of the sporocarps (Pal and Thapa, 1979). Oyster mushrooms should be harvested before spores are released (Jandaik, 1976). The fruit bodies should be harvested by twisting them, so that broken pieces are not left out on the bed surfaces, which may often cause bacterial infections and rotting of the beds. Fructification and harvest of flush can be controlled more simply by temperature—air exchange and air humidity. The composition of air and development of fruiting body are directly related. With a greater exchange of air, the amount of fruiting body for surface unit will be reduced. Further, more insufficient ventilation and low light exposure induce bunched growth. It has also been pointed out that in *Pleurotus* spp., about five flushes could be harvested and the maximum yield was obtained from the first harvest (flush) and thereafter reduced gradually in each flush (Sharma and Jandaik, 1981). When a sawdust substrate is used, the harvested surface may be scraped lightly to expose a new surface for the next flush. As long as the substrate appears white mushrooms will continue to form under adequate environmental conditions, when it appears colorless, soft and shrunken it is time to remove the bags from the house (Quimio *et al.*, 1995).

Constraints

Cultivation of oyster mushrooms has gained much importance in the country especially in southern States as a self employment venture for the rural sector. Even though a lot of technological advancements have been achieved in its production its commercialization has not yet been fulfilled. The researchers only can't bring this into a reality but the extension machinery of the State Governments and financial Institutions has to play their own roles to bring this into a real employment venture among the needy people. The following are also some important lacunae existing in this field.

1. Lack of attention of the medical field in its exploitation as necessary health food.
2. Lack of proper awarenes among people regarding its uncomparable nutritive and medicinal values.
3. Lack of manufacture of value added products to reach in the hands of common people as ready made food items from the local markets.
4. Supply of quality spawn in sufficient quantity has not yet been achieved and no approved standard are available for determining the quality of spawn.
5. Lack of established centers for supply of processed/pasteurized substrates for cultivators in local areas.

Future prospects

Viewing the last one few years of fast growth in the production of oyster mushrooms in the country there is no ground for any suspicion to its future prospects. More number of high yielding strains has to be developed suited to each and eery locality of the country. So also the cost of production has to be brought down so that the market price will also be reduced which will attract more and more poor people also as its consumers. Fruitful market strategies also has to be worked out suited to various regions of the country so as to bring more people in mushroom growing to a steady income generating programme.

References

Ahlawat, O.P. and Vijay, B. 2003. Present status and future scope of microbial inoculants in mushroom cultivation technology. In: Current Vistas in Mushroom Biology and Production. Proceedings of IIIrd Indian Mushroom Conferene (Eds.) Upadhyay. R.C.. Singh. S.K. and Rai. R.D. pp. 61-74.

Babu. K.M. and Nair. R.K. 1991. Mushroom cultivation on oil palm factory wastes. Indian Mushrooms. *Proc. Natl. Symp. Mushrooms*. pp. 104-108.

Bahukhandi, D. 1990. Effect of various treatments on paddy straw on yield of some cultivated species of *Pleurotus. Indian Phytopath.* **43**: 471-472.

Bahukhandi. D. and Munjal. R.L. 1990. *Indian Phytopath.* **43**: 70-73.

Balakrishnan, B. 1994. Improvement ont he techniques for the cultivaion and preservation of tropical species of mushrooms. Ph.D. Thesis. Kerala. Agri. Univ.. Kerala, pp. 207.

Balakrishnan, B. and Nair, M.C. 1997. Development on the biotechnology of oyster mushrooms. In: Advances in Mushroom Biology and Production (eds.) R.D. Rai, B.L. Dhar and R.N. Verma, Mushroom Society of India, Solan, pp. 83-89.

Bano. Z. and Nagarajan. N. 1976. The cultivation of mushrooms (*Pleurotus flabellatus*) on paddy straw packed in polythene bags with vents. *Indian Food Packer*. **30**: 52-57.

Bano, Z. and Rajarathnam, S. 1982. Studies on the cultivation of *Pleurotus sajor - caju. Mushroom J.*, **115**: 243-245.

Bano, Z. and Srivastava. H.C. 1962. Studies on the cultivation of *Pleurotus* species on paddy straw. *Food Sci.*. **11**: 36-38.

Bano. Z., Rajarathnam. S. and Nagarajan. N. 1978. Some aspects on the cultivation of *Pleurotus* in India. *Mush. Sci.*, (Part II): 597-607.

Beig. G.M. and Jandaik. C.L. 1989. Artificial cultivation of *Pleurotus cystidiosus. Mush. Sci.*, **12**(2): 67-71.

Bhaskaran. T.L., Sivaprakasam. K. and Kandaswamy. T.K. 1978. Compact bag method—A new method of increasing the yield of *Pleurotus sajor-caju. Indian J. Mush.*, **4**: 10-12.

Block. S.S. 1965. Carvage composting for mushroom production. *Appl. Microbiol.*, **13**: 5-9.

Block. S.S., Tsao. G. and Hom. L.H. 1958. Production of mushrooms from saw dust. *J. Agric. Food Chem.*, **6**: 923-927.

Chandrasekhar. B.S.. Veena. S. and Kulkarni. J.H. 2001. Cultivation trials of *Pleurotus sajor-caju* on sodium hydroxide pretreated sugarcane by products. *Mushroom Research*. **10**(1): 27-30.

Chang. S.T. and Miles. P.G. 1985. Mushroom Technology—Lecture given by Prof. Chan at the Second National Biology Symposium held in Bangi. Malaysia from 12-14 November (1985).

Das. T.L., Sharma. R. and Sing, B. 1988. Utilization of weeds and other waste products for spawn and production of oyster mushroom. *Weed abstract.* **37**: 504.

Dhar, B.L. 1976. Japanese method of cultivation of wood inhabiting mushroom. *Indian J. Mush.*, **2**(1): 26-32.

Easwaramoorthy, S.. Ramaraj. B. and Shanmugam. N. 1983. Studies on some factors influencing yield of oyster mushroom in Coimbature. Paper presented at symposium on Science and Technology of edible Fungi. CSIR. Regiona Research Laboratory. Srinagar.

Eger. G.. Eden. G. and Wissing. E. 1976. *Pleurotus*. breeding potential of a new cultivated mushroom. *Theoretical and Applied Genetics*. **47**: 155-163.

Eugenio, C.P. and Anderson. N.A. 1968. The genetics and cultivation of *Pleurotus ostreatus*. *Mycologia*. **60**: 627-634.

Ghose, T.K. and Ghose. P. 1978. Bioconservation of cellulosic substance. *J. Appl. Chem. Biotechnol.* **28**: 309-20.

Ghosh. N. and Chakravarty, D.K. 1991. Studies on evolving new strains of *Pleurotus sajor-caju* by selective dikaryotisation. *Adv. Mushroom. Sci.* pp. 28 (abstr.).

Gupta. V.K. and Langer. P.N. 1988. *Pleurotus florida* for upgrading the nutritive value of wheat straw. *Biological wastes*. **23**: 57-64.

Jandaik. C.L. 1974. Artificial cultivation of *Pleurotus saor-caju*. *Mushroom J.*, **22**: 405.

Jandaik. C.L. 1976. Commercial cultivation of *Pleurotus sajor-caju*. *Mushroom J.*, **22**: 405.

Jandaik, C.L. 1997. In Proceedings of the Indian mushroom conference. Advances in Mushroom biology and production. (eds). Rai. R.D.. Dhar. B.L. and Verma. R.N.. Indian. pp. 181-182.

Jandaik. C.L. and Kapoor. J.N. 1974. Studies on cultivation of *Pleurotus sajor-caju*. *Mush. Sci.*, **9**(1): 667-72.

Jandaik. C.L. and Kapoor. J.N. 1974. *Pleurotus sajor-caju* (Fr.) Singer from India. *Indian J. Mushrooms*. **1**: 1-2.

Khandar. R.R.. Vaishnav. M.V.. Akbari. L.F. and Andhania. J.H. 1991. Effect of various plant substrates on sporophore production of *Pleurotus sajor-caju*. *Indian Mushrooms*. pp. 112-13.

Martonffy. B. 1987. Certified cultivars. Two oyster mushrooms. *Kerteszet es Szoleszet*. **366**: 6.

Mathew. J.R.. Kothandaraman and Thresiamma. K.J. 1991. Cultivation of oyster mushrooms on rubber processing factory waste—A possible solid waste utilization method Indian Mushrooms. *Proc. Natl. Symp. Mushrooms*. pp. 97-99.

Mehta, K.B. and Jandaik. C.L. 1989. Cultivation of *Pleurotus sapidus* in India. *Mush. Sci.*, **12**(2): 179-185.

Pahil. V.S. 1997. In: Proceedings of the Indian mushroom conference. Advances in mushroom biology and production. (eds). Rai. R.D.. Dhar. B.L. and Verma. R.N. India. pp. 91-104.

Pal. J. and Thapa. C.D. 1979. Cultivation of dhingri (*Pleurotus sajor-caju*) made easy. *Indian J. Mushrooms*. **5**(182): 17-20.

Pande, M. and Tewari. R.P. 1989. Cultivation of *Pleurotus sapidus* in India. *Mush. Sci.*, **12**(2): 179-185.

Puri. Y.N.. Rehill. P.S. and Balwanth Singh. 1981. Cultivation trials of *Pleurotus fossulatus*. *Mush. J.*, **102**: 209-214.

Purkayastha. R.P. and Jana. K.K. 1983. A step forward for the commercial cultivation of *Pleurotus eous*. *Mush. J.*, **2**: 20-25.

Quimio, T.H.. Royse. D.J. and Menini. U.G. 1995. Technical guidelines for mushroom growing in the tropics. International book distributing Co. (Publishing Division) Chaman Studio Building. 2nd Floor, Charbagh, Lucknow - 226004. U.P. (India). p. 155.

Ram. R.C. and Thakur. D. 2005. Evaluation of new substrate preparation techniques for oyster mushroom cultivation. *Mushroom Research*. **14**(1): 37-39.

Rangad. C.O. and Jandaik, C.L. 1977. Nitrogen fixation by *Pleurotus* species. *Indian J. Mushrooms*. **3**(1): 9-12.

Rao. C.S. 1991. Evaluation of substrates and supplements for the oyster mushroom cultivation in the state of Goa. Indian Mushrooms-Proc. Natl. Symp. Mushrooms. pp. 93-95.

Rengaswamy. G.. Kandaswamy. T.K. and Ramaswamy. K. 1975. *Pleurotus sajor-caju* a protein rich nitrogen fixing mushroom fungus. *Curr. Sci.*, **44**: 403-404.

Shanel. L. 1970. Effect of carbon dioxide on the growth of wood decaying fungi. *Folia. Biol.*, **11**: 670-72.

Shanmugam. N. 1986. Biology and cultivation of *Pleurotus* spp. in India. In: Beneficial Fungi and their utilization (eds.) M.C. Nair and Balakrishnan. S., Scientific Publishers, Jodhpur. pp. 17-24.

Sharma. A.D. 1983. Effect of cropping environment on some quality parameters of oyster mushroom. *Indian J. Mushrooms*. **9**: 1-6.

Sharma, A.D. and Jandaik, C.L. 1981. Yield potential and economics of *Pleurotus* cultivation in wheat straw under Solan conditions. *Indian J. Mushrooms*. 7(1&2): 11-13.

Sharma, A.D. and Jandaik, C.L. 1985. Cultural requirement of two isolates of *Pleurotus eryngii. Indian J. Mush.* **11**: 20-26.

Sing, R.P. and Dwivedi, R.R. 1991. Standardization of substrate for production of *Pleurotus sajor-caju* (Fr.) Sing. Indian Mushrooms. *Proc. Natl. Symp.* Mushrooms. pp. 102-103.

Sivaprakasam, K. and Kandaswamy, T.K. 1982. Spawn composition (effects) on the sporophore yield of oyster mushroom. *Madras Agric. J.*, pp. 70-117.

Sivaprakasam, K. and Ramaraj, B. 1991. Studies on some factors influencing the yield of oyster mushroom. Indian Mushrooms-Proc. Natl. Symp. mushrooms. pp. 127-32.

Sivaprakasam, K., Bhaskaran, T.L. and Kandaswamy, T.K. 1979. Mushroom industry and its potential in Tamil Nadu. *The Farm Science*. **4**: 21-27.

Sohi, H.S. 1988. Mushroom culture in India-recent research findings. *Indian Phytopath.* **41**(3): 313-326.

Srivastava, H.C. and Bano, Z. 1970. Nutrition requirement of *Pleurotus flabellatus. Appl. Microbiol.*, **19**: 166-69.

Thilagavathy, D.R., Kumuthakavally and Shanmugam, S. 1991. Study of oystr mushroom cultivation on various substrates. Indian mushrooms-Proc. Natl. Symp. Mushrooms. pp. 86-88.

Upadhyay, R.C. and Sohi, H.S. 1988. Apple pomace, a good substrate for the cultivation of edible mushrooms. *Curr. Sci.*, **57**: 1189-90.

Vijay, B. and Sohi, H.S. 1987. Cultivation of oyster mushroom, *Pleurotus sajor-caju* on chemically sterilized wheat straw. *Mush. J. Tropics*. 7: 67-75.

Vijay, B. and Upadhyay, R.C. 1989. Chicken manure as a new nitrogen supplement in oyster mushroom cultivation. *Indian J. Mycol. Plant Pathol.*, **19**: 297-98. Weed abstract. **37**: 504.

Zadrazil, F. 1975. Influence of CO2 concentration on the mycelium growth of three *Pleurotus* species. *European J. Appl. Microbiol.* **1**: 327-35.

Zadrazil, F. 1978. Cultivatin of *Pleurotus*. In: The Biology and Cultivation of Edible Mushrooms. (eds.) S.T. Chang and W.A. Hays. Academic Press, New York. pp. 521-538.

Zadrazil, F. and Kurtzman, J.R. 1982. The biology of *Pleurotus* cultivation int eh tropics. In: Tropical mushrooms. Biological nature and cultivation methods (S.T. Chang and T.S. Quimio, Eds.). The Chinese University Press, Hongkong. p. 493.

Zadrazil, F. and Schneidereit, M. 1972. Die Grundlagen fur die inkultumahone einer nicht cultivation *Pleurotus. Art. Champignon*. **12**: 25-32.

Soil Microbiology and Soil Biotechnology : Vol. 2
Editor : S. Kannaiyan
Associated Publishing Co., New Delhi, 2010, pp. 233-250

13

Biodiversity of Oyster, Paddy Straw and Milky Mushrooms

T. Marimuthu[1], A.S. Krishnamoorthy[2] and Nakkeeran[3]

Introduction

The fungi are the most important groups in terrestrial ecosystems, having a biomass second only to plants and species diversity second only to the invertebrates. They have been known from fossil record as far back in time as the Silurian period, 408 to 438 million years ago in the Paleozoic era. Fungal diversity had increased by the Pennsylvanian period (286 to 320 million years ago) including Basidiomycetes and Ascomycetes; mushrooms have been part of the fungal diversity for around 300 million years. Globally 1.5 million fungal species have been reported, of which approximately 41,000 are mushroom fungi. In India about 850 species of mushrooms are known to occur. Humans of early civilizations (Greeks, Egyptians, Romans, Chinese and Mexicans) appreciated mushroom as a delicacy, knew something about their therapeutic value. China in AD 600 cultivated for the first time the *Auricularia auricula* on wood logs, others such as *Flammulina velutipes* (AD 800) and *Lentinula edodes* (AD 1000) were subsequently cultivated. But major breakthrough came in 1600 AD when France cultivated *Agaricus bisporus* on composted substrate. *A. bisporus*, commonly known as champignon or the button mushroom increased in popularity in the Western world and it became number one mushroom produced in the world. However, mushroom species like *Pleurotus* and *L. edodes* are popular and commercially produced in Asia and have made inroads into Western markets. It has been estimated that *L. edodes* may become number one and most popular in the world by 2010.

Cultivated mushrooms are now one of the important and valued agricultural products worldwide. The total world production of edible and medicinal mushrooms was estimated to exceed 6 million metric tons, with a value of about $6 to 30 billion. The bioconversion of lignocellulosic biomass by the mushroom industry to food and other valued products is already a significant contribution to the management of agricultural and industrial wastes at various levels. It is predicted that this contribution will enormously increase in the near future and

1,2. Professors. 3. Associate Professor. Department of Plant Pathology, Tamil Nadu Agricultural University, Coimbatore - 641 003

contribute to a "non-green revolution". It is said that the fungal biomass present in one ha of pasture land would be approximately 4,000 kg. Despite the importance of fungi, exploration of fungal biodiversity has only just begun and we have only described 5% of those fungi thought to exist; otherwise our knowledge on fungal biodiversity and conservation is limited. Like all organisms on the planet, fungi are being endangered and driven to extinction by human activity. Loss of diversity occurs at ecosystem, species and genetic levels. Mushrooms not only provide nutritious, protein rich food but some species produce medicinally effective products.

It is generally understood that the composition and type of plant species in a terrestrial ecosystem primarily decides the productivity and sustainability of that particular ecosystem. More interestingly, recent findings reveal that the plant biodiversity may be primarily regulated by the diversity of fungi associated with them. In other words, fungal biodiversity would indirectly control both ecosystem productivity and variability. Naturally this process would reduce the competition and enhance the overall biodiversity by promoting coexistence.

Monitoring the fungal biodiversity in a given environment and acquiring knowledge about the fungal community of an ecosystem are important assets, as fungi are considered ecological indicators of perturbation within the environment (Hawksworth, 1991). Conservation of mushroom genetic resources has high stakes in agriculture, food chain, animal feed, chemicals and pharmaceuticals. Globally limited attention has been given to the conservation and exploration of basidio-macromycetes because of the difficulty in culturing and long term preservation. Agro-ecological conditions that prevail in India favour rich mushroom biodiversity and offer great scope for industrial exploration. In spite of this fact, till date no sincere and organized efforts have been made to conserve the germplasm of various species of fleshy fungi.

World production of *Pleurotus* and *Volvariella*

World edible mushroom production has risen by 230 per cent in 2002 (2,961,493 t) from 1994 (2,064,079 t) (FAOSTAT, 2002). The production of *Pleurotus* and *Volvariella* is approximately 15-18 % of the total mushroom production. China is the leading producer of both oyster and paddy straw mushroom (Table 1).

Table 1: World production scenario of some edible mushrooms

Species	Fresh weight (x 1,000 t)				Increase
	1986	Per cent	1997	Per cent	Per cent
Agaricus bisporus	1,227	56.2	1,956	31.8	59.4
Lentinula edodes	314	14.4	1,564	25.4	398.1
Pleurotus spp.	169	7.7	876	14.2	418.3
Volvariella volvacea	178	8.2	181	3.0	1.7
Total	2,182	100.0	6,158	100.0	182.2

Source: Chang (1999)

The genus *Pleurotus*

The genus *Pleurotus* is commonly called oyster mushroom and was described first by Fries (1874). It is an important genus of edible basidiomycetes, especially in the tropics and subtropics. More than 1000 species of *Pleurotus* have been described throughout the world, in more than 25 related and/or confused genera. However, only approximately 50 valid species are recognized in *Pleurotus*. The macro and micro morphocharacters, as well as on the color of the basidioma, the spore print and the type of hyphal system, all of which play an important role in the taxonomic classification of the genus. Out of the 20 species of *Pleurotus* reported from Mexico, only seven seem to be valid taxa. *P. ostreatus* is the most studied, but with numerous taxonomic problems (Guzman, 2000).

Morphological variants

Based on the species diversity Singer (1975) classified the genus into five sections *viz.*,

Section 1. Lepiotarii

Type species: *P. dryinus* (Pers. ex.Fr.) Kummer. Internal tissues of the gills are thin walled at first, becoming sclerified with age; stipe and veil present. Cultivated species belonging to this group includes *P. corticatus.*

Section 2. *Pleurotus*

Type species include *P. ostreatus* (Jacq. ex.Fr.) Kummer. Internal tissues of gills are sclerified early. Gill edges are smooth. Basidiomata is normally coloured. Veil is not present. The cultivated species include *P. ostreatus*, *P. sapidus* and *P. eryngii.*

Section 3. Calyptrati

Type species: *P. calyptratus* (Lindb. in Fr.) Sacc. Veil very thin, not completely covering the cap; very short stem; internal tissue of the gills is irregular. Many hyphae are thick walled. In this section no edible and cultivated species has been reported.

Section 4. Lentodiellum

Type species: *Panus concavus* Berk. Tissues of basidiocarp are thick walled; normally caps are cinnamon brown or buff or rust coloured or with no pigment; sclerotia are not produced; veil and annulus (thick) are present. The cultivated species include *P.* (*Lentinus*) *sajor-caju.*

Section 5. Tuber-regium

Type species include *P. tuber-regium* (Fr.) Sing. The characters more or less resemble to that of section 4, but the species are more coloured. Conspicuous large sclerotia are produced. The cultivated species include *P. tuber-regium*. However, *P. cystidiosus* Miller cannot be placed in any of these sections based on the characters or similarities proposed. The genus *P. cystidiosus* described for the first time by Miller (1969) is unique in the sense that it is a dimorphic species. A nearly identical species to this group is *P. abalonus.*

The more the genus is studied the more discrete variation is observed. *P. sajor-caju* reported by Pegler (1977) had a veil and an annulus whereas; an isolate of the same species (ITCC 1725) collected by Jandaik and Kapoor (1976) did not possess veil or annulus. This kind of morphological diversity is wide spread in *Pleuotus* species. *P. ostreatus* and *P. pulmonarius* show more morphological similarity. However, *P. pulmonarius* prefers to grow in high altitudes mostly on coniferous trees; while, *P. ostreatus* grows on hard wood trees in the middle elevation and foothills (Stametes, 2000).

Diverse host preference

The genus is also known to exhibit diverse host preference. *P. citrinopileatus* was isolated from mango tree (Sivaprakasam *et al.*, 1986); *P. platypus* from silk cotton tree (Marimuthu *et al.*, 1991) and *P. salmoneo-stramineus* var. *roseus* Corner from a coconut tree (Krishnamoorthy *et al.*, 1997). The *P. eryngii* species-complex includes populations of choice edible mushrooms, growing in the greater Mediterranean area in close association with different genera of plants of the family Apiaceae. Their distinct host-specialization served as the principal criterion for the discrimination of several taxa; however, the genetic relationships among the various *P. eryngii* ecotypes remain ambiguous. In a study, 46 *Pleurotus* strains with a wide range of geographical origins were isolated from *Eryngium* spp., *Ferula communis*, *Cachrys ferulacea*, *Thapsia garganica* and *Elaeoselinum asclepium* sub sp. *asclepium* were subjected to isozyme and random amplified polymorphic DNA-PCR (RAPD) analysis. The 16 enzyme activities tested were controlled by 28 loci, 11 of which were monomorphic. Host-exclusive zymograms for the *Aph* (acid phosphatase) and *Phe-1* (dopa-phenoloxidase) loci were obtained from *Pleurotus* strains associated with *C. ferulacea*. Allele frequencies, genetic diversity and mean diversity were high for isolates from *Eryngium* spp. and *F. communis*. In RAPD analysis, the use of five primers allowed the production of 45 (out of 48) polymorphic bands, while four molecular markers specific for the identification of *Pleurotus* strains growing on *E. asclepium* sub sp. *asclepium* and *C. ferulacea* were obtained. The *Pleurotus* strains produced 35 distinct electrophoretic types and 42 RAPD patterns, which independently permitted the separation of the fungal populations into five clusters in accordance with their host-specificity. In addition, the evaluation of the principal ecological and morphological characters provided further evidence for discriminating between *P. nebrodensis* growing on *C. ferulacea* and the rest of the host-associated populations. All Mediterranean *Pleurotus* populations growing on umbellifers seem to have recently diverged through a sympatric speciation process that is based on both intrinsic reproductive barriers and extrinsic eco-geographical factors (Zervakis *et al.*, 2001).

Colour diversity

The sporophores of *Pleurotus* appear in varied colours including white, blue, gray, brown, gold and pink. The pink oyster mushroom varieties like *P. djamor*, *P. eous* and *P. salmoneo-stramineus* var. *roseus* represent a large complex of tropical varieties. *P. flabellatus*, *P. ostreato-roseus* and *P. salmoneo-stramineus* are included within *P. djamor* species complex and distributed throughout the world. The pink colour usually fades with maturity. The spore print of *P. ostreatus* is known as white by some workers; while, others indicate that it is violet. According to Moser (1978), *P. citrinopileatus* is often considered to be a variety of *P. cornucopiae*. However, Singer (1986) described *P. cornucopiae* var. *cornucopiae* as having a tawny brown cap whereas, *P. citrinopileatus* reported by Sivaprakasam *et al.*, (1986) was pure white and the specimen used by Stametes (2000) had a brilliant yellow pileus. Geographically *P. citrinopileatus* is limited to Asia and *P. cornucopiae* mostly appears in Europe. Unless comparative DNA techniques are employed and inter-fertility studies are undertaken, the confusion of *Pleurotus* taxa will continue to remain.

Bio-geographical variation

Understanding the pattern of genetic diversity of species is fundamental to address questions concerned evolutionary process and conservation strategies. The interest on the genetic diversity has recently increased as this will help in broadening knowledge on cultivated species and help in identifying qualitative and quantitative differences and to make any desirable improvements genetically. *P. ostreatus,* a huge complex of sub species, varieties and strains, distributed throughout the temperate and tropical forests of the world is considered to be a synonym of *P. ostreatus*. *P. florida* prefers warmer climate compared to *P. ostreatus*. Vilgalys *et al.*, (1993) reported that *P. florida* is interfertile with *P. ostreatus*. The geographic distribution of *P. tuber regium* includes most of the equatorial Africa, India, Srilanka, South East Asia, North Australia and Southern Pacific. In India, occurrence of *P. cornucopiae, P. flabellatus*, *P. eous*, *P. opuntiae* and *P. fossulatus* have been reported to occur in different geographical locations. Correlations between biogeography, phylogeny and reproductive incompatibility suggest ancient speciation within the *Pleurotus*. *P. eryngii* complex is distributed in temperate Mediterranean region, in South Germany, Hungary, Slovakia and Central Asia and in subtropical regions, in Southern Europe, North Africa and Central Asia. Four varieties belong to this complex, two of which have recently been classified as a separate species: *P. eryngii* and *P. ferulae*. These two taxa could be distinguished based on shape, color and diameter of the pileus and size of the basidiospores. *P. eryngii* colonizes roots of *Eryngium campestre* while *P. ferulae* grows mostly on *F. communis*. *P. eryngii* is mainly saprophytic while *P. ferulae* is pathogenic (Urbanelli *et al.*, 2002). Studies conducted on genetic differentiation and reproductive isolations of *P. eryngii* and *P. ferulae* populations have revealed that they are two distinct entities. The industrial production of these two taxa under the name *P. eryngii* is well developed in Italy. Urbanelli *et al.* (2003) used allozymes and PCR finger printing to estimate the genetic diversity of Italian populations of *P. eryngii* and *P. ferulae*. They observed heterozygote deficiency, which is a character to reveal inbreeding, was found within some populations of the two species at few loci; however, the inbreeding coefficients were not high, which indicated that a mechanism restrain the inbreeding act at the local level. Estimates of genetic differentiation indicated a pattern of greater variation within, rather than between populations. These results were supported by Analysis of Molecular Variance (AMOVA) analysis, which attributed a low proportion of the total genetic variation to a large geographical scale divergence and indicated that most of the genetic diversity was because of differences within populations. This distribution pattern of genetic variation of *P. eryngii* and *P. ferulae* populations was attributed to the high level gene flow, by efficient basidiospore dispersal and out crossing mechanisms, which restrain inbreeding within populations. Lewinsohn *et al.* (2001) studied the genetic diversity in populations of the *P. eryngii* complex in Israel. The levels of polymorphism and gene diversity were significantly different between populations. Sixty eight per cent of the RAPD diversity was within the populations and 32 per cent between populations. Inter-specific population genetic distance showed a positive association with geographic distance. There was also a strong positive association between high polymorphism and the aridity index. The coefficient of determination of polymorphism and gene diversity was explained by climatic variables linked to temperature and humidity. Their study showed the validity of the 'environmental theory of genetic diversity' hypothesis within *P. eryngii* populations in Israel.

Molecular analysis and phylogenetic relationships

Convincingly, molecular analysis have drastically changed the views on fungal evolution and their phylogenetic relationships. Analysis of ribosomal RNA genes, mitochondrial and nuclear ribosomal RNA genes have all proved valuable for mapping the inter-relationship between different taxa. Morphogenesis on an evolutionary time scale based on the analysis of 18S sub units of ribosomal RNA genes have been proved in several cases. Such studies have clearly indicated the time scale for the split of the ascomycete and the basidiomycete lineages as 400 million years which correspond well with the earliest fossil evidence.

Over millions of years the basidiomycetes themselves have undergone several metamorphoses. Evidence from the 18S rRNA genes has now been used in conjunction with other morphological characters to erect a system of basidiomycete classes that more closely reflects their natural evolutionary relationships. The origin and evolution of the gilled mushrooms has also been examined using mitochondrial and nuclear ribosomal RNA genes. This analysis has shown that similar gilled mushroom morphologies have evolved independently at least six times, from diverse progenitors. Most gilled forms fall into one of these lineages, called the "euagarics". While the euagarics contain familiar mushrooms such as *Agaricus, Pleurotus, Volvariella, Coprinus, Amanita* and *Cortinarius*, there are also some poroid, coral and bird's nest fungi in this group. Again, the proposed age of the euagaric is the early to mid Cretaceous (roughly 100 million years) corresponds well between the molecular data and fossil record.

In an interesting study, genetic relationships of *P. tuber-regium* were investigated by Isikhuemhen *et al.* (2000). Among several populations of *P. tuber-regium* from Nigeria, Papua New Guinea and New Caledonia, intrastock mating compatibility studies using progeny from two collections demonstrated a tetrapolar mating system for *P. tuber-regium*. Interstock matings among the geographically distinct populations were compatible. All isolates were found to be inter-sterile with tester strains of other *Pleurotus* species, showing that *P. tuber-regium* represents a unique inter-sterility group in *Pleurotus*. Nucleotide sequences of the ITS region of the rDNA gene were determined for 30 isolates and used to infer phylogenetic structure of populations. Phylogenetic analysis showed that African and Australasian-Paci/Ec isolates form at least two distinct evolutionary lineages. Higher genetic divergence was observed among ITS sequences from the Australasian-Paci/Ec region than among African isolates, which suggests a possible origin of *P. tuber-regium* in the Australasian-Paci/Ec region.

A worldwide geographical sampling of Coremio *Pleurotus* taxa and nucleotide sequence data from the internal transcribed spacer of the nuclear rRNA genes (ITS) were used by Zervakis *et al.*, (2004) to produce a molecular phylogeny for the group. Inter-fertility mating studies were also conducted. Both ITS phylogeny and mating data supported the distinction between *P. australis* (a species apparently endemic to New Zealand and Australia) and *P. cystidiosus sensu lato*. Within *P. cystidiosus*, ITS phylogeny showed a deep split between Old and New World isolates and clearly distinguished four distinct clades that strongly correspond to the geographical origin of the strains. In the Old World, one clade is composed of isolates from Europe and Africa, and one clade is composed of isolates from Asia (including collections from Hawaii). In the New World, one clade is restricted to isolates from Mexico; one clade includes all the North America isolates, one collection from Japan and one collection from South Africa. Mating data revealed a high level of inter-fertility among strains of *P. cystidiosus*, except that isolates from Mexico were nearly fully inter-sterile with the other collections. Nucleotide sequence divergence in the ITS1–5.8S rDNA–ITS2 regions among inter-compatible *P. cystidiosus* collections was very high (0–6.9%) in comparison to that reported in other biological species of basidiomycetes (0–3 %), indicating significant genetic divergence between geographically isolated populations of the *P. cystidiosus* group. The phylogenetic species concept, as well as molecular, mating and geographical evidence, was used to recognize five species in the subgenus Coremio *Pleurotus*: *P. australis* (in New Zealand and Australia), *P. abalonus* (in Asia and Hawaii), *P. fuscosquamulosus* (in Africa and Europe), *P. smithii* (in Mexico) and *P. cystidiosus* sensu stricto (in North America). However, geographical boundaries between these species are not strict, as rare events of long distance dispersal have occurred. Comparative study of the V4, V6 and V9 domains of the mitochondrial small-subunit (SSU) rRNA was conducted by Gonzalez and Labarere (2000) to evaluate the use of these sequences to investigate phylogenetic relatedness within the genus *Pleurotus*. The PCR products encompassing these regions from 48 isolates belonging to 16 *Pleurotus* species were sequenced and compared. From this comparison, the length and sequence of the three domains were found to be constant within a species. Significant inter-species variations due to insertion/deletion events were found, in most cases occurring in regions not directly involved in the maintenance of the standard SSU rRNA secondary structure. Phylogenetic analysis based upon these mitochondrial sequences was in agreement with relationships previously established by morphological descriptions and with previous studies based upon the nuclear genome or isozymes; moreover such analysis resolved some ambiguities in earlier analysis. It was confirmed that *P. ostreatus*, *P. florida*, *P. pulmonarius* and *P. sajor-caju* represent a single species. The phylogenetic analysis also made it possible to assess the relative positions of *P. rattenburyi*, *P. lampas*, *P. sapidus*, *P. colombinus* and *P. eryngii*. The results clearly showed that sequences of the V4, V6 and V9 domains of the mitochondrial SSU rRNA could provide good markers for use in the taxonomy and phylogeny of species of Basidiomycota. Because of their nucleotide conservation, the major advantage of these species-specific markers was the possibility to study only one isolate from each species to determine phylogenetic relatedness.

Bioremediation with *Pleurotus*

Pleurotus, the saprotrophic macrofungi that are indigenous to subtropical and tropical regions, can play key and critical role in water sequestration, nutrient recycling, animal and human nutrition, soil health, bioremediation of waste materials and ecosystem function, particularly in water-constrained environments. The biotechnological potential of *Pleurotus* spp. and those of related genera, has been exploited to enhance the digestibility of animal feeds to reduce the incidence of cancers, to synthesize fine chemicals, for bioremediation and as part of a strategy to produce fuel ethanol from wood (Kashangura *et al.*, 2006). Bioremediation has become an attractive alternative to physicochemical methods of remediation of polluted sites. White rot fungi (WRF) are increasingly being investigated and used in bioremediation, because of their ability to degrade an extremely diverse range of very persistent or toxic environmental pollutants.

Polycyclic aromatic and heteroaromatic compounds (PACs) & PAH

Pollution of soils with PACs from various uses of oil products has important economic and toxicological implications for the polluted areas. Hestbjerg *et al.* (2002) described the effects of WRF, *P. ostreatus* on PACs degradation in a field scale experiment. The refuse from commercial production of *P. ostreatus* were used as amendment and was found to enhance PACs degradation. Eggent and Sveum (1999) showed that PAH degradation in aged creosote contaminated soil was effective when *P. ostreatus* was used to degrade the same. The soil temperature and pre-treatment affect the degradation. Wolter *et al.* (1997) studied the effect of *P. florida* on PAH which mineralized 53% [^{14}C]pyrene, 25% [^{14}C]benzo[a]anthracene and 39% [^{14}C]benzo[a]pyrene to $^{14}CO_2$ in the presence of eight unlabeled PAH (50/7g applied) within 15 weeks. *P. florida* degraded more than 40% of the wheat straw amended with PAH.

Cadmium

Cihangir and Saglam (1999) showed that the biomass of *P. sajor-caju* could adsorb the heavy metal cadmium (Cd^{2+}); freeze-dried biomass was more effective than oven dried and non-metabolizing live biomass in the pH range of 4.5 to 6.0. Crude oil polluted soil Isikhuemhen *et al.* (2003) examined the ability of *P. tuber-regium* to ameliorate crude oil polluted soil. They showed that bioremediation of soil contaminated with crude oil was possible, especially when the fungus had been allowed to establish and fully colonize the substrate mixed with the soil. There were significant improvements in per cent germination, plant height and root elongation values of *Vigna unguiculata,* when seeds were planted after 14 days of soil treatment. At 1 to 5% crude oil pollution, the per cent germination values were comparable with the values in control plants. Also, at the highest level of crude oil pollution (15%), there was about 25% improvement in germination value over the 0 day treatment. The plant growth parameters and biomass accumulation increased beside the decreased total hydrocarbon content.

DDE (Dichlorodiphenyldichloroethylene) - phytotoxicities and genotoxicity

P. pulmonarius, which secretes extracellular ligninolytic enzymes causing non-specific cleavage of lignin and organopollutants, remediated DDE spiked to soil. In 5 weeks, 78% of 10 mg kg^{-1} DDE was biodegraded, and the fungal-treated soil reduced acute toxicity on *Triticum aestivum* using the seed germination test (Gong *et al.*, 2006).

Dioxins and furans (polychlorinated dibenzo-p-dioxins and polychlorinated dibenzofurans - PCDD/Fs.)

A. campestris IZ 1877, *Phanerochaete chrysosporium* ATCC 24725, *Pleurotus* sp. PSC 94/03, *P. tailandia, P. sajor-caju, Pleurotus* sp. CCB068 and *P. sajor-caju* 020 presented very good growth in solid medium containing 2.5 g fly ash/plate. All strains produced lignin-peroxidase on the 3rd day of cultivation, but the best activities were produced on the 42nd day of cultivation. Laccase was detected in the supernatants of *Pleurotus* CCB 068 throughout the period of test; the greatest activity being on 42nd day of cultivation (20 U/L). All the other strains produced lower laccase activities. The highest manganese-peroxidase (MnP) activities were produced after 3 days for *Pleurotus* CCB068 (18 U/L) and 49 days for *P. chrysosporium* (18 U/L). Laccase and MnP were the predominant enzymes present in the cultures media and are suggested to be involved in the degradation of PCDD/Fs (Magar *et al.*, 2003).

Biomolecules from *Pleurotus*

A polycyclic aromatic compound pleurotin has been isolated and identified from *P. griseus* which has also antibiotic properties (Grandjean and Huls, 1974). *P. flabellatus, P. ostreatus, P. citrinopileatus, P. pulmonaris, P. sapidus, P. cornucopiae, P. ostreatus* Z-15 and *Pleurotus* sp. HK-35 produced 4-hydroxy 2-methyl benzaldehyde after 10 days of inoculation in culture medium. The maximum amount was synthesised by *P. citirinopileatus, P. pulmonarius* and *P. flabellatus* (Verma *et al.*, 2003). *P. mutilis* produced pleuromutulin, which is a diterpene compound possessing antibiotic properties against gram-positive bacteria and mycoplasmas (Kavanagh *et al.*, 1951). Two antitumor polysaccharides have been isolated from *P. ostreatus* by Yoshioka *et al.* (1975). Both antitumor compounds consisted primarily of glucose. One of the compounds was found to have a β-(1-3) glucan accounting for 69% of the weight, with 13% galactose, 6% mannose and 13% uronic acids. The galactose, mannose, and uronic acids apparently were side groups on the glucan.

The genus *Volvariella*

Volvariella species is known as paddy straw mushroom or Chinese mushroom. It is also called "fukurotake" in Japan and "choku" in China. Chinese were the first to cultivate this mushroom almost 300 years ago. Migrating overseas Chinese introduced this mushroom to other South East Asian countries (Chang, 1977). It is also grown in other countries like Mexico, Puerto Rico, India, Mauritius and Madagascar. Well suited to the tropical and sub-tropical climate of Asia, *Volvariella,* the fast growing mushroom, prefers to grow at high temperature and high humidity. Optimum temperature for the production of paddy straw mushroom is 30-35°C. The fruiting bodies are formed within 10-12 days of spawning. *Volvariella* has more than one hundred species, sub species and varieties. Previously the genus was known as *Volvaria*. Shaffer (1957) published the only monograph of this species and described about 23 species. In India, *V. bombycina* and *V. diplasia* had been elaborately studied, while, the rest of the world concentrated more on *V. volvacea*. *V. esculenta* is yet another species reported from different parts of the world. The difference between the species is apparently colour. *V. diplasia* is white, while *V. volvacea* is dark gray. *V. bombycina* is white to yellow. Tu and Cheng (1979) proved that an isolate of *Volvariella* was white at low light intensity as compared to dark brown mushrooms at high light intensity. Hence, colour is not a constant character to distinguish the species. *V. bombycina* and *V. volvacea* can be differentiated based on the habitat of occurrence. *V. bombycina* is found on woods while, *V. volvacea* is found on soil or compost.

The genus is known for its peculiar taste and medicinal properties. Outdoor cultivation of this mushroom is practiced in China, Philippines, Taiwan and other countries. Recently, at Tamil Nadu Agricultural University, Coimbatore an outdoor production method of paddy straw mushroom as inter crop in maize, sugarcane and banana fields has been developed (Krishnamoorthy *et al.,* 2005). This type of cropping system helps in permaculture, soil amelioration, addition of organic nitrogen, weed control and all the most eco-restoration and *in situ* organic recycling. In indoor production systems, a variety of substrates like composted cotton wastes, oil palm wastes, corn cobs and sugarcane bagasse are used in addition to paddy straw for growing *Volvariella.*

The genus *Calocybe*

Calocybe is the genera with large, pale, tricholomatoid basidiomata in Tricholomataceae. Its mature basidia contain siderophilous granulations; clamp connections are present on all hyphae. It is mainly a grassland species, saprophytic in nature, never obligate and some times ectomycorrhizal. It is temperate to tropical and belongs to the tribus Lyophylleae (Pegler *et al.* 1998). In addition to wide adaptability the genus *Calocybe* is known for its prolonged shelf life, rapid growth on a variety of substrates and increased productivity as compared to any other cultivated mushroom.

Biodiversity of *Calocybe* in relation to other mushroom fungi

Manjula (1983) reported 538 species belonging to 115 genera of 20 families from India. Verma *et al.*, (2003) conservatively estimated that the fungal species available in India would range from 23,000 to 24,000. Out of the total number of fungi, almost half of the species would be fleshy fungi. They further reported that a total number of 912 species from Agaricales, Tramellales and Auriculariales belonging to 121 genera and 16 families have actually been identified in India.

Only limited literature is available on the genus *Calocybe*. Approximately, 20 species have been from different parts of the world (Singer, 1948; Purkayastha and Chandra, 1974; Pegler, 1983 and Krishnamoorthy *et al.,* 1998) of which only two species *viz. Calocybe indica* P&C and *C. gambosum* (Fries) Singer are known to be edible. Baroni *et al.*, (1999) reported the rare find of a beautiful lilac coloured *Calocybe cyanea* (Singer ex Redhead & Singer) from Puerto Rico though the edibility was yet to be confirmed.

C. gambosum (Fries) Singer. (St. George's Mushroom) an edible and excellent fungus which typically fruits by St. George's Day, 23 April, is therefore reported to be a spring to summer species (Singer, 1948). It grows in lawn, field, road side grassland and also in woodland clearings. It is characterized by white cap 5-12 cm, convex with incurved margin, sometimes faintly yellowish or faintly pinkish. Gills are close and white. Stipe 3 -10 cm tall, 1-3 cm thick, cylindric or thicker towards base and white. Smell and taste strong. Spore print white - creamy and ellipsoid and spores 5-6 x 3.3 mm. Basidia with siderophilous granulation. Clamps present.

The basidiomata of *C. rubra* Rick in Singer ex Redhead and Singer is uniformly purplish red, lamellae dark reddish and crowded, spores short ellipsoid, 3-5 x 2-4 mm, cheilocystidia hyaline, cylindric and septate (Singer, 1948). *C. cyanea* Singer is a striking species with bright violaceous pileus, yellow lamellae and stipe, and the delicate violaceous fibrils which can produce contrasting costae or a reticulum over the bright yellow apex of the stipe. Basidiospores white in deposit, smooth, slightly thick-walled, wall siderophilic and cyanophilic. Basidia 14.6 - 23.5 x 4-6.4 mm, two or four sterigmate, narrowly clavate, thin walled, with obvious siderophilous and cyanophilous character. Trauma is hyaline with parallel hyphae and a distinct mediostratum composed of inflated cells of 8-18 mm diameter and a laterostratum composed of cylindric cells, four to six mm in diameter. Sub-hymenium is made up of pseudoparenchymatous cells, 1.6 - 3.2 mm in diameter. Pileus context composed of hyaline, cylindrical or slightly inflated hyphae, 4 - 14 mm diameter. Pileipellis clavate or sphaeroid or pyriform (Baroni *et al.*, 1999).

C. cyanello Singer ex Redhead and Singer has lamellae white, stipe bluish and pileipellis (Singer, 1948). *C. bipigmentata* Singer has pileus convex-papillate, dark reddish brown, glabrous or slightly punctate, approximately 7 mm broad; lamellae ochraceous, narrow, crowded; stipe ochraceous; cheilocystidia cylindric or fusoid. *C. alneti* Singer has pileus convex-depressed, light fulvous with a pale greyish brown disc, approximately 10 mm broad; lamellae yellow, rather broad, close; stipe fuscous or fuscous yellow; cheilocystidia absent. *C. coniceps* Singer has pileus pale cream with a greyish brown disc, convex-papillate; lamellae white, adnexed; stipe pale sordid. *C. atropapillata* Singer has pileus bluish grey with a fuliginous disc, convex with prominent papilla, 13-18 mm broad; lamellae white, subdecurrent and stipe fuliginous (Singer, 1977). *C. cyanocephala* (Pat.) Pegler has lamellae with violaceous or lilac hues. *C. eborina* Pegler has basidiomata uniformly white or ivory, spores globose, 5-6.5 mm in diameter (Pegler, 1983).

C. eborina reported by Pegler (1983) is also a pure white species whereas, *C. cyanocephala* (Pat.) Pegler and *C. cyanea* Singer ex Redhead & Singer contain violet or yellow pigments in the pileus, stipe and in the lamellae. Using DNA sequences *C. cyanea* has been closely allied to *C. carnea*. *C. cyanea* appears to be more closely related to the temperate species *C. onychina* (Fr.) Donk, typically found in conifers forests in Europe and North America. Even though they are clearly different species, these two taxa share several similarities of macroscopic and microscopic features.

C. bipigmentata Singer has dark reddish brown pileus; *C. alneti* Singer has convex-depressed pileus with a pale greyish brown disc and the stipe appears fuscous yellow; *C. coniceps* Singer has pale cream pileus with a greyish brown disc; *C. atropapillata* Singer has bluish grey pileus; *C. cyanocephala* (Pat.) Pegler has lamellae with violaceous or lilac hues (Singer, 1948). Appearance of *C. cyanea* from a xerophytic habitat in Mexico indicates the wide adaptability of the genus.

Calocybe indica P&C (milky mushroom), the Indian introduction

Milky mushroom, *C. indica* P&C from India was first described by Purkayastha and Chandra (1974). Its milky white color and robust nature are appealing to all. Hence, it is known as "milky mushroom" or "Dudhi chhata" and its identity and edibility were confirmed by Purkayastha and Chandra (1974). *C. indica* has been placed under *Calocybe* - Sect I *Calocybe* (Guttatae) (Fr.) Sing. based on the following characters. The hymenophoral trauma is regular but for slight divergence below the hymenium, presence of filamentous pileopellis, carminophilic basidia, thin walled hyaline spores, lack of highly differentiated cystidia although several cylindric units with granular contents are present. Although Purkayastha (1976) could induce fruit bodies in culture of *C. indica* very limited attempts were made on its cultivation until 1998 (Purkayastha and Nayak, 1981; Doshi *et al.*, 1989 and Pandey and Tewari, 1993).

Natural occurrence of this species in Tamil Nadu was reported by Krishnamoorthy (1995). Krishnamoorthy *et al.*, (1998) studied the exact cultural requirements and perfected the techniques for commercial cultivation of milky mushroom for the first time. One particular strain of *Calocybe indica* collected from a sugarcane field near Coimbatore, Tamil Nadu was more efficient with 142 per cent bio-efficiency and subsequently in 1998 released as APK 2 milky mushroom variety by TNAU, Coimbatore. Some of the edible species of *Calocybe* are listed in Table 2.

Table 2: Edibility of some *Calocybe* species

Name	Colour	Edibility
C. rubra	Red, violet, lilac, yellow or bluish gray	Not known
C. gambosum	Creamy white	Edible
C. cyanella	Yellow or white; stipe bluish	Not known
C. cyanea	Yellow with violet fibrils	Not known
C. indica	Milky white	Edible
C. bipigmentata	Yellowish or reddish brown; some times deep violaceous, lilac or bluish gray	Not known
C. alneti	Pale grayish brown, fuscous or fuscous yellow	Not known
C. coniceps	Pale creamy white with a grayish brown disc	Not known
C. atropapillata	Violaceous, lilac or bluish gray	Not known
C. cyanocephala	Violaceous or lilac or with lilac hues	Not known
C. eborina	White or ivory	Not known
C. carnea	Pale to grayish brown	Not known
C. onychina	Violaceous or lilac	Not known

During 2002 – 2005 twenty one wild isolates namely, WCi-1 to WCi-21 were collected and evaluated for their yield potential and quality parameters (Table-2). Among the isolates three selections *viz.*, WCi 2, WCi 6 and WCi 19 have shown the best performance. Interestingly, majority of the wild isolates collected by the authors were found to be associated with finer roots of *Cocos nucifera* (L.) and the rest were collected from sugarcane and banana fields and barren land (Table-3).

Table 3: Wild isolates of Milky mushroom collected during 2002-2005

Code No.	Collected from	Habit	Soil type
WCi 1	Erode	Sugarcane field	Black loam soil
WCi 2	Mullai Nagar	Coconut tree	Black loam soil
WCi 3	Nanjundapuram	Coconut tree	Red loamy soil
WCi 4	Anaikatti	Near a thorny Bush	Clay soil
WCi 5	Bhavanisagar	Banana field	Black loam soil
WCi 6	Gobichettipalayam	Coconut tree	Red loam soil
WCi 7	Glass House, TNAU	Coconut tree	Black loam soil
WCi 8	Vadavalli	Coconut tree	Black soil
WCi 9	Thondamuthur	Coconut tree	Red loamy soil
WCi 10	Pollachi	Palmyrah palm	Red sandy soil
WCi 11	Veerakeralam	Rain tree	Huaman soil
WCi 12	Kalaramani	Coconut tree	Black loam soil
WCi 13	R.S. Puram	Lawn	Mixed soil
WCi 14	Thukkanaickenpalayam	Banana field	Black soil
WCi 15	Narasimmanaickenpalayam	Near a bush	Red sandy soil
WCi 16	Mevani	Sugarcane field	Red loamy soil
WCi 17	Karupparayan Kovil	Coconut tree	Red loam soil
WCi 18	Narasimmanaickenpalayam	Coconut tree	Red soil
WCi 19	Thukkanaickenpalayam	Coconut tree	Red sandy soil
WCi 20	Coimbatore	Near tamarind tree	Black soil
WCi 21	Thukkanaickenpalayam	Palmyrah tree	Black loam soil

Strain variation based on protein profile

Biochemical markers like protein profiles, isozymes, serological properties and DNA probes offer great scope for the study of genetic variability and may be useful in strain improvement programme. Gel electrophoresis of soluble proteins of *C. indica* fruit bodies at developmental stages revealed the presence of 20, 18 and 16 different protein bands in button, stipe and well differentiated sporophores of milky mushroom respectively (Purkayastha and Nayak, 1981). Kalpana *et al.*, (2005), studied the variations in protein profile of six wild *Calocybe* isolates and compared the results with that of the ruling variety APK 2 through SDS-PAGE. Wide variations were observed in the protein profile of different isolates. The isolate WCi 5 showed very high protein profile as compared to other isolates. High molecular weight protein bands of 68 kDa and 97 kDa were conspicuous in WCi 5; whereas, the other isolates including the var. APK2 did not express these proteins. Difference in the protein profiles of six isolates were subjected to cluster analysis using NTSYS-PC program version 2.02. The similarity index (SI) values were computed as a ratio of number of similar bands to the total number of bands in pair wise comparison of the isolates. The genetic similarity index calculated from the protein profile data was used to estimate the genetic relatedness between the isolates. Based on the genetic distance dendrogram was drawn out. Among the isolates WCi 6 showed only 24 per cent SI with other isolates; WCi 2 and WCi 5 showed 100 per cent SI; the variety APK 2 and WCi 4 expressed 81 per cent SI. Remaining two wild isolates namely, WCi 1 and WCi 3 had 60 per cent and 41 per cent SI with the other counterparts. The results further indicate that the wild isolates WCi 4 and the var. APK 2 close relatives as compared to other wild isolates. Interestingly, in dual culture plates perfect hyphal anastomosis of APK 2 with WCi 2 and WCi 4 was observed. Interstrainal fusion and anastomosis had been confirmed by large number of clamp connections at the point of contact. In the preliminary experiments the fusions namely, CWC 2 and CWC 4 showed superior performance in yield and morphological characters as compared to their parents.

Many fungi are endangered

Because of their prominent fruiting bodies and long history of use, the decline of species in the macrofungi is more obvious than microscopic forms. Macrofungi act as indicators of losses of fungal diversity and as a whole, ecosystem degradation. Hawksworth (1991) reported that only 17% of the known species and 1% of the world's total estimated fungi are conserved in the form of cultures; totalling to about 2,54,000 strains representing 11,500 species. Based on the published fungal names, roughly 70,000 species have been described. The described species however, represent only a fraction of those in existence. In well-studied systems there are about 6 times as many fungal species as that of plant species. Since the number of known plant species is approximately 2.7 lakhs; the total number of fungal species may be more than15 lakhs. But, due to our inadequate efforts to conserve the fungal wealth, today we face extinction, more or less, comparable to that of historical extinctions which we understood only through fossil records. Certain anthropogenic factors; loss of natural habitats; fragmentation of habitats; environmental pollution: co-extinctions due to loss of plant species; illegitimate use of pesticides and fertilizers; expansion of monocultures: over-collection and loss of genetic diversity are the real reasons. Even though a little more than 1, 00, 000 fleshy fungi are known, the genome of even the best described species like *Agaricus bisporus* has not been fully understood and explored. Certainly there are species, genes and fungal products awaiting discovery which otherwise will be lost forever if immediate attention is not paid to conserve them. Rehabilitation of ecosystems would help to preserve the coexisting fungal biodiversity (Hawksworth, 1991).

Conservation of fleshy fungi

Ex situ, preservation in established culture banks is the most practical approach for the conservation of fungi. However, the major limitation of this approach would be the uncertainty of how much of the genetic diversity of each species is preserved. In addition, obligate fungi can not be cultured and preserved. The cost of establishment and sustainable maintenance of culture collection centre or gene bank will be very heavy. In poorly maintained gene banks with limited staff strength and funding long-term safety of such cultures is a big question.

Contrarily preservation of fungi *in situ* is also problematic. Here again a full inventory or a database of fungi of any particular geographical location is not available. Practically, to undertake such work lot of funding and networking is needed. Unless this task is completed it would be very difficult to choose areas for conservation. With all difficulties even if the task is achieved the diversity of coexisting fauna and flora also need thorough investigation. Pollutants and global warming pose greater threat for *in situ* preservation. Establishing mushroom orchards is not beyond our imagination. Permaculture approach is also advisable (Stametes, 2003). However, this type of approach should ensure the survival of the greatest diversity of fungal species. Due to the overall global loss of habitat such as tropical rainforest it has been estimated that over 3,50,000 species of fungi will become extinct within the next 25 years (Hawksworth 1990). Already 14 years have passed and by this time who knows, we might have lost at least 50% of the estimate.

Globally, World Federation of Culture Conservation (WFCC) coordinates, supports and promotes networking and exchange of microbial genetic resources. It also helps to build a data base through World Data Centre on Micro-organisms. The species of fungi of varying numbers are stored at various culture collection centers like American Type Culture Collection (ATCC) (27,000), International Mycological Institute (IMI), U.K (3,60,000 dried and 18,000 live), Central Bureau Voor Scimmelcultures (CBS), Netherlands (29,000), International Bank of Edible Saprophytic Mushrooms (IBESM), Italy (434 commercial and Wild strains of mushrooms especially *Pleurotus* spp) and Global Bank of Mushroom Genetic Resources (GBMGR), France (3,000 strains of 80 species of mushrooms) are conserved (Verma *et al.*, 2003).

India, is one of the most diverse agro-ecological regions in the Globe with a rich biodiversity of flora and fauna. With the establishment of Advanced Centre for Biosystematics at Indian Agricultural Research Institute in the year 1993 some efforts have been initiated for the conservation of fungi including edible mushroom species. Due to the efforts taken up by the Indian Council of Agricultural Research through All India Co-ordinated Mushroom Improvement Project with the main centre at National Research Centre for Mushroom (NRCM), Solan, Himachal Pradesh and nine coordinating centres distributed in the entire length and breadth of the country take up serious attempts to conserve edible mushroom cultures. Presently, NRCM, Solan maintains 122 strains of *Agaricus* spp; 114 collections of oyster mushroom species and a few isolates of *Volvariella, Lentinulla* and *Auricularia* in its gene bank. Similarly, Tamil Nadu Agricultural University, Coimbatore has released six oyster mushroom varieties, one milky mushroom variety and two button mushroom varieties for commercial cultivation.

Conclusion

Biodiversity of mushrooms especially, tropical mushrooms provides wide scope for exploitation. Scientific inquest must be made immediately to understand the full secretes of the beautiful and fascinating fleshy fungi to rediscover the current imbalances in mycology and mycotechnology. Fast decline in the habitat and ecosystem is visualized throughout the globe due to population explosion, threat to ethical and moral values and loopholes in governance and law enforcement. As a result, our plants and animals breathe badly; our reptiles forgot to crawl and our birds lost their wings; the sixth sensed animal lost its sense to conserve Mother Nature. Considering the species diversity and increasing demand for different kinds of tropical mushrooms including oyster, milky, paddy straw and Reishi mushrooms in the global market for human medicine and newer biomolecules, continuous and sustained efforts are needed to conserve these fungi. Molecular biological approaches especially DNA analysis would give the best solution to describe and create a database for these fungal species. Detailed sequence information is of immense value to identify and recover useful genes. Gene sequencing would also reveal the biochemistry and physiology of fungi. Genetic markers are useful for rapid distinction of fungal population.

References

Baroni, T. J., N. W. Legon, R. Vilgalys, and D. J. Lodge. 1999. *Calocybe cyanea* – a rare and beautiful agaric is discovered in Puerto Rico. *Mycologist*, **13**: 7-10.

Chang, ST. 1977. The origin and early development of straw mushroom cultivation. *Econ. Bot.*, **31**: 374-76.

Chang, ST. 1999. World production of edible and medicinal mushrooms in 1997 with emphasis on *Lentinus edodes* (Berk.) Sing. in China. *Int. J. Med. Mush.*, **1**: 291-301.

Cihangir, N. and N. Saglam. 1999. Removal of cadmium by *Pleurotus sajor-caju* basidiomycetes. *Acta Biotechnologica*, **19**(2): 171-177.

Doshi, A., N. Sidana and B.P. Chakravarti. 1989. Cultivation of summer mushroom.*Calocybe indica* (P&C) in Rajasthan. *Mush. Sci.*, **12**: 395-400.

Eggent, T. and P. Sveum. 1999. Decontamination of aged creosote polluted soil : the influence of temperature, white rot fungus Pleurotus ostreatus, and pretreatment . *International biodeterioration & biodegradation*, (*Int. biodeterior. biodegrad.*), **43**(3): 125-133 .

FAOSTAT. 2002. Agriculture database of Food and Agriculture Organization.

Fries, E.M. 1874. Hymenomycetes Europaei sive Epicriseos Systematis. Mycologici. Ed. Alt. Ipsala, Berling. p. 755.

Gong, J., K. Chan and S. Chiu. 2006. Toxicities of DDE on Wheat and Bioremediation of DDE by Fungus *Pleurotus pulmonarius*. *Human and Ecological Risk Assessment*, **12**(2): 21-235.

Gonzalez, P. and J .Labarère. 2000. Phylogenetic relationships of *Pleurotus* species according to the sequence and secondary structure of the mitochondrial small-subunit rRNA V4, V6 and V9 domains. *Microbiology*, **146**: 209-221.

Grandjean, J. and R. Huls. 1974. Structure de la Pleurotine: Une Benzoquinone Extraite de *Pleurotus griseus*. *Tetrahedron Lett.*, **22**: 1893-1895.

Guzman, G. 2000. Genus *Pleurotus* (Jacq. Fr.) P.Kumm. (Agaricomycetideae): diversity, taxonomic problems, and cultural and traditional medicinal uses. *International Journal of Medicinal Mushrooms*, **2**(2):

Hawksworth, D. L. 1990. Guidelines for the Establishment and Operation of Collections of Cultures of Microorganisms. Campinas SP, Brazil: World Federation of Culture Collections.

Hawksworth, D. L. 1991. The fungal dimension of biodiversity magnitude and significance and conservation. *Mycol. Res.*, **95**: 641-655.

Hestbjerg, H., P.A. Willumsen, M.C.O. Andersen & C. S. Jacobsen. 2002. Field scale bioremediation of tar contaminated soil with commercial mushroom Pleurotus ostreatus refuse. Ecology and Environment, 59. In: Oil and Hydrocarbon Spills III: Modelling, Analysis and Control Ed. C.A. BREBBIA, p. 480.

Isikhuemhen O. S., G.O. Anoliefo, O.I. Oghale. 2003. Bioremediation of crude oil polluted soil by the white rot fungus, Pleurotus tuber-regium (Fr.) Sing. *Environ Sci Pollut Res Int.*, **10**(2): 108-12.

Isikhuemhen, O. S., Jean-marc moncalvo, Fantisek nerud and Rytas vilgalys. 2000. Mating compatibility and phylogeography in *Pleurotus tuber-regium*. *Mycol. Res.*, **104**(6): 732-737.

Jandaik, C.L. and Kapoor, J.N. 1976. Studies on the cultivation of *Pleurotus sajor-caju* (Fr.) Singer. *Mushroom Sci.*, **9**(1): 667-672.

Kalpana, K.. A. S. Krishnamoorthy and S. Nakkeeran.2005. Exploration of the diversity of milky mushroom, *Calocybe indica* (P&C). In: *Advances in Fungal Diversity and Host Pathogen Interaction* (ed. B.F.Rodrigues, H.N.Gour. D.J.Bhat and N.Kamat).Goa University, Goa. India. pp. 97-108.

Kashangura, C., J. E. Hallsworth and A. Y. Mswaka. 2006. Phenotypic diversity amongst strains of *Pleurotus* sajor-caju: implications for cultivation in arid environments. *Mycol. Res.*, **110**(3): 312-317.

Kavanagh, F., A. Hervey and W. J. Robbins. 1951. Antibiotic substances from Basidiomyces. VIII. *Pleurotus mutilis* and *Pleurotus passeckerianus*. *Proc. Natl. Acad. Sci. U.S.A.* **37**: 570-574, 1951.

Krishnamoorthy, A. S. 1995. Studies on the cultivation of milky mushroom. *Calocybe indica* P&C. Ph.D. Thesis, Tamil Nadu Agricultural University, Coimbatore, India, p. 124.

Krishnamoorthy, A.S., G. Thiribhuvanama., K. Shanthi and T. Marimuthu. 2005. Out door System of Cultivation of Paddy straw Mushroom as Inter crop in Maize field. *Mushroom Res.*, **14**: 9-12.

Krishnamoorthy. A.S., M. Muthusamy, T. Marimuthu, V. Narasimhan and A. Muthusankaranarayanan. 1998. *APK 2 milky mushroom – Extension Bulletin*, Regional Research Station (TNAU), Aruppukottai. Tamil Nadu, India.

Krishnamoorthy, A. S., T. Marimuthu., S. Mohan and R. Jeyarajan.1997. *Pleurotus salmoneo stramineus* an edible oyster mushroom from conconut stump. *Mushroom Res.*, **6**: 79-80.

Lewinsohn, D., E. Nevo, S. P. Wasser, Y. Hadar and A. Beharav. 2001. Genetic diversity in populations of the *Pleurotus* eryngii complex in Israel. *Mycological Research*, **105**(8): 941-951.

Magar. V. S., M. F. Reis-Gorny, L. R. dos, Durrant, and M. E. Kelley. 2003. Screening fungi for their possible ability to degrade dioxins and furans. In: *In situ* and on-site bioremediation. *Proceedings of the Seventh International In Situ and On-Site Bioremediation Symposium*, Orlando, Florida, USA, 2-5 June, 2003-2004.

Manjula, B. 1983. A revised list of agaricoid and boletoid mushrooms from India and Nepal. *In: Proc. Indian Acad. Sci., (Plant Sci.)*. **92**: 81-213.

Marimuthu. T., A. S.Krishnamoorthy and R. Jeyarajan.1991. *Pleurotus platypus*, an edible mushroom from silk cotton tree. In: *Indian Mushrooms* (eds. M.C. Nair *et al.*), *Proc.Natl.Symp.on Mushrooms*, KAU, Thiruvananthapuram. pp. 66-68.

Miller Jr, O.K. 1969. A new species of *Pleurotus* with a coremioid imperfect stage. *Mycologia*, **61**: 887-93.

Moser, M. 1978. Key to Agarics and Boleti. Roger Philips, London.

Pandey, M. and R. P. Tewari. 1993. Milky mushroom grown on paddy straw. *The Hindu*.

Pegler, D. N. 1977. Pleurotus (Agaricales) [Fungi] in India, Nepal and Pakistan. *Kew-Bull.*, **31**(3): 501-510.

Pegler, D. N. 1983. *Agaric flora of the lesser Antilles. Kew Bulletin Additional Series IX.* **10**: 1-668.

Pegler, D. N., D. J. Lodge, and K. K. Nakasone. 1998. The pantropical genus. *Macrocybe* gen. nov. *Mycologia*, **90**(3): 494-504.

Purkayastha, R. P. and A. Chandra. 1974. New species of edible mushroom from India. *Trans. Br. mycol. Soc.*, **62**: 415-418.

Purkayastha, R.P. and D. Nayak. 1981. Development of cultivation method and analysis of protein of promising edible mushroom, *Calocybe indica* P&C. *Mush. Sci.*, **11**: 697-713.

Shaffer, R. L. 1957. *Volvariella* in North America. *Mycologia*, **49**: 545-579.

Singer, R. 1948. New and interesting species of Basidiomycetes. II. pap. *Mica. Acad. Sci. Arts*, **32**: 103-150.

Singer, R. 1975. The Agaricales in Modern Taxonomy. Vaduz.: J. Cramer.

Singer, R. 1986. The Agaricales in Modern Taxonomy 4th Edn. Koeltz cientific Books. Koenigstein, Germany.

Singer, R. 1997. Keys for the identification of the species of Agaricales. *I. Sydowia*, **30**: 193-279.

Sivaprakasam, K., F., S. Rajan, and R. Jeyarajan. 1986. *Pleurotus citrinopileatus* – Anew edible mushroom. *Curr. Sci.*, **55**: 1203.

Stametes, P. 2000. *Growing edible and Gourmet mushrooms.* Ten Speed Press. Berkeley, California. p-574.

Stametes, P. 2003. Potentiation of cell-mediated host defense using fruit bodies and mycelia of medicinal mushrooms. *Int. J. Med. Mushr.*, **5**: 179-191.

Tu, C.C. and S. Cheng. 1979. A new white variety of *Volvariella volvacea* (Bull. ex.Fr.) Sing. *Mushroom Sci.*, **10**(2): 185-90.

Urbanelli, S., C. Fanelli, A. A. A. Fabbri, V. DellaRosa, L. Maddau, F. Marras and M. Reverberi. 2002. Taxonomic studies on two taxa of the *Pleurotus eryngii* complex: *Pleurotus eryngii* var. *eryngi* and *P.eryngii* var. *ferulae* (Quel.:Fr.). *Biol J. Linnean Soc.*, **75**: 125-136.

Urbanelli, S., V. DellaRosa , C. Fanelli , A. A. Fabbri and M. Reverberi. 2003. Genetic diversity and population structure of the Italian fungi belonging to the taxa *Pleurotus eryngii* (DC.:Fr.) Queland P.ferulae (DC.:Fr.) Quel. *Heredity*. **90**: 253-259.

Verma, R.N., R.C. Upadhyay., M.C. Yadav., S.K. Singh and B.L. Dhar.2003. Genetic resources of commercial mushrooms, their conservation, characterization and improvement. In: Current Vistas in Mushroom Biology and Production. (Eds.) Uphadhyay, R.C., Singh, S.K. and Rai, R.D. Mushroom Soc. of India. pp. 1-9.

Vilgalys, R., A.Smith., B.L.Sun. and O.K.Miller. 1993. Interfertility groups in *Pleurotus ostreatus* complex. *Canadian Journal of Botany*, **71**: 113-128.

Wolter, M., F. Zadrazil, R. Martens and M. Bahadir. 1997. Degradation of eight highly condensed polycyclic aromatic hydrocarbons by Pleurotus sp. Florida in solid wheat straw substrate. *Applied Microbiology and Biotechnology*, **48**(3): 398-404.

Yoshioka, Y., M. Emori, T. Ikekawa and F. Fukuoka. 1975. Isolation, purification, and structure of components from acidic polysaccharides of *Pleurotus ostreatus* (Fr.) Quél. *Carbohydr. Res.*, **43**(2): 305-320.

Zervakis G.I.., Giuseppe Venturella and Kalliopi Papadopoulou. 2001. Genetic polymorphism and taxonomic infrastructure of the *Pleurotus eryngii* species-complex as determined by RAPD analysis, isozyme profiles and ecomorphological characters. *Microbiology*, **147**: 3183-3194.

Zervakis,G.I., Jean-Marc Moncalvo and Rytas Vilgalys. 2004. Molecular phylogeny, biogeography and speciation of the mushroom species *Pleurotus cystidiosus* and allied taxa. *Microbiology*, **150**: 715-726.